Calculated Risk

Greed, Politics, and the Westray Tragedy

DEAN JOBB

94 95 96 97 98 99 6 5 4 3 2 1

Nimbus Publishing Limited
PO Box 9301, Station A
Halifax, NS B3K 5N5
(902) 455-4286

Design: GDA, Halifax
Cover photo: Canapress Photo Service (Ryan Remiorz)
Printed and bound in Canada

Canadian Cataloguing in Publication Data
Jobb, Dean, 1958-
Calculated risk
Includes index.
ISBN 1-55109-070-8
1. Westray Mine Disaster, Plymouth, N.S., 1992. 2. Coal mine accidents—Nova Scotia—Plymouth (Pictou) I. Title.
TN806.C32N684 1994 363.11' 9622334'0971613 C94-950214-6

CONTENTS

PREFACE

THE GREY SKY THREATENED RAIN as a fresh crew of miners descended into the Westray mine in Plymouth, Nova Scotia, on the evening of May 8, 1992. At 5:18 the following morning, a powerful explosion killed all twenty-six men working underground. It was Canada's worst mining disaster since 1958, when a massive cave-in, or "bump," at another Nova Scotia colliery, in the hard-luck town of Springhill, claimed the lives of seventy-five miners. The survivors included eight men plucked from the depths after a nine-day ordeal. Despite the horrendous loss of life, the Springhill bump came to be regarded less as a disaster than as a miracle of survival.

In the tense days following the Westray explosion, relatives and friends of the twenty-six trapped men prayed for another miracle. As rescue workers gingerly picked their way through the damaged tunnels, people across Canada watched the drama unfold on their television screens, hopeful that some survivors would be found. But there was no Westray miracle. After five days, the rescue effort was called off; the mine yielded fifteen victims, and the bodies of the remaining eleven were left underground. They were the latest men to fall prey to Pictou County's disaster-prone coalfield, where explosions have claimed the lives of two hundred and seventy-two miners since the mid-nineteenth century.

There were no survivors, but there were plenty of questions—and plenty of allegations that Westray had been a disaster waiting to happen. Although the mine had been in operation a scant eight months, it had a long and controversial history. It was the focus of a bitter political battle that inflamed regional rivalries within Nova Scotia. The mine was built to take coal from a seam with a reputation as one of the gassiest and most dangerous in the world. There were repeated warnings, in public and private, that the project faced serious technical and safety hurdles, and might well prove a financial failure. In the headlong rush to bring Westray into production, the warnings were either ignored or dismissed as partisan troublemaking. The mine opened in the fall of 1991, just as the politicians and the mine officials promised it would, but it never lived up to the high expectations of its promoters. Safety took a back seat as the company desperately tried to fulfil its coal-supply contracts. It was a recipe for disaster. The explosion led to charges of manslaughter and criminal negligence causing death against the mine's Toronto-based developer, Curragh, and two former Westray managers. It is one of the few cases in Canada in which mine fatalities have prompted a criminal prosecution.

The Westray explosion and the criminal charges are a black mark on an industry that employs ten thousand people and pumps $3 billion into the Canadian economy each year. Despite the popular image of coal mining as a dangerous job, the industry boasts a better safety record than logging, construction, or manufacturing. Most of Canada's coal comes from surface mines–huge pits carved out of the earth to reach coal deposits. Working a coal seam hundreds of metres below the surface of the earth is a riskier proposition. The roof can come tumbling down, or methane can reach explosive levels in the enclosed atmosphere. Dust from the coal is combustible, creating another explosion hazard. Every mine is different, presenting its own blend of rock formations and obstacles to be overcome. Since geology and mine engineering are inexact sciences, mining methods and approaches must change to meet the conditions encountered as tunnels advance. There are only four underground operations left in Canada–one each in British Columbia and Alberta, two on Nova Scotia's Cape Breton Island.

If underground coal mining is a risky business, the development of the Westray mine was a calculated risk. Engineers and consultants, with few exceptions, were confident that modern mining methods and equipment would prevent a repetition of past explosions. The federal and Nova Scotia governments gambled roughly $100 million in taxpayers' money on the project, without ensuring the province's mining regulations and inspection staff were up to the challenge. Westray's operators focused their energies on producing the precious coal needed to fulfil corporate and political agendas. Westray's miners also took chances. Men went underground day after day, knowing the mine was unsafe. They worked in an atmosphere laced with methane, and trudged though the thick layer of coal dust piling up on the mine floor. They had families to feed, mortgages to pay, and little hope of finding other work. But there was a crucial difference; while politicians and promoters gambled with money, Westray's miners gambled with their lives.

The statute books of Nova Scotia and other jurisdictions are filled with regulations designed to minimize the risks of mining coal. But safety laws and regulations are only as good as the miners who observe them, the companies that obey them, and the government officials who enforce them. And they must be overhauled on a regular basis to keep pace with new mining methods. That system of checks and balances broke down at Westray, with fatal consequences. Disasters like the Westray explosion serve as a crucible for the development of improved safety laws. One of the tragic ironies of the mining business is that improvements in working conditions are often measured in human lives. And it has taken the deaths of twenty-six men to force yet another hard look at the way coal mines are operated.

There are heroes in the story that follows. Miners from Westray and other mines in Nova Scotia and New Brunswick risked their lives in the futile search for survivors. A handful of federal bureaucrats fought against the generous aid package sought by Westray's owners, enduring the wrath of their political masters. A few politicians and journalists went public with tough questions about cave-ins and other safety problems. A couple of miners blew the whistle on Westray, only to have government officials ignore their complaints. And in the tumultuous months that followed the explosion, relatives and friends of the men who died had the courage to demand a public accounting of what happened on the morning of May 9, 1992.

And there is no shortage of victims. The disaster was one of the factors that cut short the political career of one of Westray's biggest supporters, the former Nova Scotia premier, Donald Cameron. It toppled Westray's parent company, Curragh, throwing hundreds of people out of work. It shattered Curragh chairman Clifford Frame's dream of building a multimillion-dollar mining empire. It left shareholders with stock certificates that were next to worthless. But politicians come and go, and businessmen and investors can lick their financial wounds and start over. The real victims are the 26 men who died, the families they left behind, and the approximately 160 Westray employees who lost their livelihoods.

Where there are victims, there are usually villains. But no single culprit was responsible for the deaths. The explosion was the culmination of a series of political, corporate, and engineering decisions that began with the project's inception a decade ago. Its origins can be traced to the dreams of politicians and mine promoters, and their blind faith that technology would overcome the formidable challenges of mining Pictou County coal. The disaster was the product of shoddy planning, unrealistic production targets, poor safety practices, outdated mining laws, and ineffective government monitoring. A long-overdue public inquiry will eventually pass judgment on the way the mine was designed, operated, and regulated, and suggest ways of preventing a similar disaster from happening in the future. The courts will decide whether the company and its managers were criminally responsible for the deaths. Regardless of these findings, the Westray disaster will stand as an indictment of a style of politics that puts more emphasis on economic development and job creation than the safety of employees.

This book is based on interviews and research conducted over a two-year period. The author reviewed thousands of pages of material—feasibility studies, reports of mine inspectors, internal government memos, financial records, and other documents. The federal and Nova Scotia governments made some of this material public, in response to the public outcry over the disaster. Much more was unearthed through access-to-information applications filed with federal and provincial departments responsible for overseeing Westray. To flesh out the picture, interviews were conducted with dozens of people who were involved in the planning, financing, monitoring, and operation of the mine. Dialogue is recorded as related by those who were privy to the conversations; all quotations are drawn verbatim from interviews, documents, and press reports. All the key players were contacted with requests for interviews, and they responded with varying degrees of cooperation. Many, most notably two former premiers, Donald Cameron and John Buchanan, discuss Westray at length for the first time within these pages. A lot of dedicated people lost their jobs when Curragh went out of business, and a few were willing to tell the inside story of the company's demise.

As this book goes to press in the summer of 1994, the criminal trial is seven months away. The public inquiry remains on hold, awaiting a court ruling on when its work can begin. More than two years after the explosion that claimed twenty-six lives and touched the hearts of an entire nation, the story of the disaster still lacks an ending. What follows is an account of the events and decisions that brought the mine into being, setting off the chain of events that reached a tragic climax on May 9, 1992. It is a story of political intrigue that leads to the highest office in the land. It is a story of corporate empire-building that embraces the rise and fall of one of Canada's most promising mining companies. It is the story of the men who died, the families they left behind, and the miners who survived to tell the tale. And it is yet another chapter in the tragic history of man's attempt to master Pictou County's unforgiving coal seams.

* * *

I AM GRATEFUL to have been part of a team of journalists at the Halifax *Chronicle-Herald* and the *Mail-Star* that was assigned the monumental task of reporting what happened at Westray, and why. Wilkie Taylor, Steve Harder, Paul MacNeill, Pam Sword, Janice Tibbetts, and photographer Len Wagg chronicled the five-day rescue effort. Brian Underhill, Brian Ward, Judy Myrden, Pat Lee, Gail Lethbridge, and Rick Conrad advanced the story in the weeks that followed. Jane Purves, the managing editor, deserves special praise for her commitment to ensuring the Westray story was covered properly, and for her conviction that in-depth stories about important public issues still have a place in the newspapers of the 1990s.

The Canadian Broadcasting Corporation in Halifax permitted me to view video tapes of scores of TV reports on Westray, making it possible to re-create some of the events described in this book. I thank the CBC's librarian, Doug Kirby, for his patience in hunting down tapes. Kevin Cox of the *Globe and Mail* provided tapes and notes that were crucial to reconstructing the hectic events of the May 1992 rescue effort. Barb Lemky, curator of the Cumberland & District Historical Society in Cumberland, British Columbia, provided information on that community's services to mark the Westray disaster. Marianne Mack, government records archivist at the Provincial Archives of Alberta in Edmonton, dug out the court file on the Brazeau Collieries case of the 1940s. Andy Thompson, a journalism student at the University of King's College who hails from Pictou County, provided copies of several useful articles and editorials from the local press.

Others helped in a variety of ways. *Herald* columnists Jim Meek and Don MacDonald shared their insights into some of the political figures involved in the Westray story. John DeMont, the Atlantic correspondent for *Maclean's,* unravelled some of the mysteries of the book trade. Toronto literary agent Dean Cooke suggested using Westray's 1991 grand opening as the starting point for the book. Dartmouth lawyer Sandra MacPherson-Duncan guided me through some of the intricacies of freedom-of-information legislation. Max Keddy of the *Herald* provided disks containing more than 1,100 articles on Westray that have appeared in the newspaper since 1988. Ian Scott, also of the *Herald,* tracked down a computer system that was up to the task of handling the mountain of information about Westray. Garry Beattie of Harris and Beattie in Halifax, a computer whiz who missed his calling as a stand-up comic, gave me a crash course in how to use it.

Most of the photographs used in this book are the work of Len Wagg, the photo editor of the *Herald;* I thank him for choosing some of his most enduring images of the disaster. Terry Eyland of the Dalhousie University Archives, working on short notice, located the photographs depicting Pictou County's mining past. The staff of the *Herald* library—Alberta Dubé, Anne-Marie White, Debbie Reid, and Anita Evans—helped track down information and photographs.

Special thanks to Dorothy Blythe, managing editor of Nimbus Publishing, for her unflagging belief that the Westray story needed to be told. Liane Heller of Nimbus edited the manuscript; her sharp eye and suggestions for improvement have made this a better book.

Westray was the subject of countless news stories before and after the explosion. This book benefited from the efforts of a number of journalists who have reported on Westray over the years. Stewart Lewis, formerly of the *Chronicle-Herald* and the *Mail-Star,* Thomson Newspapers correspondent Betsy Chambers, and Linden MacIntyre of CBC-TV's *the fifth estate* recognized early on that the mine's development raised important political, business, and safety issues. Since the explosion, Kevin Cox, Atlantic correspondent for the *Globe and Mail,* Stephen Thorne of the Canadian Press, Toronto freelance journalist Stevie Cameron, Robert Mason Lee of the *Ottawa Citizen,* Halifax freelancer Richard Starr, and Andrew

Mitrovica, formerly of *the fifth estate* and now with CTV, have helped ensure the Westray story gained the national profile it deserved.

In the course of writing Calculated Risk, the following articles and books were of invaluable assistance: *The Pictonian Colliers,* by James M. Cameron; *Coal in Our Blood: 200 Years of Coal Mining in Nova Scotia's Pictou County,* by Judith Hoegg Ryan; *The Roman Empire: The Unauthorized Life and Times of Stephen Roman,* by Paul McKay; *The Westray Tragedy: A Miner's Story,* by Shaun Comish; "Diary of a Draegerman," by Nancy Robb, in *Occupational Health and Safety Canada,* July/August 1992; "The Fault Line," by Jennifer Wells, *Report on Business Magazine,* December 1992; "The Westray File," *Maclean's,* April 19, 1993; "Death by Consensus: The Westray Mine Story," by Harry Glasbeek and Eric Tucker, *New Solutions,* Summer 1993; and "Burying Westray," by Stevie Cameron and Andrew Mitrovica, *Saturday Night,* May 1994.

Finally, I would like to thank all those who agreed to be interviewed for this book. For the widows of those who died, and the miners who survived the blast, it meant reliving the most traumatic moments of their lives. They did so in the belief that the Westray story should not be swept under the rug, that twenty-six men should not die in vain, and that the public should not have to wait years for courts and inquiries to pass judgment on what happened at Westray. I hope this book gives them some of the answers they seek.

Dean Jobb
Halifax, Nova Scotia
July 8, 1994.

COUNTDOWN TO DISASTER

"JOBS, JOBS, JOBS"

HERE WOULD BE NO LOBSTER or smoked salmon for the men of B-crew working almost half a kilometre underground at the Westray mine. There would be no free booze or music, either. But at least they would be spared the speeches of the politicians and the corporate bigwigs, who would drone on about their long struggle to make the mine a reality. And there was one other consolation for the miners stuck on the back shift; on this soggy and unseasonably brisk morning in September 1991, the tunnels being burrowed into the glistening coal of the Foord seam were about the only dry place left in Pictou County, Nova Scotia.

A heavy rain was drumming on the steel roof of the mine's entrance as tractors carrying B-crew—two dozen grimy, weary miners—emerged from underground about eight o'clock in the morning. The Westray mine squatted on the edge of a wide river valley, its coal-storage silos—a pair of drab, concrete-grey cylinders, each topped with a box-shaped cupola of bright blue—towered over the neighbouring houses and farms. Despite the rain and the morning gloom, the place had a decidedly festive air. Perched on a small knoll at the southern edge of the mine site were two large striped tents, one blue and white, the other yellow and white. It was as if the circus had come to the village of Plymouth and set up shop next door. Within hours, those tents would echo with live music and speeches, as the Westray mine was officially opened. But no-one from B-crew would be on hand to take part in the celebrations.

Eugene Johnson and the other miners showered, changed into street clothes, and sprinted through the driving rain to their cars and pick-up trucks. After a long shift underground, they were eager for breakfast and bed. The mine was worked in twelve-hour shifts by four crews, designated A, B, C, and D, and scheduled for four days on, four days off. B-crew had to be back at eight o'clock in the evening for the start of their next shift, and that left no time for the luxury of free drinks and a fancy luncheon.

Mine manager Gerald Phillips had invited all two hundred employees and their families to the opening ceremonies on September 11, 1991, a Wednesday.

But the company refused to give the men of B-crew time off to attend. For Johnson, it was just another example of their rotten luck. What was it Phillips had said back in May, after a large section of one of the main tunnels collapsed? That night, Johnson and the other miners had come running up the slope scared half to death, sure that the whole mine was about to collapse on their heads. But the next morning, when the men complained, Phillips responded with curses and ridicule. "This goddamned mine will go," he sneered in his thick British accent, "with or without B-crew." The official opening was also going to go ahead without them.

Johnson was thirty-three, with a tuft of reddish hair and a thin moustache. His lanky six-foot-one frame had earned him the nickname "Stringbean," but most people shortened that to "String" or "Beaner." At nineteen, he had gone to work in the Drummond colliery in his home town, Westville, just a short drive from Plymouth. It was back-breaking work, digging coal the old-fashioned way, with pick and shovel and blasting powder. But Johnson loved it; mining coal was in his blood. Four generations of Johnson men had toiled in the mines of Pictou County, a swath of Nova Scotia's northern coast as rich in coal as in Scottish tradition. But Johnson knew all too well that digging coal could be a deadly business; his great-grandfather and grandfather had died working in local mines.

When Westray revived coal mining in Pictou County, Eugene Johnson jumped at the chance to get back underground. [Photo courtesy of Donna Johnson]

About fifty guys worked at Drummond. It was a close-knit bunch; many of the younger miners had grown up together in Westville, and shared the same passions in their off-hours—hunting, fishing, playing hockey. Johnson would often tell his wife, Donna, how much he liked the hard work. But mostly he enjoyed the camaraderie, the bonds that developed from working in close quarters underground. Soon after he and Donna were married, in 1980, Johnson bought an acoustic guitar. He was shy about playing for people at first, but as his technique improved, he would sometimes pull it out at parties and sing his favourite country tunes. He even penned a couple of his own songs. In "Westville Miners," Johnson gushed with pride about working "down in the mine, where few men dare go." The chorus said it all:

We are the Westville miners so tall and so proud,
I'll say it again and we'll say it aloud,
We are the Westville miners so tall and so proud,
And that's what I think and I'll tell you right now."

But for Johnson and his fellow miners in Westville, underground mining in Pictou County came to an abrupt end in 1984, when a fire closed the Drummond pit. Johnson switched to surface work, driving a truck at the Pioneer Coal Company's adjacent strip mine. He toughed it out for seven years, all the while longing to get back underground. So the Westray project was a dream come true. Curragh Resources of Toronto, the mine's developer, promised fifteen years of work, possibly double that. The colliery, in the village of Plymouth, just across the East River from the once-bustling mining town of Stellarton, tapped the last major block of coal in Pictou County. At least forty-five million tonnes of low-sulphur coal remained in the thick Foord seam, which was as much as eight metres from top to bottom in the area targeted for mining. The main market for that cache of black gold was on Westray's doorstep—two coal-fired electrical generating plants a few kilometres away, in Trenton.

It was long-term, stable work at good pay in a region where a job—any job—was tough to find. A coal miner could make $60,000 a year, even without working overtime. When Westray's tunnelling crews reached coal in March 1991, and hiring began in earnest, Johnson was one of the first taken on. He even agreed to take a pay cut in order to work there—$12.90 per hour to start, compared with $15.00 per hour at the open pit. Johnson had his miner's papers, but started out working on the conveyor used to carry waste rock and coal to the surface. He had a lot to learn; digging coal by hand at Drummond was a far cry from running the complicated mining machinery used underground at Westray. The 65-tonne continuous miner—a low-slung mechanical dinosaur, its three-metre-wide drum bristling with steel teeth—could gouge out 15 tonnes of coal in half a minute. At Drummond, they had been lucky to dig 150 tonnes a day.

New mining technology was not the only surprise in store for Johnson. He had been at Westray a little over a month when B-crew narrowly escaped a major cave-in, the one that led to the showdown with mine manager Phillips. Nothing like it had ever happened at Drummond; Johnson could recall three, maybe four minor roof-falls in his six years there. But he had enough experience to know when the roof was about to give way. Crews were carving out the main slopes, two parallel tunnels, each more than a kilometre long and sloping downward to tap the coal seam at a depth of about two hundred metres. It was the night of May 23, 1991, and several miners were working in one of the main tunnels. Johnson was the first to spot the signs of an imminent collapse; the roof about five metres above his head was moving slightly ("creeping," miners call it), causing particles of rock to fall like a gentle rain. He warned the other men, and they retreated to the safety of the nearest crosscut, a passage connecting the main tunnels.

Glyn Jones, the supervisor in charge of the shift, showed up and ordered them into the tunnel and back to work. Johnson and the other miners refused, insisting that the roof was about to collapse.

"You're crazy," one miner protested. "We'll get killed down there."

"Get down there and work," Jones shot back.

As they argued, a thirty-metre section of the tunnel below the crosscut collapsed with a deafening roar. Everyone ran for their lives towards the surface, stopping only when it became obvious that the cave-in was over. Gingerly working their way back down the slope, they found the tunnel choked with rock, piled two metres high in places. The area where Johnson's crew had been working only minutes before was engulfed in debris; a boom truck—a flatbed vehicle used to haul timbers, roof supports, and other materials—was buried up to the level of the driver's cab.

Before leaving the mine site the next morning, the entire shift demanded a meeting with Phillips to discuss the incident. Westray workers, like an increasing number of miners across Canada, did not have a union to go to bat for them on safety issues; less than 30 per cent of the country's miners were unionized by the late 1980s. So the men had to stand up for themselves as best they could. They gathered in a conference room in the mine's administration building. The events leading up to the cave-in were described to Phillips, and Johnson expected Jones to get his knuckles rapped. But Jones and Phillips went back a long way—both were British-born, and they had worked together at other mines. Johnson listened in disbelief as Phillips turned his wrath not on Jones, but on the miners, calling them "a no-good bunch of bastards." It was during that tirade that Phillips declared the mine would succeed, with or without B-crew. To Johnson, the message was loud and clear: do what you're told or find another job. By the time word of the cave-in hit the press a couple of months later, Phillips wondered aloud what all the fuss was about. The collapse had been anticipated, he told one newspaper, and the miners simply resupported the roof and got back to work.

The cave-in, and Phillips's hard-nosed response, left Johnson rattled. For the next week, Donna Johnson noticed that her husband was quieter than usual. She thought little of it at the time—he usually spared her the details of what was happening at the mine. Even when he was still at Drummond, he remembered that she always got worried if he was late getting home after his shift, thinking the worst. But this time, Johnson was just as tight-lipped around his buddy Alex Ryan, who worked a different shift underground at Westray. Johnson and Ryan had laboured side by side at Drummond, and hunted and fished together every chance they got. After his shift, Johnson often dropped by Ryan's place, or phoned to tell him what was happening with B-crew. But it was a long time before he levelled with Ryan about the big cave-in and the tongue-lashing that B-crew had taken from Phillips. Truth be told, he admitted to Ryan, he was thinking about giving up the work he loved.

Soon after the cave-in, Johnson asked to be reassigned to surface work; there were plenty of safer jobs in the coal-washing plant or the supply warehouse. But management refused to allow the move, saying Westray could ill-afford to take a qualified miner like Johnson out of the pit. Johnson decided to stay put, but only for the moment. The mortgage on his house in Westville, a tidy bungalow trimmed with cedar, which he shared with Donna and his two school-age sons, would be paid off in August 1992. He would tough it out for another year. If working conditions and management's attitude did not improve by then, Johnson would leave Westray and take his chances on finding work elsewhere.

* * *

THE DRUMMOND COLLIERY, where Eugene Johnson got hooked on coal mining, was the last gasp of an industry that had been the mainstay of Pictou County's economy for more than a century and a half. By 1900, coal was king in Nova Scotia—twenty-nine collieries fuelled the locomotives, steam engines, and factories that were the key to prosperity in the industrial age. And Pictou County was blessed with one of the richest coalfields in the province, covering an area of more than one hundred square kilometres in the heart of the county. The field contained scores of coal seams, stacked one on top of the other and separated by bands of rock much like the layers of a cake. One seam, the Foord, was twelve metres thick in places, and was reputedly the thickest band of coal in the world. Pictou County's early settlers were quick to exploit the wealth that lay beneath their feet. The first collieries were opened in the early 1800s; by the turn of the century, a quarter of the county's population, some twelve thousand people, depended on a half-dozen coal mines for their livelihood.

While nature was generous in bestowing coal on the county, geology was not as kind. The Pictou field is fractured by faults—major shifts of ground that cut across the seams, the product of millions of years of geological stress and upheaval. To take the cake analogy one step further, it was as if the Pictou coalfield had been jammed into an undersized pan, breaking up the original pattern of the layers. This all added up to headaches for miners and mine engineers. Faults weakened the surrounding rock, making it tough to hold up the mine roof, and shifted the bands of coal at crazy angles, so that it became difficult to follow the seam.

Other problems were caused by the nature of the coal itself. Pictou coal is prone to spontaneous combustion; pockets of underground coal can smoulder and burst into flame when exposed to air during mining. Another hazard was methane, an invisible gas released naturally when coal is dug. If allowed to collect in the enclosed atmosphere of a mine, the gas can reach explosive concentrations. Methane is found in all coal mines, but Pictou County earned a reputation for having some of the gassiest collieries anywhere. The worst seam by far was the Foord; its thickness meant more coal was exposed during mining, releasing more methane.

It added up to a recipe for disaster. In roughly a century and a half of intensive mining, Pictou County's collieries were plagued by dozens of explosions and scores of underground fires. The quest for coal exacted a human price; by the 1950s, 246 miners, some of them boys as young as ten, had died in explosions. About 400 more people were killed in other mine accidents, bringing the death toll to roughly the number of men Pictou County lost in both world wars. Disasters in Pictou and other coalfields forced the Nova Scotia government to begin enacting safety laws in the mid-1800s; for example, miners were required for the first time to be certified. Each new explosion brought new standards for underground coal mining, but that offered little solace to the families of the men who paid with their lives for improved safety regulations.

The best way to ward off a methane explosion was to flush out the gas before it could reach explosive levels. As ventilation systems improved, fresh air could be circulated throughout a mine's workings, keeping methane levels within safe limits. But forcing air into mines also drove out the moisture, creating a new problem. Dust accumulated on the floor or tunnel walls as coal was dug and broken up for delivery to the surface. When picks and shovels gave way to modern machinery, more coal could be dug, and even more dust was produced. The coal dust was combustible and, if touched off by a methane fire or explosion, could turn a mine into a roaring inferno within seconds. Coal dust, methane, or a combination of the two were the cause of all Pictou County's mine explosions. To keep pace, safety laws were devised that required mine operators to water down the tunnels, or to spread copious amounts of powdered limestone to render the dust inert. The limestone, known to miners as stone dust, proved its worth, limiting the extent of explosions and saving scores of lives after it was introduced in the Pictou mines in the 1920s. Mining regulations in Nova Scotia and elsewhere now make this procedure a requirement.

Methane and coal dust could spell death for miners; for their bosses, an explosion could mean financial disaster. Mining Pictou County's troublesome seams was a risky and expensive proposition, making it tough to compete with larger collieries in Springhill and on Cape Breton Island, Nova Scotia's other coal-producing areas. When oil and gas dethroned coal as the king of fuels after World War II, no coal-mining job was safe anywhere in Nova Scotia. Pictou's money-losing mines were the first victims of the new energy order. A way of life disappeared: men who had followed their fathers and grandfathers into the pits scrambled to find other work in an area beset by high unemployment. Their sons, more often than not, packed their bags and headed west in search of jobs. By the 1970s, the Drummond mine, a throwback to the nineteenth century—horses were still being used to haul coal—was the only colliery still in operation. But by 1984, when the mine closed, the groundwork was being laid for a new project that would put Pictou County back in the coal business.

* * *

IN THE TEN MONTHS he had worked at Westray, Mike Wrice had never seen the workings as clean and shipshape as they were on September 11, 1991, the day of the grand opening. Housekeeping was usually the last thing on the bosses' minds; it was always push forward, get to the coal, and worry about the tidying-up later. "A goddamn pigpen" was how Alex Ryan would later describe the place. Wrice assumed the mine was being decked out in its Sunday best, above and below ground, for the dignitaries and the bigshots from Curragh's head office who would be on hand for the ceremonies. For days, men had been taken off their regular duties and told to haul out scraps of pipe, torn sections of ventilation tubing, discarded oil cans, and all the other garbage strewn about the tunnels. Even the continuous miners and other machines displayed on the surface had been given a fresh coat of paint. Unknown to Wrice, there was a second reason for the housecleaning effort: officials from the Nova Scotia Department of Labour had inspected the mine a week earlier and told mine manager Gerald Phillips to clean up the mess.

Wrice, a Cape Bretoner with a decade of hard-rock mining under his belt, normally worked with the trainees, showing them how to install roof supports and operate tractors and loaders. But in preparation for the opening, he spent a shift dropping off pallets stacked with bags of stone dust at spots throughout the mine. Stone dust—or rock dust, as it's sometimes called—is spread over the roof, walls, and floors to suppress the fine black coal dust. Under Nova Scotia's Coal Mines Regulation Act, at least twenty bags of stone dust must be stored in every working section of a coal mine; the act stipulates that after stone dust is spread, the dust in all areas of the mine must contain less than 35-per-cent combustible coal. Westray had been tunnelling through the Foord seam for months, digging the roadways that would be needed to put the mine into production, but it was the first time Wrice had seen stone dust in the mine.

Wrice was no expert on the combustibility of coal dust, and he was in no position to tell Westray how to run a coal mine. Although he grew up in the coal-mining town of Glace Bay, he didn't set foot in a mine until he left Nova Scotia at the age of twenty-five to find work. Wrice's experience was limited to hard-rock mining in Northern Ontario—gold and copper in Timmins, uranium in Elliot Lake. But the experienced miners on his shift, men who had been in the coal business for years, said again and again that things were not being done right at Westray. Ray Savidge, an underground surveyor with decades of mining experience in Britain, did not mince words. "Everything I was ever taught in the mines from day one," he told Wrice, "they're going against here."

Such talk only added to Wrice's uneasiness. His specialty was development work—driving access tunnels and shoring up the roof for the production crews. That's how he ended up at Westray in the fall of 1990, working for Canadian Mine Development, the contractor hired to dig the mine's main tunnels. He was laid off about six months later, when Westray took over the final phases of tunnel development. Wrice snagged a job at the Westminer Canada lead-zinc operation in Gays River, about fifty kilometres northwest of Halifax. But within weeks, flooding shut down that mine, and Wrice was again out of work. He got back on at Westray in June 1991, but he never felt good about the place; Westray had been implementing its own ideas about how to support the mine roof. Canadian Mine Development had driven the first nine hundred metres of the main tunnels, securing the roof with wire screening, which was bolted into place with metre-long metal rods drilled into the rock. Then the entire surface of the roof was coated with shotcrete—cement sprayed on with a special machine to seal the roof and prevent moisture from weakening the rock. When Westray took over tunnel development in March 1991, its crews skipped the shotcreting step and left the roof exposed. The rock above the Foord seam was shale—flaky and particularly susceptible to the elements.

It had come as no surprise to Wrice when a large, unsealed section of the main tunnel collapsed in late May, narrowly missing Eugene Johnson and his co-workers. Wrice was at Gays River at the time, but the cave-in was no secret. Moreover, it became a political football in July 1991, when the incident was brought up on the floor of the Nova Scotia legislature in Halifax. This was more ammunition for Bernie Boudreau, a Liberal politician from Cape Breton who had fought the Westray development from the outset. Boudreau caught the government flat-footed—Labour Minister Leroy Legere, whose department was responsible for mine inspection, was forced to admit he knew nothing about the roof-fall. To deflect criticism, Legere accused the opposition Liberals of playing politics. "They are not concerned about the safety of the miners," he contended. "As long as they can find blame, they are satisfied." The following day, Legere announced that Department of Labour inspectors had investigated the collapse, and Westray had retained engineering consultants to find ways of tackling the roof problem.

There had been several cave-ins in the three months since Wrice's return to Westray in June; none had happened on his shift, but the roof conditions were wearing on his nerves. He was already thinking of quitting when he went underground with the day shift at eight o'clock on the morning of September 11, relieving B-crew. Just before noon, word came down for the men to stop work for the day and join in the festivities. Wrice was collared for one last duty—he was told to take a group of about six visitors on an underground tour. One of the main tunnels was used to get men and equipment in and out of the mine; the other contained the conveyor line for hauling coal. Wrice drove them down the travel slope on one of the converted farm tractors Westray used to transport miners to and from the workings. It was like driving into a highway tunnel under a river—only this one kept descending, for more than a kilometre, to a point about 350 metres below the surface. They stopped at the first area to be mined, the southwest section. The visitors were full of questions, but Wrice was in no mood for talking up Westray. He told them to tag along with John Bates, a mine foreman in his mid-fifties who, like many of the men calling the shots at Westray, was a Brit who had come to Nova Scotia from the coal mines of Alberta. The group walked a short distance into the southwest section, where the coal seam was being cut into a grid of interconnected tunnels. As soon as they returned to the surface, Wrice headed off for a shower and a free lunch. The rain was still streaming down, but it felt good to be back above ground.

* * *

WESTRAY OWED its existence to the whims of the global energy market, the prodding of mining promoters, and the power of politicians. After years of mine closings and layoffs, Nova Scotia's coal industry staged a remarkable comeback in the 1970s. When the Arab embargo put oil prices through the roof, Nova Scotia's coal reserves became a cheap and secure source of energy. The province's Crown-owned electrical utility, the Nova Scotia Power Corporation, began converting its oil-fired generating plants to burn coal. The beneficiary of the burgeoning local market was the Cape Breton Development Corporation, better known as Devco, which operated the few collieries still in operation on Cape Breton Island. Devco was also a Crown corporation, this one set up by the federal government in the late 1960s, when private industry pulled out of Nova Scotia's ailing coal business. Devco was supposed to phase out the collieries and create new industries for displaced miners; thanks to the global energy crisis, it was back in the coal business.

So was Pictou County. In the early 1980s, a major Canadian oil and gas producer, Suncor, staked out an unmined block of the Foord seam near Stellarton. By

1985, Suncor was drawing up the plans for a colliery, a multimillion-dollar project that promised to bring hundreds of jobs to an area where they were sorely needed. The Nova Scotia government did its part, dangling the carrot of a long-term contract to supply a provincial generating station in nearby Trenton. But that plant was already burning Devco coal, and Cape Bretoners viewed the Suncor proposal as a threat to their mining jobs. For two regions hungry for investment and work, the Pictou mine was a test of political muscle, pitting solidly Conservative Pictou County against mainly Liberal Cape Breton.

Suncor's plans sounded like an attempt to reinvent the wheel. Despite Pictou County's sorry record of disasters and explosions, engineers and consultants felt the Foord seam could be mined safely. The seam's reputation for gassiness and spontaneous combustion was discounted; laboratory tests suggested that the coal had a low methane content, easily handled by modern ventilation systems. A lot of money was spent mapping the fault lines that crossed the seam, so the tunnels would avoid areas vulnerable to cave-ins. To manage the project, Suncor hired a young, ambitious British expatriate—Gerald Phillips. Phillips came on board with years of coal experience, and some unwelcome baggage. During a stint as underground manager at an Alberta colliery, he raised the ire of union officials with his cavalier attitude towards their safety concerns. Phillips left the mine in 1980, a matter of weeks before a major cave-in killed four workers. His name surfaced a few times at a subsequent public inquiry into the deaths, but his reputation emerged unscathed.

From Suncor's point of view, all the pieces were falling into place. But then, the vagaries of the market intervened. By 1986, oil prices were on the way down; coal was losing its appeal, and petroleum producers like Suncor were losing money. Suncor backed off, and tried to entice another firm, Placer Development of Vancouver, to share the cost of developing the mine. Placer commissioned a new set of feasibility studies but abandoned the project in mid-1987. Unwilling to tackle the project alone, Suncor then put its Pictou County coal rights and mining plans on the auction block. The Nova Scotia government was desperate to find a new developer; the provincial power authority was building a second power plant at Trenton, providing a captive market for Pictou County coal. The environment, the hottest political issue of the day, was also a consideration. Burning Cape Breton coal produced sulphur emissions, which in turn produced the acid rain that was fouling lakes and rivers. Coal from the Foord seam had a lower sulphur content, offering a cheap way to reduce acid-rain emissions.

The jobs and other benefits of mining Pictou coal were not lost on the politicians. One of the mine's biggest boosters was Donald Cameron, then a backbencher in the Conservative government of Nova Scotia Premier John Buchanan.

Cameron, whose Pictou East riding bordered the site of the proposed mine, was a self-styled reformer who had waged a lonely campaign against political patronage. He rose to the post of industry minister and replaced Buchanan as premier in 1991, all the while promoting the economic and environmental benefits of his native county's coal. Cameron's ally in Ottawa was Elmer MacKay, the member of Parliament for Central Nova and a federal cabinet minister. MacKay, a steadfast supporter of Brian Mulroney, had temporarily relinquished his seat to give the Tory leader a toehold in the House of Commons. Cameron and MacKay may have been the most vocal proponents of the mine, but they were not alone. Municipal politicians, business leaders, the local papers, people longing for stable jobs—they all jumped on the bandwagon. After watching from the sidelines as the federal government poured hundreds of millions of dollars into Cape Breton's mines through Devco, Pictou County was poised to cash in on the coal industry's revival.

There was just one hitch: Nova Scotia had to entice a mining firm to pick up where Suncor and Placer had left off. The province's salvation came in the form of Clifford Frame, a Toronto mining promoter whose political connections were rivalled only by his ambition to turn his company, Curragh Resources, into a major player on the international mining scene. Frame arrived in Nova Scotia with a chequered past. He cut his teeth at Denison Mines, a firm that grew fat on government uranium contracts but was stingy about miners' safety, particularly when it came to ventilation. Frame was Denison's golden boy until he was tapped to develop the massive government-backed Quintette coal project in British Columbia in the early 1980s. Quintette turned out to be a financial disaster, plagued by technical glitches and high production costs. Denison and the federal and B.C. governments lost millions; Frame took the fall and was fired.

The setback was temporary. In 1985, Frame took over an abandoned lead-zinc mine in the Yukon, gathered up as much government money as he could, and cashed in when metal prices began to rise. The Faro mine became a cash cow, reaping huge profits and turning his new company, Curragh, into one of Canada's leading exporters. By 1987, when Frame got wind of a coal property up for sale in Nova Scotia, the Quintette fiasco was a fading memory and the upstart Curragh looked like the answer to Pictou County's dreams. Robert Coates, a veteran Tory MP who was a friend of both Frame and Buchanan, did the introductions. Frame snapped up the Suncor rights, set up a subsidiary called Westray Coal, and waded back into the coal business.

Frame was giving Nova Scotia's Tories their mine, but on his terms. And that meant a healthy investment from the public purse. The provincial government made an offer to help put Westray's financing in place: a $12-million loan and a fifteen-year contract to supply the Nova Scotia Power Corporation with 700,000 tonnes per year, and a promise to buy another 275,000 tonnes of the coal that Westray could not sell each year. But Frame wanted more. He asked the federal government for money to ease interest payments on his bank loan, plus a loan guarantee that would put Ottawa on the hook if the project failed. That request hit a brick wall within the federal bureaucracy; officials argued that the project was too profitable, and warned it would take a bite out of Devco's coal sales. Questions were also raised about Westray's engineering plans and the quality of Foord-seam coal. In the headlong rush to get Westray up and running, those questions were left unanswered.

The impasse continued for almost two years, stalling construction and putting the project far behind schedule. Cameron and MacKay backed Westray to the hilt, accusing Devco sympathizers within the federal civil service of trying to scuttle the new mine. The battle was waged all the way to the Prime Minister's Office, and some of Brian Mulroney's closest aides were enlisted to negotiate a deal the politicians and the bureaucrats could live with. By the fall of 1990, the deal was done. Ottawa guaranteed the bulk of the $100-million bank loan needed to finance construction, and threw in $8.7 million to reduce interest payments on the loan. The package was less than Frame had hoped for, but it still gave him the coal mine he wanted—with minimal risk to the Curragh treasury.

But the long funding wrangle also sealed Westray's fate. Construction began in earnest in the fall of 1990, less than a year before the mine was slated to begin shipping coal to Trenton 6, the new power plant NSPC was building nearby. The task of ramming Westray into production fell to Gerald Phillips, who was hired away from Suncor to see the project through. To make up for lost time, Phillips scrapped the original mine plan and diverted the main tunnels. The gambit enabled tunnelling crews to tap the Foord seam earlier, but created new headaches. The revised route was crisscrossed by faults, leaving the roof unstable and increasing the risk of cave-ins. To make matters worse, the first coal found was of poor quality, high in sulphur and "ash"—the impurities left behind when coal is burned.

Westray was a project driven by political and business deadlines. The safety of the men working in the mine took a back seat to the push for production. The task of enforcing safety laws at Westray was in the hands of the Nova Scotia Department of Labour, which was ill-equipped to monitor a modern coal mine. By the mid-1980s, Devco operated two of the province's three collieries, and those mines fell under the jurisdiction of federal inspectors. Nova Scotia's labour department was understaffed, and its mine inspectors were armed with a set of safety rules—the Coal Mines Regulation Act—that was thirty years out of date. Westray was having its own expertise problems; miners with little or no experience in coal mining were pressed into service to get the project up to speed. It was a risky approach to mining a seam with a deadly history of explosion and fire.

As the politicians and promoters descended on Plymouth to christen the Westray mine in September 1991, trouble was brewing underground. Coal production was lagging far behind schedule; a measly 250 tonnes had been delivered to NSPC by the last week of August, and the new Trenton power plant was slated to go into service in October. There was no way Westray could reach its full production level—nearly 3,000 tonnes of coal per day—in time. There had been two major cave-ins in the previous four months, and both the company and the Department of Labour were trying to figure out a better way to stabilize the roof. Bigger safety problems loomed ahead; as Westray reached deeper and deeper into the Foord seam, it would have to deal with methane and dust, the twin menaces of all coal mines. Time would tell if the consultants had been right when they downplayed the gas hazard. Time would tell if the company was prepared to keep the coal dust at bay.

* * *

DOWN AT WESTRAY'S main gate, off East River Road, the mine was under siege on its official opening day. About one hundred people clutching placards and umbrellas had braved the downpour and were marking the grand opening with a picket line. As a half-dozen Royal Canadian Mounted Police officers looked on, the protesters slowed traffic headed onto the mine property. "Hey, hey, ho, ho," they chanted, "the Tories must go." There were rumours that the prime minister himself, Brian Mulroney, was coming to the mine's opening, and the demonstrators, mainly federal civil servants who were in the second day of a nationwide strike, had descended on Westray to vent their anger. The people of Pictou County had the dubious distinction of giving Mulroney his start in federal politics, electing him as Central Nova's MP after Elmer MacKay graciously stepped aside in 1983. (MacKay got his seat back in 1984, when Mulroney moved on to claim the Manicouagan riding in his native Quebec.) Mulroney had led the Conservatives to power a year later, promising "jobs, jobs, jobs." But now, after seven years of Tory rule, unemployment remained high and labour was feeling the pinch of layoffs and wage freezes.

The ranks of the strikers were bolstered by other unionized workers from the local area. The heavily subsidized mine was to provide more than 250 badly needed jobs once it reached full production, but employees of Pictou County firms hit hard by the recession wanted to make sure they were not overlooked in the hoopla surrounding the official opening. Employees at Trenton Works Lavalin wanted a financial package from the Nova Scotia government to assist the railcar plant; shipyard workers at Pictou Industries were looking for a federal contract to build a car ferry. Jim Cameron, president of the New Glasgow and District Labour Council, summed up the local mood to a reporter: "There's more people who think Elvis is alive than like Mulroney."

But Mulroney was not on the guest list. The demonstrators had to be content with harassing two other powerful Tories who had been instrumental in the mine's development. Nova Scotia Premier Donald Cameron and federal Public Works Minister Elmer MacKay, who were eager to bask in the glow of Westray's success, ran the gauntlet at the mine gate. For the strikers trying to rain on Westray's parade, a federal cabinet minister was a good substitute for the no-show Mulroney. When MacKay drove up to the gate in his Dodge pick-up, he faced a mob of chanting, placard-waving protesters. MacKay glared at them coldly, then drove on.

The more than four hundred people on hand for the opening ceremonies were only slightly more comfortable than those picketing in the rain. It was cold and damp inside the tents; water collected in puddles at the sides, then seeped in under the walls, turning the ground to ankle-deep mud in places. Tables covered in white cloth stretched the length of one tent; the buffet was set up in the other tent, a few metres away. When lunch was served, about noon, people had to duck outside in the heavy rain to collect food and drinks. The crowd included mayors and councillors from the neighbouring towns, and executives of the Nova Scotia Power Corporation. Some Devco officials were on hand, even though relations between the rival coal-producers were far from cordial.

There was no head table. Cameron and MacKay sat at the end of one of the long dining tables, just in front of the stage where they would deliver their speeches. Clifford Frame, chairman and chief executive officer of Curragh Resources, sat off to one side with his back to the wall of the tent. Seated on his right, looking out of place at what amounted to a Tory love-in, was John Turner, the former Liberal prime minister. Turner, back in law practice in Toronto after being drubbed by Mulroney in two federal elections, was a member of Curragh's board of directors and chairman of the audit committee.

Frame, the beefy fifty-eight-year-old promoter whose ability to attract government money had made this day possible, had the privilege of speaking first. He was introduced as having been twice named Canada's mining man of the year. The honour, bestowed in 1982 and 1987 by the industry's weekly newspaper, the *Northern Miner*, had been in recognition of his leading role in two previous projects—Quintette and the revitalized Faro mine. Like Westray, both projects had relied heavily on public money. While Quintette had proved to be an embarrassment for all concerned, Faro remained a success story. With Westray, Frame was looking to go two-for-three at the plate. And maybe, just maybe, he could get the Quintette monkey off his back for good.

Dressed in a dark blue suit with a fat, expensive Monte Cristo—his favourite cigar—tucked in the breast pocket, Frame stepped up to the podium. To mark the occasion, he sported a wool necktie in the blue, green, and yellow Nova Scotia tartan. He began with an explanation of how Curragh, a miner of base metals in Western Canada and one of the world's largest producers of lead and zinc, had come to develop an underground coal mine at the other end of the country. The answer, he boasted, was having the right political connections. In the late 1980s, Robert Coates had, as he put it, "dragged" him off to a meeting with then-premier Buchanan, who was looking for a firm willing to mine Pictou County coal.

Any reluctance Frame may have displayed at the outset evaporated as soon as the politicians opened the public purse. The new colliery, Frame now declared, was "great for Curragh, great for Pictou County, great for Nova Scotia, great for Canada." Frame had high hopes for his new mine; it would end Curragh's dependence on lead and zinc, the main products of its Faro mine. Base-metal prices on the world market had fallen by as much as one-third during 1991, leaving Curragh splattered with red ink. After turning a $32-million profit in 1990, when prices were still respectable, the company was headed for the first operating loss in its six-year existence, and it was a hefty one—$98.3 million. Westray, with the market and the price for its coal guaranteed into the next century, offered welcome relief from the boom-and-bust cycle of base metals.

Thanks to Frame's astute deal-making, there was little doubt (around those luncheon tables, at least) that Westray would be great for Curragh. The company stood to make a bundle. NSPC would pay at least $60 per tonne for Westray coal, and an additional $14 per tonne for premium-grade coal, higher in heat content and lower in impurities. Even if Westray only supplied lower-grade coal, selling 700,000 tonnes per year to the utility would bring in $42 million. The annual revenue would be closer to $50 million if Westray reached its target of shipping more than 50 per cent of its output as premium coal. The firm's business plan projected operating costs of about $35 million per year, leaving a surplus in the range of $7 million to $15 million. The deal was a whole lot sweeter, thanks to the generosity of the Nova Scotia government. Curragh and the province had struck a "take-or-pay" deal, committing the Nova Scotia government to buy up to 275,000 tonnes of unsold Westray coal every year for fifteen years, regardless of whether it was brought to the surface. The government agreed to pay the higher, premium price, on that coal, assuring Westray a windfall from the provincial treasury of up to $20 million per year if buyers failed to materialize. The only catch was that Westray had to repay all the money paid out under the "take or pay," without interest, at the end of the fifteen years.

Frame spared his audience the mundane details of the Westray deal, choosing instead to lavish praise on others. Westray had been developed, he said, through the efforts of the "top-notch people on his team." One was Marvin Pelley, Curragh's president for corporate development and projects. Pelley, who had worked with Frame at Quintette, had helped hammer out the aid packages with the politicians. Gerald Phillips, vice president and general manager of Westray Coal, was singled out for special praise. Despite the long wait for money from Ottawa, the mine had been brought into production $11.5 million under budget. Frame dubbed Phillips "the miracle man" in recognition of his herculean effort. That was stretching it, to say the least. Westray was months behind its production targets, and the struggle to prop up the roof was pushing up operating costs. Frame ignored the problems in his speech, which ended with a tip of the hat to "the one individual without whose help the mine wouldn't have been possible—the Honourable Donald Cameron."

That was the premier's cue. Even though he had won the long battle to bring Westray to Pictou County, Cameron had some old scores to settle with opposition politicians. "This is a project that is a symbol of what happens far too often in our province, in our country, when politics gets in the way of good things happening," he said. Westray, "a truly good project for the people," had been brought to fruition despite "silly hurdles." Then the premier launched into his oft-repeated explanation of why the mine made economic and environmental sense. Under a federal-provincial agreement to reduce acid rain, the Nova Scotia Power Corporation was required to cut sulphur emissions from its coal-burning generating stations. Burning Westray's low-sulphur coal at the province's Trenton 6 power plant, slated to start up within weeks, would cut emissions by 17,000 tonnes alone. The alternative was building a $200-million scrubbing plant to clean emissions from Trenton 6, which would have cost another $16 million per year to operate. "And what did this cost the taxpayers of Nova Scotia?" asked the silver-haired Cameron, deeply tanned after his first summer as premier. "A loan of $12 million with standard interest rates. And there is an indication that loan will be repaid ahead of time."

As if following a script, Phillips took the floor and picked up Cameron's train of thought. Westray, he confirmed, was not getting a free ride from Nova Scotia taxpayers; at 11.75 per cent, the interest rate on the provincial government's loan represented the highest level of repayment of all the assistance packages for the $124-million mine. Given the heady praise he had just heard from the boss, Phillips would have been forgiven if he had chosen to brag a bit. Instead, he lavished praise on his employees and on Westray's contractors. "At times we felt we couldn't do it, but we managed to pull it together." How well Phillips and his cohorts had pulled the project together, and at what cost for the mine's future viability, remained to be seen. Westray, Phillips promised, would be a good corporate citizen; the company sponsored little-league baseball and gave generously to Plymouth's community centre. "We want the operation to blend into this lovely community," he said. After less than a minute at the podium, he called on the final speaker, MacKay, to take over.

"This is Cliff Frame's day and Curragh's day," declared MacKay. But like Cameron, he was not about to let anyone forget that politicians, Tory politicians, had been instrumental in bringing Westray to Pictou County. "There were complications and there were closures because of bureaucratic problems, to which Donald Cameron has alluded," MacKay said. "Maybe this is inevitable. I don't think it is. I think that it was caused to a large extent by bureaucratic inertia, resisting an idea that came from the top down instead of the bottom up." The well-timed announcement of the Westray project had helped get the Tories through the Nova Scotia and federal elections in the fall of 1988; now, three years later, with the Tories low in the polls and new elections on the horizon, Westray was still good politics. Mindful of the protesters at the mine gate, MacKay put in a plug for his boss, Brian Mulroney. "Without his support," he said, "it's doubtful the bureaucratic inertia could have been overcome."

"This is Cliff Frame's day," federal cabinet minister Elmer MacKay declared, as the Westray mine was christened on September 11, 1991. The Curragh chairman, cigar and hard hat in hand, was among the dignitaries on hand for the opening ceremonies. [Photo by Steve Harder, courtesy of the *Chronicle-Herald* and the *Mail-Star*]

There was only one more item on the agenda. Frame, his blue eyes aglow with pride, broke into a broad smile as he stood on the stage, a pair of scissors at the ready. Cameron and Phillips held one end of a long ribbon, MacKay and Pelley the other. After a pause to allow photographers to record the historic moment, Frame delivered the ceremonial snip. Canada's newest coal mine was officially in business.

* * *

MIKE WRICE helped himself to the buffet, but it was the music that really interested him. The Rankin Family, an up-and-coming band from Cape Breton, was booked for the opening. Wrice had never heard them play be-

Declaring the mine "great for Canada," Clifford Frame cut the ribbon to officially open Westray. He was flanked, from left, by mine manager Gerald Phillips, Premier Donald Cameron, federal Public Works Minister Elmer MacKay, and Curragh executive Marvin Pelley. [Photo by Dave Glennen, New Glasgow *Evening News*]

fore. He endured the speeches and the ribbon-cutting, nearly choking when Frame called Phillips "the miracle man." Then, thankfully, the Rankins took over, filling the main tent with haunting Gaelic harmonies and the cry of the fiddle.

The rain outside was letting up as Wrice and another miner listened to the band. After a while John Turner walked over and, to Wrice's surprise, the former prime minister struck up a conversation. "It's quite the operation," he said of Westray. "It's really something." Then he mentioned that his grandfather had been a coal miner in Pictou County, but had ended up working in collieries in British Columbia.

"I find it so hard to believe that he worked in areas where you had to crawl on your stomach," Turner said.

Wrice was too fed up with Westray and worrying about cave-ins to hold his tongue, even to a Curragh director. Besides, he would be gone within a month—he was quitting. "Yeah," Wrice replied, without missing a beat, "the only difference now is that the ground's got a little farther to come down before it hits you." Then he let out a chuckle and walked away.

Amid all the back-slapping and the rhetoric, despite all the predictions and promises made that day about Westray's bright future, it was the only time anyone had said anything about whether the mine was safe.

"EVE OF DESTRUCTION"

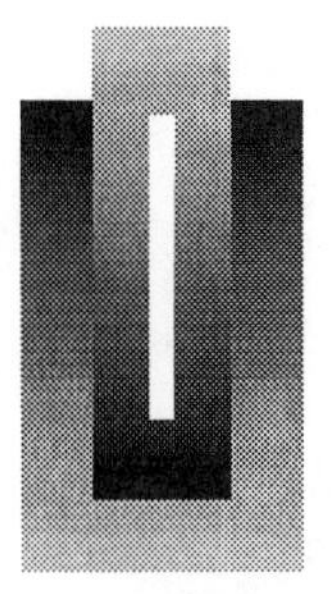

IN THE DREAM, the dreaded call always came in the night, waking him from a deep sleep. Then, before he could reach for the receiver, the telephone exploded in a flash of light.

Carl Guptill was not the kind to put much stock in dreams. Heavy-set, with a gold stud in one earlobe and shoulder-length brown hair spilling out from under a ball cap, he looked more like a castoff from a heavy-metal band than a man about to turn forty. Guptill was a doer, not a dreamer; over the years he had operated fishing boats in New Brunswick and filled in as lighthouse keeper on Nova Scotia's east coast. He had worked in the mines for the past five years, and that was why he was having recurring nightmares about exploding phones. Any day now, he was certain, the call would be real. And the voice on the other end of the line would announce that his worst nightmare had come true—that the mine had blown up.

Guptill had barely lasted two months at Westray. He had been hired in mid-November 1991, six months after being laid off from Westminer Canada's mine in Gays River. Until then, he had only worked in hard-rock mines. Guptill loved the challenge and the excitement of working underground; he was the first to admit that, in a sick sort of way, he thrived on the danger. It was a rush. But that did not mean he was cavalier about his work. He had put a lot of extra time and effort into learning how to mine safely, and had his mine rescue certification. Safety, he thought, was the way of the future in mining, a sure ticket to getting ahead. By the time Gays River closed down, Guptill was captain of a mine rescue team and chairman of the safety committee, and he held the management-level post of shift supervisor.

Guptill stressed safety during his interview with Roger Parry, the underground manager, saying he considered it the number-one priority in mining. For most Westray miners, Parry, a gruff, overweight Brit who was pushing fifty, was the bane of their existence. He was always turning up underground, chewing out the men for not working fast enough or not doing things his way, even when his way defied common sense. Parry's style was to couch his orders in a torrent of profan-

Roger Parry, a tough-talking Yorkshire native, was Westray's underground manager. Most miners there considered him the bane of their existence. [Photo by Darren Pittman, courtesy of the *Chronicle-Herald* and the *Mail-Star*]

ity—"dog-fuckers" was one of his more-original terms for the men under his command. Other Parry gems were lost in his thick Yorkshire accent; many of the miners could barely make out a word he said. They got even by christening him "the Pig." But in an interview with a new recruit, Parry was on his best behaviour. Westray, he assured Guptill, also considered safety to be a priority. It was a perfunctory interview, five minutes at most, but Guptill got the impression that his safety training had clinched the job.

The rush Guptill got from working underground soon turned to fear. Training was almost non-existent; he was shown a few videos, mostly promotional flicks extolling the virtues of Curragh Resources. Trainees were supposed to work with certified miners for the first six months but, within days, Guptill and other new recruits were left on their own to figure out how to put up steel I-beams to support the roof. On his fifth shift, the power went out underground for almost five hours, shutting down the ventilation fans. The continuous miners were electrically powered, so mining came to a stop. Guptill and four other miners were operating a scooptram, a diesel-powered front-end loader, and using the bucket as a platform to reach the roof. They were told to keep the machine running, even though it was using up oxygen and pumping out poisonous carbon monoxide. By the time the power came back on and ventilation was restored, the men were groggy and disoriented from breathing the foul air.

After that, Guptill refused to work in conditions he considered unsafe, either because of poor ventilation or lack of roof support. He was relegated to carrying supplies to the men working on the roof or at the coalface, the exposed seam of coal in a mine. Guptill considered this "dirt work," and he felt the company was trying to force him to quit. As luck would have it, his thirteenth shift underground in early December was his last. The light on Guptill's mining helmet was out, but the shift boss, Angus MacNeil, ordered him to keep working. Guptill was trying to move a heavy steel arch when he tripped in the darkness. The steel beam flipped and landed on his back, rupturing a disc. Guptill spent three days in hospital, then called the Nova Scotia Department of Labour in Halifax.

Claude White, the province's director of mine safety, arranged a meeting at the provincial Department of Mines building in Stellarton. In mid-December 1991, Guptill sat down with Albert McLean, a veteran mine inspector based in Glace Bay; John V. Smith, responsible for the inspection of electrical and mechanical equipment used in mines; and Fred Doucette, who was in charge of mine rescue for the province. Guptill launched into a litany of complaints about lax safety practices at Westray. He told them about working without proper ventilation, lack of training, use of acetylene torches underground, the company's failure to report cave-ins, working twelve-hour shifts without a lunch break, the coating of coal dust building up throughout the mine, and men being forced to work in high levels of methane.

The dust and methane were hazards unique to coal mining; although his background was in hard-rock mining, Guptill knew from his rescue training that they were dangerous. Methane is the biggest safety threat in a coal mine: the colourless, odourless gas is explosive in concentrations of 5 per cent to 15 per cent in the atmosphere of a mine. And the Foord seam, with its thickly exposed coal, was especially prone to the dangers of methane. Nova Scotia's Coal Mines Regulation Act requires that electricity be shut off underground when methane reaches a level of 1.25 per cent, and that mining be brought to a halt; workers must be withdrawn immediately from any part of a mine where the gas level exceeds 2.5 per cent.

Westray relied on several types of sensors to detect methane. The continuous miners were equipped with methanometers designed to shut down the machine once the gas reached 1.25 per cent. Underground bosses used hand-held detectors that gave methane readings up to 5 per cent—the explosive threshold. Finally, remote sensors were installed in the mine to continuously monitor gas levels and display the results on a computer screen on the surface. The computer system was calibrated to trigger a signal at a methane content of 0.25 per cent, well within government guidelines.

Even without detection equipment, Westray miners knew when there was too much gas. Dizziness, a numbness in the tongue, shaky knees—at one time or another, most of them had felt the symptoms of working in high methane. Methane is lighter than oxygen, and collects at the mine roof; several miners passed out as they worked near the roof, putting up arches and timbers. But work was never called off, and the mine was never cleared. If the men complained and the foreman's hand-held detector showed the methane was too high, they were told to keep working while the area was ventilated. If the methanometer on the continuous miner was broken, the machine was kept in service until it was repaired; sometimes the detectors were altered so that the machine would continue to operate in an atmosphere containing methane concentrations as high as 1.5 per cent.

The Department of Labour officials wrote down all of Guptill's allegations. He was the first Westray employee to come forward and lodge a formal complaint of unsafe conditions, they said, even though the department's own inspection reports had noted some of the problems he outlined. Guptill was left with the impression that the mine would be closed down, maybe even that same day. But weeks passed, and he heard nothing. Finally, McLean called and told Guptill to meet him again, this time at a room in the Heather Hotel in Stellarton. The meeting was cloak-and-dagger stuff. There was a television set in the room, and it was blaring so loudly that Guptill could barely make out what McLean was saying. The department was not prepared to act on the complaints, McLean said, but he offered to put in a good word for Guptill, maybe even get his job back. "The bottom line is," Guptill recalls McLean saying, "if you don't go back there and go underground, you're going to be walking the street. You take the choice." Guptill could barely believe what he was hearing; he even turned down the TV at one point, but McLean jacked up the volume again, so no-one could overhear the conversation. Guptill knew McLean from his days at the Gays River mine, and when he looked him in the eye, he saw a defeated man who was trying to justify what he had been told to say. Guptill was stunned; he knew he had lost, and that it would only be a matter of time before he was fired. When his separation papers did come through, Westray branded him "unsuitable for underground mining."

That was back in January 1992. It was already mid-April; the spring weather had been miserable, so cold and wet that the yellowed lawns and fields were just beginning to take on a tinge of green. Four months gone, and it would be many more before Guptill saw a penny of workers compensation for his injury. He figured his mining days were over, and that he would be blacklisted for blowing the whistle to the Labour Department. But it made little difference; his back was so bad that he could barely lift twenty pounds, let alone work underground. Guptill still kept in touch with the men at the mine, so it was no surprise when Roy Feltmate called early one Saturday morning in April to invite Guptill and his girlfriend over. Nothing special—B-crew had just gotten off shift, and some of the guys had gathered at Feltmate's house in Stellarton for a few drinks.

Guptill and Feltmate had been good friends for years. A big man with dark hair and a thick moustache, Feltmate was as comfortable in a mine as he was driving stock cars at the local speedway. He was thirty-three, almost seven years younger than Guptill, but he had more mining experience. The two men had worked together at Gays River and, before that, at the Forest Hill gold mine in Guysborough County, about an hour's drive east of Stellarton. Feltmate was the reason Guptill ended up at Westray after Gays River closed down. Feltmate, with two kids to support, was eager to get another job; Guptill, with no dependents and a girlfriend who worked, just went along for the drive when Feltmate applied at Westray, but decided to put in an application anyway. Feltmate was hired in November 1991, Guptill a few weeks later. Guptill worked with B-crew during his stint underground; Feltmate was on another shift, but switched to B-crew not long after Guptill was fired.

So Guptill needed no introduction to the men gathered around Feltmate's kitchen table: Randy House, one of the handful of Newfoundlanders at Westray; Robbie Fraser, a twenty-nine-year-old mechanic from Westville; and Mike MacKay, another Pictou County boy. Guptill had worked with them all, but he knew MacKay best. Thirty-eight years old and the father of two young daughters, MacKay had a big smile and a heart to match; he was the kind of person who would meet strangers on the street and end up inviting them home for dinner. He had a passion for motorcycles and dearly wanted to get his hands on a Harley-Davidson. Guptill owned a Harley that he stored in Feltmate's shed; when MacKay saw it, and felt the surge of power as he revved the engine, he was hooked. MacKay hoped to be able to afford a Harley by the summer, when he, Guptill, and Feltmate planned to head over to Newfoundland for four or five days.

But Guptill had not been invited over to discuss Harleys or vacation plans. The mood was tense, and the talk around the table was about how dangerous things were in the mine. A couple of weeks earlier, B-crew workers had returned from their days off to find that a good chunk of the area then being mined, the Southwest One section, had been sealed off and abandoned because of cave-ins. Then there was the methane; it was so bad on one shift that Feltmate and another miner blacked out. A union drive was under way, and there was debate around the table about whether putting in a union would make a difference. The United Mine Workers of America had tried in the fall of 1991 but could not recruit enough workers; many harboured a distrust of the UMW because it represented Devco miners, and Cape Bretoners wanted nothing better than to see Westray shut down for good. This time, the United Steelworkers of America was taking a run at unionizing Westray. The largest mining union in Canada, the Steelworkers had already organized Curragh's Faro miners–and a lot of the men at Westray were signing on.

A union could demand safer working conditions, but only if there was still a mine to work. Westray's coal production was still lagging far behind projections, and there were rumours it would close if output did not improve by the summer. The company, desperate to make up the shortfall, got government approval in early 1992 to dig up to 100,000 tonnes of the surface coal at a site near Stellarton called the Wimpey pit. Feltmate hoped that was his ticket out, and was trying to get on with the contractor working the strip mine. Other miners were fed up and thinking of quitting, but the chances of finding another job in a county where almost one in five people was out of work—let alone one that paid as much money—were bleak. For those with children, and those who had mortgages to pay, it was a gut-wrenching choice.

The consensus in Feltmate's kitchen was that Westray was a disaster waiting to happen, that miners were going to die. There would be a major cave-in or, worse, some kind of explosion; it was only a matter of when and where. And that was why they had summoned Guptill. These B-crew workers had a request. MacKay did most of the talking. "Carl," he said solemnly, "if we die in that mine, you go public. You go public as soon as she blows, and you tell the world what you know. Do it for our widows."

There was a brief silence. "It's going to happen," MacKay went on. "There's a 25-per-cent chance that it's going to happen to us." Four shifts worked the mine, so the odds were one in four that B-crew would be underground when all hell broke loose. "If it does, it must never be laid to rest," said MacKay. "I don't want to die and nothing be done to the bastards that are going to be to blame."

Guptill, who was already being tormented in his sleep by images of Westray's impending doom, promised to do as they asked.

* * *

ON THE WAY HOME after working the back shift, Glenn Martin often stopped for a cup of tea at his parents' house near Thorburn, a few kilometres east of Stellarton. It was still early in the morning, but as long as the curtains were open, he knew they were awake. His father, Albert, was retired after more than twenty years at Scott Maritimes, the giant pulp mill that filled the air for miles around with the rotten-egg stench of sulphur. Before that, Bert Martin had worked for sixteen years in the nearby McBean colliery, which closed in 1972. Glenn, who had been at Westray since September 1991, represented the fourth generation of Martin men to work in the coal mines.

As the tea steeped, Glenn would unwind by talking about work—routine stuff, like how much coal they had dug the night before. The odd time, he would report that so-and-so had been injured in some minor accident; or he would mention the coal dust, how it would billow up so thick that he could hardly see. Bert had no experience with modern mining equipment, and he wasn't surprised that the big machines were stirring up the dust. At McBean, a couple of men had been assigned to do nothing but spread stone dust; he assumed Westray was doing the same. Then again, Westray did a lot of things that seemed at odds with Bert's coal-mining experience. Everything had to be done in a hurry; the bosses were always cursing and raving at the men. Once, Glenn tried to get his father underground at Westray to take a look around. When the idea was canned, Bert figured it was because he had mining experience.

But when Glenn was with his brothers, things were different. The Martin boys—Glenn had four elder brothers—lived within a few kilometres of each other. They fished and hunted together, but Glenn was the one who knew the woods best; he seemed to have a knack for bagging deer. If the baseball game was on TV, they would gather at one of their houses to split a case of beer or share a bottle. That's when Glenn would open up about Westray. Cave-ins were so frequent, he said, that some of the miners joked about getting together a pool on who would be the first to be killed under a pile of rock. And the coal dust was so thick on the mine floor that you probably couldn't find a tool you dropped there. Chris Martin, who was two years older than Glenn, had no idea that the dust was dangerous; nor did Glenn mention it as a hazard. But Chris was left with the impression that his brother was scared by what he was seeing underground at Westray.

Glenn Martin did not scare easily. And he certainly had no fear of hard work. He could do more work alone than all his brothers combined, Bert would sometimes say with a chuckle. After Glenn bought an old house in Thorburn, he dug a basement by hand and poured the foundation himself. The one thing he had been unable to tackle was school work. He had dropped out in junior high, barely able to read or write because of an undiagnosed learning disability. With little education, Glenn had trouble finding steady work: he cut firewood, carried lumber at a sawmill, and worked off and on at the railcar plant in Trenton. Sometimes he drew unemployment insurance benefits, but he never liked the idea of being paid to do nothing. He was too restless, too determined to get ahead—so determined, in fact, that he went to a tutor in the evenings for two years until he finally learned to read.

Glenn was thirty-five and out of work when Westray opened, and he was eager to get a mining job. He went over to Plymouth a couple of times for interviews, but got nowhere. But Glenn's luck changed after his girlfriend, Sheila Dykstra, happened to run into a woman she knew. The woman worked in Elmer MacKay's Central Nova constituency office, and Sheila happened to mention that Glenn was trying to get on at the mine. A little political clout went a long way at Westray; the next day, Glenn got a phone call telling him to report for work.

As a trainee, Glenn started on the conveyor belt, keeping it clean and adding new sections to it as mining progressed. He graduated to putting up roof supports, then to second operator on the continuous miner—the second operator keeps the machine's water-hose and electric cables from tangling as it moves. Sometimes the operator would let Glenn take the controls briefly, so he could get a feel for the mechanical miner. The shifts were long, but Glenn liked having four days off; it gave him a chance to get away to the family cottage, where he could hunt or roam through the woods. But it was poor compensation for the fear he felt each time he went underground.

After eight months working underground at Westray, Glenn Martin was ready to quit. All he wanted was a few more paycheques—enough money to put siding on his house and pave the driveway. [Photo courtesy of Chris Martin]

By the late spring of 1992, Martin wanted out. As much as he hated being on pogey, it was better than taking his chances at Westray. He would work just a little longer, he told his brother Chris, until he had saved enough money to put siding on his house and pave the driveway. Then he would quit.

* * *

THE NAMES were put in a hat to see who would get the free trip to Montreal. They all knew the John T. Ryan award was a sham, but Westray wanted a miner to accept the trophy. The Canadian Institute of Mining, Metallurgy, and Petroleum gave the annual award to the mine with the best safety record in Canada, and although Westray had not even been in operation a year, it still won the 1992 title. And now, one of the miners was going to have to go up on a stage, shake a few hands, and pretend the company deserved the honour.

The award was based in part on the number of employees injured on the job who claimed workers compensation benefits—and that was how Westray reached the top of the heap. Plenty of miners were being hurt in minor accidents in the mine, but they were kept on the payroll and reassigned to surface work while they recovered from their injuries. Westray called it "light duty," but the term was a misnomer—the men usually ended up in the warehouse, carrying roof supports and other heavy supplies destined for use underground. After Alex Ryan tore most of the ligaments in his wrist, he spent more than a week in the warehouse; despite his injury, he ended up lifting 35-kilogram steel plates. Most of those hurt at Westray were happy to get full pay, and not have to wait for compensation payments. For the company, playing with the numbers gave the mine an enviable safety record, at least on paper—4.24 injuries for every 200,000 hours worked. "It's a record we're pretty pleased with," mine manager Gerald Phillips crowed to an interviewer after the award was announced.

Eugene Johnson, too, knew all about "light duty." A couple of months earlier, after a slight injury, he had worked a few shifts in the warehouse. When his name was drawn from the hat, Johnson just shook his head and refused to go. "That award," he told Alex Ryan, "is a farce." But Johnson's fellow miners told him to go anyway; Westray was picking up the tab, and he would be crazy not to take Donna away for four days at a posh Montreal hotel.

It turned out to be the time of their lives. The trophy was going to be presented at a banquet, so Donna scoured local stores in search of a dress, and even took the two-hour drive to Halifax. She finally found the perfect one, but it was more than three hundred dollars, and she thought that was too much money. Eugene insisted she buy it, and got himself a light grey suit and a dapper paisley tie. They flew to Montreal at the end of April and stayed at the Queen Elizabeth, in the heart of the city. Phillips, and his wife, Catherine, had the room across the hall, and the couples did some sightseeing together. Phillips even snagged tickets to a playoff game at the Forum between the Canadiens and the Hartford Whalers. The Johnsons were die-hard Toronto Maple Leafs fans, but the game was a highlight of the trip; Eugene, normally reserved in public, surprised his wife by joining in as fans did "the wave." At the awards banquet, Eugene and Donna met Clifford Frame, who had flown in from Toronto for the occasion.

The Johnsons got back to their home in Westville on April 29. Eugene, who was now driving the twelve-metre-long shuttle cars that ferried coal from the continuous miner to the conveyor belt, worked the next day. It was the last shift in the four-day rotation for B-crew, so Eugene got another four days off. He got some mileage out of his encounter with Frame, teasing the guys by saying he was going fishing with his new buddy, Cliff. Eugene returned to work on May 5 for the back shift. B-crew would work the next four nights, until 8:00 a.m. on Saturday, May 9.

* * *

GERALD PHILLIPS was back in his office at the mine site on April 29, but he probably wished he were still in Montreal. Three officials from the Department of Labour dropped in first thing in the morning for an inspection. Westray miners usually could tell when government inspectors were about to visit; that was one of the few times that stone dust was spread around the workings to neutralize the ever-present coal dust. But Phillips had been away for several days, and no-one, it seemed, had taken the usual precautions.

Albert McLean normally toured the mine alone, but this time he was accompanied by his boss, Claude White, and by Fred Doucette, the department's mine rescue official. It was almost the same group that had given Carl Guptill's complaints short shrift just a few months earlier. The only one missing was John V. Smith, the inspector of electrical and mechanical equipment; he had taken the day off to drive to Halifax for a medical appointment. Roger Parry accompanied the trio underground to the southwest section, then even deeper, into the North Mains, an area about 1.5 kilometres from the surface which was just coming into production. Methane levels were within legal limits; McLean measured 0.75 per cent. He was less than complimentary about the roof in the southwest section—an area plagued by cave-ins—describing its condition as "fair." But that was nothing compared to the build-up of coal dust. Westray miners had seen it up to half a metre deep on the mine floor; McLean, in his inspection report, made no comment on the amount of coal dust he saw on April 29, but what he saw was enough to prompt quick action.

After the tour, McLean met with Phillips and Parry and ordered them to deal with the coal dust or face charges under provincial mining laws. He handed them four written directives citing violations of the Coal Mines Regulation Act and the Occupational Health and Safety Act. Two of the orders were to be carried out immediately: Westray was to spread stone dust to prevent a coal-dust explosion, as required by law, and to ensure that at least twenty bags of stone dust were stored in each area being mined. McLean gave Westray until May 15 to comply with the remaining two orders: develop a plan for spreading stone dust and set up a system for sampling dust in the mine to make sure its coal content was not higher than the level stipulated by law, 35 per cent. If the orders were ignored and charges were laid, Westray could be fined up to $10,000 on each count; if Phillips, Parry, or any other company official were singled out in the charges, they faced the fine and up to a year in jail. To make sure the men in the pit knew what was up, McLean followed the standard procedure and posted the orders at the mine site.

The threat of prosecution was a serious step, but McLean could have taken more-drastic measures. Under the Coal Mines Regulation Act, he had the power to close down Westray on the spot. An inspector had the right to stop work and order all miners to the surface if he considered a mine, or any part of it, to be unsafe. McLean, however, opted to let Westray stay in business, under orders to take immediate steps to eliminate the coal-dust hazard. That was not the only break for Westray; the two-week grace period for devising schemes for dealing with coal dust was nothing short of generous. McLean and his superiors had been hounding the Westray manager for stone-dusting and dust-sampling plans since September, when the mine went into production; eight months had passed, and Phillips had yet to implement procedures for spreading the dust.

The problem was simple: at Westray, management and miners paid little heed to the danger lurking right before their eyes. Airborne dust danced in the beams of light coming from miners' helmets as they worked, and black grime coated their faces. Tractors and shuttle cars left deep tire ruts in the black powder that settled on the mine floor. Shaun Comish, an experienced hard-rock miner, likened it to stepping on the soft, powdery surface of the moon. Some of the men wrapped tape around the legs of their dark-blue coveralls to keep dust from filling their boots.

The continuous miners were equipped with spray nozzles to water down the dust as coal was dug, and parts of the mine were equipped with a spray system to keep the tunnels damp. But that was not enough. Time and time again, miners asked their bosses to put on extra crews to do nothing but spread rock dust. Management's response was an offer to pay overtime to anyone willing to stay after their shift to do the work. The priority, as always, was to dig coal. But after a gruelling twelve-hour stint underground, few miners were interested in putting in extra hours. Besides, the mining machines advanced so quickly through the seam that a few hours of stone dusting would make little difference; Westray needed a systematic approach, the kind the mine inspector had been demanding for months. Complaints and grumbling gave way to complacency; coal dust seemed less of a hazard each day the miners emerged unscathed at the end of a shift. Most of the men knew the dust was explosive, but few realized it had the potential to wreck an entire mine. Their biggest worry was the menace they could not see—methane.

The April 29 ultimatum did nothing to change conditions underground. By some accounts, up to six hundred bags of stone dust went into the mine after the order was posted. But on May 5, when miner Gordon Walsh went home after his final shift in the four-day rotation, the area he was working still lacked the required twenty bags of stone dust. How much of that dust was actually spread around the workings is also unclear. The company collected four dust samples from the roof and floor of the mine's roadways on May 8; the samples, analyzed later at an independent laboratory, were found to contain as much as 67-per-cent coal dust. Not one sample fell within the 35-per-cent limit for combustible coal dust, as stipulated by the Coal Mines Regulation Act.

There was no follow-up inspection to ensure that the April 29 orders were carried out. During the first week of May, McLean and other Department of Labour officials were preoccupied with a fire at the Evans colliery in Cape Breton. McLean could have checked the mine again on May 6, when he was in Plymouth to administer tests to men seeking their miner's papers and to help plan the provincial mine-rescue competition Westray was to host in June. There is no record of McLean conducting an underground inspection that day.

* * *

THE LAST THING Gerald Phillips needed was flak from the Department of Labour over safety; head office was breathing down his neck about Westray's lacklustre production record. Curragh was throwing a lot of money into a bottomless pit and getting precious little in return. From the official opening in September 1991 to the end of March 1992, Westray's coal sales were $7.3 million. But the mine cost almost twice as much—$13.3 million—to operate for those seven months. The figures for March were the best yet, thanks to the timely addition of surface coal from the nearby Wimpey pit. But Westray was still losing $18 on every tonne of coal sold. Clifford Frame's "miracle man" was desperate to pull off another miracle; Phillips needed coal, lots of coal, and he needed it right away.

For months, cave-ins had been the biggest safety issue at Westray, not coal dust. Even the officials of the Department of Labour were concentrating on the roof-control problem; that was one reason it took McLean until late April to finally get tough about Phillips's failure to devise a stone-dusting plan. The cave-ins were also bad for business. Clearing debris from the tunnels and putting up additional roof supports was costly and time-consuming. Too much time was spent working on the roof, and not enough on digging coal. From March 1991 to the end of the year, Westray shipped only 13,000 tonnes to the Trenton power station, all of it lower-priced coal with a high ash content. That was the total output for 1991; the contract with Nova Scotia Power called for delivery of almost 60,000 tonnes every month. To make up the difference, the Nova Scotia government gave Westray the green light in January 1992 to take coal from the Wimpey pit.

Coal from the surface nearly doubled production figures for February and March, when Westray shipped a respectable 50,000 tonnes. And it was far cheaper to dig, putting much-needed cash in Westray's pocket. But at the end of March, a new problem cropped up underground. A large portion of the southwest section had to be abandoned, leaving behind thousands of tonnes of coal. Westray used the room-and-pillar method of mining; tunnels were cut at intervals through the seam, creating large blocks of coal that served as pillars to hold up the roof. Once the tunnels were laid out in a grid pattern, crews dug more coal from the pillars as they retreated from the area. In the southwest section, the tunnels were driven too close together, leaving pillars as thin as five metres. They began to collapse under the weight of the roof long before the coal was extracted, giving off high levels of methane. Plywood barriers had to be erected to close off the section and seal up the gas.

The loss of the southwest section was a severe blow; the area had been producing up to 3,000 tonnes of coal per day. Westray was forced to rely even more on the open pit until other areas of the mine could be brought into production. In April, the company asked the provincial government for a permit to mine 200,000 tonnes per year at Wimpey. But the Town of Stellarton and landowners bordering the site were pressing for a lengthy environmental review of the proposal. A decision was expected by the middle of May. Underground, work continued on two fronts. Tunnels were driven to the west of the abandoned area, and a new section, Southwest Two, was created. Deeper into the mine, the North Mains were being developed, but even there the roof refused to stay up. Tunnelling crews were fighting a rearguard action; the roof over roadways and intersections deteriorated quickly, forcing workers to go back to install additional steel arches for support.

By the middle of April 1992, the top brass at Curragh decided Westray was broken, and it was time to fix it. Phillips had spent a lot of time commuting to and from Toronto during the spring of 1992; word around the mine was that his bosses were chewing him out over the shoddy production record. Now one of his superiors descended on Plymouth to try to sort out the Westray mess. The task fell to Colin Benner, who had been in charge of Curragh's lead and zinc mines out west, until the beginning of April, when he was promoted and became president of operations for the entire company. One of Benner's first moves was to create the Mine Planning Task Force, a seven-member team of outside consultants and officials from Westray and Curragh, to tackle the mine's safety and production woes. Phillips headed the group, but the direction came from Curragh's executive suites in Toronto. "This is an extremely important project," Benner wrote in a memo setting up the task force, "which requires the highest level of priority in the company."

The goal, in Benner's view, was threefold: to develop "a safe mine plan, an achievable mine plan, and an economic mine plan." To reach these objectives, the Mine Planning Task Force was to start from scratch. Benner wanted Westray to rethink roof-support systems and the way blocks of coal were accessed and cut into pillars, and he wanted the new approach to be based on past experience and visits to other mines. He stressed the need for a systematic, coordinated method of tackling the mine's problems; clearly, he felt Westray's managers had not been using such a method. The group was to meet at least once a week at the mine site, starting April 15. Benner planned to fly to Nova Scotia for as many sessions as possible; if he was unable to attend a meeting, the minutes were to be forwarded to him.

That was not the only memo Benner wrote to Frame and his colleagues at head office. He looked over Westray's production figures, and took a guess at how much coal would be brought to the surface in the coming months. His conclusion was that Westray would fall short of the production level required under the loan agreement with the Bank of Nova Scotia. Unless there was a dramatic turnaround by October 1992, the bank's deadline, Curragh would be penalized millions of dollars. The one ray of hope was that the bank might be persuaded to count coal from the Wimpey pit, which might be enough to push Westray's total production figures over the hump—assuming, of course, that the Nova Scotia government allowed Westray to strip-mine more coal. And assuming that the task force could straighten out the crisis at the underground mine.

* * *

CLIFFORD FRAME, a white carnation pinned to the label of his dark business suit, was upbeat. It was the first Tuesday in May, and Frame, the man with the controlling interest in Curragh Resources, was getting together with his fellow shareholders for their yearly pep talk. But Curragh's investors, the people counting on Frame's bravado and mining savvy to make money for them, had little reason to rejoice as they gathered at Toronto's Royal York Hotel on May 5. Curragh shares had gone on the market at $11.87 when the firm went public in 1990; when the Toronto Stock Exchange opened on the day of the meeting, those shares were trading at $4.50. In a press release accompanying the latest financial statements, the firm accentuated the positive: Curragh had lost only $6 million in the first quarter of 1992, a vast improvement over the $63-million loss posted during the last three months of 1991.

Everyone on hand for the annual meeting in the Upper Canada Room of the Royal York knew why Curragh was taking a beating. The stubborn recession meant fewer cars and trucks and tractors and other vehicles were being built and sold. Lead for batteries and zinc for galvanizing metal were in less demand, which meant lower prices for both metals, and a heap of trouble for a company that produced little else. Zinc, which peaked at 87 cents per pound in 1989—a year Curragh cleared a cool $60 million—commanded an average price of 51 cents per pound in 1991. Low prices, combined with a higher Canadian dollar that year, meant misery for any mining company reliant on export markets. A ten-week strike at Faro, Curragh's main lead and zinc operation, north of Whitehorse, reduced base-metal production for the year, adding insult to injury. To cut costs, Curragh froze salaries at the management level, trimmed the payroll by 20 per cent, and cut back on exploration. One of the few to escape the penny-pinching was the chairman and chief executive officer; Frame earned $190,200 in bonuses in 1991, on top of his $520,000 salary.

But this was a day for talking about the future, not dwelling on the past. "As most of you appreciate," Frame told the shareholders, "forward progress is hardly ever painless." Curragh had made money during March and April, he noted, and metal prices were creeping upward. To cash in when the recession finally lifted, the company was concentrating on having lots of lead and zinc on hand. Since Faro was nearly depleted, Curragh was trying to get the federal government to guarantee a $40-million loan to develop an adjacent ore deposit. Sä Dena Hes, a new mine near the Yukon–British Columbia border, had come on stream in 1991; further down the road, Curragh had high hopes for the remote Stronsay site in northern B.C., and its fifty million tonnes of lead-zinc ore.

The company had invested $60 million in the property, but estimated it would take a lot more money—$200 million—to bring it into production. It would make Curragh the world's fourth-largest producer of lead and zinc concentrates. After the meeting, Frame told reporters he was considering selling up to 35 per cent of Stronsay, either to another mining firm or through a public share offering.

Then there was the Westray coal mine, thirteen hundred kilometres to the east. Shunted aside like some poor waif, the colliery was almost overlooked in Frame's grand strategy for Curragh's revival. Just as Stronsay's star was on the rise, Westray's was growing ever dimmer in the Curragh firmament. In an effort to raise cash, Frame had put Westray up for sale back in November 1991, seeking a buyer or at least a partner. Curragh's glossy annual report for 1991, its cover graced with a photograph of two Westray miners examining the huge cutting head of a continuous miner, presented a brave face; the mine, shareholders were assured, had overcome its growing pains and was now in full production. But the report also noted that Westray's losses had contributed to Curragh's poor showing for the year. Shortly before the meeting, Frame admitted he still had a soft spot for Westray. "I like that project and have very mixed feelings about selling it," he told the *Northern Miner Magazine*. But there is little room for sentiment in business; the mine, Frame said, was still on the market.

As he emerged from the shareholders meeting, Frame had reason to believe he was seeing the light at the end of the tunnel. He even put through a resolution changing the company's name from Curragh Resources to Curragh; the change was minor, but dropping the word *resources* hammered home the idea that Curragh was a mineral producer and no longer a collection of mineral properties waiting to be developed. "Everyone forgets that on May 8, 1992, we were in the top forty exporters in the country," a Curragh executive noted later. "It was a company full of piss and vinegar." Frame, the man who had built a major Canadian-owned mining company from scratch, was once again on the rebound. Like a cheerleader on the sidelines, the *Northern Miner Magazine* was heading to press with a sixty-four-page issue devoted entirely to Curragh and its promising future. A few days after the shareholders meeting, Frame and his wife, Catherine, boarded a plane in Toronto for the thirteen-hour flight to Tokyo, where Frame would attend a meeting of a Canada-Japan trade association. The date was Friday, May 8. As the plane took off for the Land of the Rising Sun, digging coal in Pictou County was the last thing on Clifford Frame's mind.

FLASHPOINT

HE HEADLINES IN THE MORNING PAPERS on May 8, 1992, screamed robbery and murder. Two men and a woman working the night shift at a McDonald's fast-food outlet near Sydney, Nova Scotia, had been gunned down in the early hours of May 7; a fourth employee, shot in the head at close range, was in hospital, clinging to life. RCMP officers were combing the Sydney area for suspects in one of Nova Scotia's worst killing sprees in years. But at eight o'clock, as miners gathered at Westray to start the day shift, the shocking events unfolding in Cape Breton, only a few hours' drive away, quickly took second place to disturbing news from underground.

A-crew was working the fourth and last shift of its rotation, and that was usually enough to put the men in a good mood. Just twelve more hours and they would have four days off, four days away from a place most of them knew was a deathtrap. The miners changed into coveralls and collected a hard hat, a battery pack for the helmet light, and a self-rescuer, a small face mask that filters out carbon monoxide and smoke produced in an explosion. Then they gathered in the deployment area of Westray's administration building for coffee and smokes before heading into the mine. As the men coming off shift passed through on their way home, the replacement crew had a chance to find out what had happened overnight. On the morning of May 8, the mood on A-crew took a definite plunge as B-crew brought them up to speed.

The mine was dusty, as usual, and methane levels were running high in the Southwest Two section. Those were conditions the miners had come to expect. But this time, there was a new twist. A continuous miner was being used in the southwest section, even though the "sniffer"—the miners' name for the methanometer—was broken. This meant the machine would not shut down if gas levels crept towards explosive levels; a spark could set off a fire or, worse, an explosion. B-crew miners who had worked the back shift in the North Mains were incensed when they reached the surface and found out about the broken methanometer, and the men starting their shift were no happier with the news. As Shaun Comish hopped on a tractor for the twenty-minute drive underground, he asked one of the foremen about the sniffer. The question was met with a shrug. "If we get killed," Comish said, half-joking, "I'll never speak to you again."

The indifferent response did not surprise Comish. After eight months at Westray, nothing surprised Comish anymore. He had put in a dozen years underground at hard-rock mines in Ontario and Nova Scotia, but he had never seen anything quite like Westray-style coal mining. He had never worked at a mine where cave-ins were almost a daily occurrence, or where the pressure to produce was so intense. He had never worked for a company that refused to install portable toilets underground, and expected the men to relieve themselves in an out-of-the-way corner. But he was working in a mine like that now.

An intense thirty-six-year-old with dark hair and moustache, Comish bore more than a passing resemblance to country music star Aaron Tippin. He had started work at Westray during the first week of September 1991, the same day as three other members of A-crew: Lenny Bonner, Steven Cyr, and Wyman Gosbee. Bonner, a barrel-chested man in his early thirties, with a booming voice and a mop of curly black hair, was Comish's best friend. They had worked side by side at several mines over the years, teaming up again at Westray after Gays River shut down. Comish stayed at Bonner's house during his four-day work week, commuting to his home near the Halifax airport for his off days. More than once, Comish called his wife, Shirley, and told her he had seen enough; but each time, he convinced himself things would get better. When the Steelworkers union came to town in the spring of 1992, Comish and Bonner jumped on the bandwagon, signing up some of their co-workers right in the parking lot at the mine site. By the first week of May, Westray workers were on the verge of getting a union—at last.

Once he got underground on May 8, Comish took over the controls of a continuous miner deep in the North Mains. But he kept thinking about the broken methanometer on the machine eating into the coal farther up the slope, in the southwest section. He was not alone. The crew in the North Mains talked of little else when they got a break for lunch. The normally soft-spoken Cyr, who was driving a shuttle car, was fed up. If things did not change, he announced, it would be his last shift at Westray. Comish was so uneasy that he left work about five o'clock in the afternoon, three hours early, making up an excuse about needing to get his car fixed.

Unknown to the men in the North Mains, the faulty methane detector was fixed about noon. This was a relief to Bonner, who spent the shift in the southwest section, operating a bolter with Gosbee. A tracked vehicle like the continuous miner, the bolter was used to drive long steel rods into the roof of newly mined areas. It was common for the bolts to give off sparks as they were inserted; Gosbee had seen it happen many times. But the bolters were not equipped with gas detectors, even though methane tended to collect at the roof. The miners recognized the hazard, but requests that management install sniffers on the bolters, like so many other complaints about safety, fell on deaf ears.

As he worked the bolter, Bonner realized it was getting hot. To his disgust, he found that the ventilation to the area had been tampered with. The vent tube, a flexible hose used to direct air through the mine, had been blocked off. This happened all too often at Westray; mining crews would rejig the ventilation to direct more air to the working face. That kept the continuous miners from shutting down, or "gassing out," as the methane level rose. Bonner reopened the tube, and demanded a methane reading when a foreman came by about fifteen minutes later. The hand-held methanometer pegged the mine's air at 3.75-per-cent methane. The gas had undoubtedly been over 5 per cent—well within the explosive range—before ventilation was restored. One spark from the bolter could have set off a lethal explosion or fire. Bonner, well aware of his luck, rewarded himself by leaving work a half-hour before the shifts changed.

Shaun Comish, left, and Lenny Bonner, experienced hard-rock miners on A-crew, were fed up with the short-cuts and unsafe practices that were rife at Westray. And they were glad to get off shift on the evening of May 8, 1992. [Photo by Len Wagg, courtesy of the *Chronicle Herald* and the *Mail-Star*]

* * *

FRIDAY NIGHT would be B-crew's last shift for four days, but Robbie Fraser was fit to be tied. He had planned to sleep all day Saturday after work, then take his son fishing on Sunday at his family's camp in the woods outside Westville. But now he had to fill in on Monday for another mechanic who was going on vacation. They didn't have the decency to tell him to his face; one of the higher-ups left a note instructing him to report for work. After he got on at Westray as an apprentice in the summer of 1991, Fraser took all the overtime he could get. But not now, not with all the problems underground. As Fraser and some of the others on B-crew had told Carl Guptill, the chances were one in four that they would be there when all hell broke loose. Working extra shifts meant taking more of a gamble.

Fraser had already planned his escape from the whole mess. He told his wife he was going back to driving truck, although it meant being away from home for days, even weeks, at a time. He was waiting until he got his next paycheque in the middle of May. It was a tough decision; Joyce was six months pregnant with their fourth child, and they were still living in an apartment. But going back on the road was a lot safer than working at Westray. Fraser's father, who had worked in the Drummond mine, kept warning him about the dangers of cave-ins and methane and coal dust. "You're crazy," he had told him more than once. "Get out of there."

Just a few more shifts, and Fraser was going to heed that advice. Before leaving home on the evening of May 8, Fraser kissed his oldest son. "Now you're the man of the house," he told Robbie, who was only five. "You've got to look after Mommy." Strange, Joyce thought; her husband never made such a fuss about going to work. He had also started wearing his wedding band, despite the risk that it would get snagged as he worked on machinery. Eugene Johnson's house was next door, and Fraser walked over to get a lift to work. The two men jumped into Johnson's car, as they had countless times before, for the ten-minute drive to Plymouth.

* * *

ROY FELTMATE left for work a half-hour early so he could drop in to see Carl Guptill. Feltmate's brother, Lloyd, was staying at Guptill's apartment in Stellarton, and Roy arranged to pick him up after he got off work at eight o'clock on Saturday morning. He also wanted to invite Carl and his girlfriend over the next day to help celebrate his daughter Amy's birthday. She would be eleven on May 9, and Roy hoped she would wait until he got home before opening her presents.

The conversation turned to the mine. Guptill tried to convince Feltmate to skip work, even though he knew Feltmate needed the money. But Feltmate scoffed at the idea of playing hooky. The sky was clouding over, and the forecast was for showers and temperatures not much above freezing. "I'd better wait until July," he chuckled. "It might be better weather."

* * *

FIVE MINUTES AFTER leaving Westray, Steven Cyr pulled into the lane of his home in Churchville, a few kilometres up the road. Born in Westville, Cyr had moved to Ontario when he was in his teens but had come back to Pictou County in the late 1980s. For a few years, he held down odd jobs—putting in foundations for houses, or cutting pulpwood. Then he got on at Westray in the fall of 1991. Cyr was twenty-eight, and he and his wife had a three-year-old daughter; he was glad to finally land a steady job.

Cyr had done well in eight months at Westray, going from a trainee with no underground experience to a certified miner with his mine rescue papers. He was still fuming over the day's events and, to top it all off, he felt like he was coming down with the flu. About half-past nine, he unplugged the telephone and went to bed. He had to be up early for a first-aid course at the Heather Hotel in Stellarton.

* * *

AS HE WENT UNDERGROUND on Friday night, Myles Gillis was looking forward to a busy weekend. Gillis, an electrician on B-crew, planned to catch a few hours of sleep after he got off work Saturday morning, then take his nine-year-old son, Christopher, fishing. After that, he had promised to help John Halloran, another electrician on his shift, who was building a house not far from his own. Gillis had been working at Westray for more than a year; like so many others, he had come to Pictou County by way of the hard-rock mines at Forest Hill and Gays River. The coal business was new to Gillis, but he knew about methane, underground fires, and other hazards from his mine-rescue training. And he did not like what he saw at Westray.

After Carl Guptill joined Westray in late 1991, Gillis would sometimes take him aside when they got a free moment underground. Westray, Gillis told Guptill, was a time bomb, and he was writing down every problem and safety lapse he saw on his shift. The two men discussed the chances of getting out alive if there was an explosion; Gillis wondered aloud to Guptill whether the self-rescuers they carried on their belts would last the twenty minutes needed to reach the surface on foot.

Gillis had not always thought of Westray as a deathtrap. When he got the job, back in March 1991, it seemed more of a dream come true. He and his wife, Isabel, were able to build a house on land they owned in Pleasant Valley, just outside Antigonish, the main town in the county next door to Pictou. Working at Forest Hills and Gays River had meant being away from home and his family for days at a time. But Plymouth was less than an hour's drive away, and that meant the

Gillises could think about putting down roots. Westray was always calling Myles in to work overtime, so there was enough money to start building that new house.

For Gillis, providing for his family was everything. An electrician who had ended up plying his trade in the mines, Gillis did not fit the popular image of the rough-and-ready miner. He was shy around new people, with a boyish face and a winning smile. He rarely drank, even a beer. He liked hunting and fishing and scuba diving, but he was more likely to be puttering around the house or helping out at a friend's place. The only time he watched TV was Sunday nights, when he and Isabel and the kids—Christopher, six-year-old Ashley, and Daniel, not yet one—would munch popcorn in front of the Disney show.

A settled home life, time for the family, a new house—everything Myles had worked so hard to achieve could be lost if he quit Westray. He felt trapped. He told Isabel some of the things he saw, about the time part of the roof collapsed moments after he walked under it. But Isabel assumed this sort of thing was part of the job, and that her husband would quit if he were in real danger. The company claimed to use state-of-the-art mining technology, and the government had sunk millions into the project; how, she reasoned, could the mine not be safe? Then, one morning around the first of May, not long after Myles's thirty-second birthday, they were talking about the problems at Westray when Myles asked his wife to make him a vow. "If I ever die in that place, Isabel, promise me you'll have a full investigation into my death." She was startled. "Surely it's not that bad," she said, not sure whether he was joking or serious. But Myles made it clear that things really were that bad. "Just you promise me," he said.

Isabel Gillis's husband Myles, an electrician at Westray, asked her to make him a promise in early May 1992. "If I ever die in that place," he said, "promise me you'll have a full investigation into my death." [Photo by Len Wagg, courtesy of the *Chronicle Herald* and *Mail-Star*]

On the night of May 8, after Myles had left for work, Isabel was surprised that all three children crawled into bed with her for the night. They rarely slept with her, even when their father was working the back shift.

* * *

It was after midnight when Gerald Phillips returned to his home in Plymouth. Westray's manager was in the midst of renovating and expanding a split-level just down the road from the mine site. Looking out the back windows, over the patio and the S-shaped in-ground swimming pool, he could see the lights that delineated the mine's crisscrossing lines of steel-sheathed conveyors. Phillips and Roger Parry had flown back to Nova Scotia late Friday, after a fact-finding mission in the United States. As members of Colin Benner's internal task force, Phillips and Parry had toured several coal mines, looking for new ways of dealing with Westray's roof problems.

Phillips was responsible for the overall operation of the Westray mine. In the words of Nova Scotia's Coal Mines Regulation Act, he had "charge, control, and supervision" of the colliery. The manager's duties included appointing other top mine officials, promptly reporting any accidents to government inspectors, personally supervising operations, and maintaining detailed, up-to-date plans of the underground workings. His immediate subordinate was the underground manager, Parry, who was required to supervise work underground each day, check the ventilation system, and ensure that roof supports and seals on mined-out areas were properly installed. It was also Parry's job to keep Phillips abreast of conditions in the mine on a daily basis.

Since Phillips and Parry had only returned to town in the early hours of May 9, Westray was the responsibility of Glyn Jones, the man in charge during that cave-in, almost a year before, which had so terrified Eugene Johnson and the other B-crew miners. Jones, a thirty-eight-year-old originally from Britain who had been mining coal since he was a teenager, held the title of assistant superintendent, a post referred to as "overman" in the archaic language of the provincial coal-mining statute. That made Jones next in line to the underground manager. The overman was essentially management's safety officer; he was responsible for ensuring compliance with all provisions of the Coal Mines Regulation Act, and for evacuating workers from any part of the mine that became unsafe.

Despite the explicit chain of command established under the act, it was not uncommon for mining operations at Westray to be left in the hands of a senior foreman, who worked underground along with the rest of the men. That was in fact the case on the night of May 8–9, as Phillips slept and the back shift churned out coal. Jones was off work, and the senior employee on shift was John Bates, a foreman who boasted four decades of experience in coal mines in Alberta and his native Britain. Bates was the man in charge.

* * *

THE PAINS WOKE JOYCE FRASER at 2:15 in the morning. Worried that she might be going into labour, she thought about calling the mine to tell Robbie to come home. No, she decided, she would wait until three o'clock, and see how she felt then. But the pains went away, and she drifted back to sleep.

* * *

AT 4:45 A.M., a message flashed on a computer screen in Westray's control room at the surface. A remote gas sensor in the North Mains indicated the presence of methane and carbon monoxide. The sensors were set to trip at relatively low levels—0.25-per-cent methane and 20 parts per million of carbon monoxide. Government regulations required mining to stop when the respective gases reached levels of 1.25 per cent and 50 parts per million. The sensor in the North Mains signalled "ALARM," but that reading was followed a second later by a "HEALTHY" message. Twenty-five seconds later, the same sensor raised a methane alarm, which was again followed, three seconds later, by a "HEALTHY" message.

The control room operator paid little heed to the alarms. He knew one of the electricians, probably John Halloran, was hooking up a sensor that had been out of service for a day and a half. It was common for sensors to send out inaccurate readings as they were reconnected; the manufacturer of the equipment even had a term for them—"voltage spikes." The almost immediate return to "HEALTHY" status confirmed that the readings had been false alarms. At the time, methane levels in the North Mains ranged between 0.16 per cent and 0.20 per cent, too low to have been the source of the readings.

* * *

ABOUT THREE HUNDRED METRES below, eleven miners were at work in the southwest section, among them Eugene Johnson, Robbie Fraser, and Myles Gillis. Roy Feltmate, Glenn Martin, Mike MacKay, John Bates, John Halloran, and ten others were down in the North Mains. It was three-quarters of the way through the shift, not too early to think about getting back to the surface—and getting away from Westray for a few days.

Then, at 5:18 a.m. on Saturday, May 9, 1992, thirty-three minutes after the false alarms from the North Mains, something went horribly wrong. There was a sudden rush of methane. The continuous miner in Southwest Two could have cut into a pocket of methane that had been trapped in the coal. The methanometer on the miner could have been out of order again, allowing the machine to run even though the gas level was rising. Or a cave-in inside the abandoned Southwest One section could have blown out the flimsy plywood seal; methane had been accumulating behind the barrier for weeks, and the ventilation system would have carried the gas straight into Southwest Two, where coal was being dug.

One spark–likely from the cutting head of the continuous miner striking a band of stone embedded in the coal seam–and the methane erupted in a flash of flame. The operator of the continuous miner, Trevor Jahn, jumped from his seat and began to run away from the coalface. Eugene Johnson and Robbie Fraser reached for their self-rescuers. Within seconds, the fireball sucked oxygen out of the mine air and flooded the area with carbon monoxide.

A wall of flame swept out of the southwest section, gaining force and momentum as it stirred up coal dust. It roared down to the North Mains and up the slope towards the surface. The coal dust was like gunpowder, fuelling the blast; the tunnels were like the barrel of a gun, channelling its destructive force in the direction of the men in the North Mains. At 53 seconds past 5:18, the control room on the surface detected a slip in the conveyor belt. Then, power to the underground was cut. The monitors fell silent.

HOPE IMPRISONED

HE OVERCAST SKY WAS BEGINNING to brighten in the east as Lauchie Scott manned the pumps at the Esso gas bar just off the Trans-Canada Highway in Blue Acres, near Stellarton. It was still a half-hour before dawn on a Saturday morning, and business was slow. Suddenly, the teenager saw a bright flash rip through the morning gloom. A bolt of blue-grey light flickered against the clouds for an instant, like the flashbulb on some giant camera. The light came from the direction of the mine, about a kilometre to the south. A split-second later, Scott heard a boom, like a huge thunderclap, roll over the gas station, shaking the ground.

Down the road at his home in Plymouth, Brad Miller was jolted awake. He sprang to his feet and ran to a window overlooking the Westray mine. A thick cloud of dust and smoke billowed out of the portal leading underground, engulfing the mine site. After ten, maybe fifteen seconds, the cloud began to dissipate in the breeze. It left one neighbouring house coated in black ash. The blast rattled windows and shook houses more than one kilometre away. One resident of Stellarton likened the sound of the blast to "two great big metal doors clanging shut in an old castle." Then, within ten minutes, the morning stillness was broken a second time, by the sirens of fire trucks, ambulances, and police cars speeding towards the mine.

* * *

DONNA JOHNSON'S phone woke her about twenty minutes to six. A woman with a British accent was on the line, asking if Eugene was home. No, she said, he's working. Johnson asked who was calling, but the woman hung up without answering. Something must be wrong at the mine, she thought. Why else would someone be looking for Eugene? She agonized over what to do for a few minutes, then dialled Joyce Fraser's number.

Fraser had not received the mysterious wake-up call, but she was just as anxious about Robbie. The women spoke briefly, then Johnson called the mine office. The man who answered brushed off her questions, telling her to go the fire hall in Plymouth. Fed up and beside herself with worry, Johnson wished she could

The force of the explosion tore metal sheets off the portal at the mine entrance, and shook homes more than a kilometre away. [Photo by Len Wagg, courtesy of the *Chronicle-Herald* and the *Mail-Star*]

crawl though the phone line and choke whomever was on the other end. Johnson and Fraser arranged for relatives to look after the children, then got a drive over to Plymouth with Eugene's mother.

They arrived about 8:00 a.m. Police officers, who had closed off the road leading to the mine, directed them to the Plymouth Fire Hall. A two-storey building set into a hillside, the hall was within a half-kilometre of the Westray site, and provided a clear view of the coal towers and conveyors. Plymouth's volunteer fire department kept its trucks in three bays on the lower level, and a large room on the upper floor was used for meetings and games. A couple of clergymen were waiting inside, but all anyone would say was that there had been an accident at the mine.

Fraser kept thinking some of the men must have been hurt, that they were already in hospital. Catherine Phillips, the mine manager's wife, dropped in. She and Johnson had gotten along well in Montreal, but she had few words of encouragement for her new friend. "It doesn't look good," Phillips whispered. The room started to fill up as wives and family of other members of B-crew arrived. Then people began talking about an explosion. As soon as she heard the word, Johnson's heart sank. It was as if a little piece of her had died. From that terrible moment on, she felt in her heart that Eugene was gone.

* * *

CHRIS MARTIN lay in bed for a couple of hours, unable to sleep. He was uneasy, but he had no idea why; it was just a feeling that something was wrong. At about six, he got ready for work; it was a Saturday, but Martin, a stocky man with a wide, bushy moustache and a gentle voice, was on shift at Scott Maritimes. The feeling was still there as he left his home in McPherson's Mills, near Thorburn, and drove to the pulp mill. On the way through New Glasgow, Martin stopped at a Tim Hortons to pick up a coffee. By now it was half-past six, and already people were talking about the explosion at the mine.

Martin got back in his car, but his mind was racing. Had Glenn been in the mine when it exploded? Had his brother even been on shift Friday night? He could call his parents to see if they knew, but that would only make them as worried as he was. Martin drove to work, but he had to know if Glenn was safe. He got permission to leave work, then called his two elder brothers. Neither knew if Glenn was at work, so they decided to try to find him. There was no answer at Glenn's house in Thorburn, and no-one was there when they drove up. They doubled back to the mine, but a policeman at the roadblock directed them to the fire hall. Glenn was missing, but Martin still clung to hope. Maybe he had not been at work when the mine exploded. Maybe, just maybe, Glenn was off in the woods somewhere, fishing.

* * *

VOLUNTEER FIREFIGHTERS, police officers, and ambulance attendants began converging on the Plymouth mine at half-past five. Staff at the Aberdeen Hospital, little more than a kilometre up the road in New Glasgow, were put on disaster alert; doctors and nurses about to go home after working the night shift stayed on. By six o'clock, attendants with Dort's Ambulance had set up a triage centre at the mine's portal.

The quick response was the result of a eerie coincidence. The local Emergency Measures Organization had staged a mock disaster the previous weekend, its largest since the 1970s. That exercise, based on the premise that a plane had crashed into a campground, leaving more than thirty people injured, helped sharpen communications among the area's fire departments, police forces, and medical services. But this was no drill. There had been a massive underground explosion, and everyone was braced for a rush of casualties as men were brought to the surface.

* * *

LENNY BONNER RACED to Plymouth about half-past seven. A neighbour had gotten him out of bed, telling him that there had been an accident at the mine, and men were trapped. His hopes were dashed as soon as he drove onto the mine site. Dozens of metal sheets had been blown off the roof of the portal leading underground; they lay scattered on the ground, as if someone had tossed a deck of cards into the air. Much of the portal's steel skeleton was exposed, offering mute testimony to the force of the explosion. Emergency vehicles were parked everywhere.

Bonner, who was certified in mine rescue, went to the administration building to get ready to go underground. He was summoned to Gerald Phillips's office, but the mine manager was about the last person he wanted to see. A few days earlier, after a row with Roger Parry, he had been sent to the surface to talk to Phillips. Bonner had used the occasion to vent his anger about the lack of rock dusting, about the high methane, about all the cave-ins. Now that the worst had happened, Bonner was fuming. "I just can't believe that it was let get this far, to actually blow up, especially when you knew about it," he told Phillips when they were once again face to face. "Look, I know you want to kick some ass right about now," Phillips replied. "If you want to kick ass, wait until this is over."

* * *

ISABEL GILLIS HAD TO REACH over her sleeping children to grab the phone.

"Hello, is Myles there?"

"No, I'm sorry, he's not home." Gillis checked the clock. It was ten minutes to eight. Who was this woman, she thought, and why was she looking for her husband at this time of the morning?

"Can you tell me if he's at work?"

Gillis, a petite woman with reddish hair cut in a bob that made her look younger than her twenty-nine years, assured the caller he was.

"There's been an accident at work," the woman said, "and we're asking people to meet at the Plymouth Fire Hall." Gillis asked directions and hung up. It was likely something minor, she thought; maybe a tractor had overturned, and Myles, with his rescue training, was helping those who were hurt. But why the cryptic phone call? Why had she been asked to go to the fire hall? The more she thought about the call, the more the fear rose within her.

Gillis asked one of her sisters to drive. When they pulled in for gas just outside Antigonish, Gillis turned on the radio. The newscast said there had been an explosion at the Westray mine, shaking houses more than one kilometre away. Gillis screamed, and burst into tears. This was nothing minor–this was her worst nightmare come true. Gillis cried for the entire forty-five-minute drive to Plymouth. Myles is all right, she told herself, as they worked their way through the roadblocks to the fire hall. People have been hurt, but Myles is all right. He knows too much about mine rescue. He's fine.

* * *

STEVEN CYR HAD SHAKEN OFF his flu symptoms and was feeling better as he got ready to head into town for his first-aid course. As his wife made toast and coffee at about eight o'clock, Cyr plugged the phone back in. It rang within minutes.

"You sure can sleep some heavy," came an exasperated voice on the other end of the line. It was Cyr's father, calling from Westville. Cyr explained that he had disconnected the phone overnight so he could get some sleep.

"Of all the times to unplug your phone," his father exclaimed. "The mine blew up."

"Fuck off," Cyr blurted out. "You're kidding." He went numb as his father told him that the radio was reporting an explosion at the mine, and that miners were trapped below.

Within minutes Cyr was roaring down the winding road to Plymouth at 140 kilometres per hour. As he rounded the last corner before the mine site, he had to swerve to avoid a Mountie manning a roadblock. Cyr slowed down long enough to shout, "Mine rescue!" Then he tore off towards the mine.

* * *

IVAN BAKER WOKE UP to the sound of someone pounding on the front door of his brick bungalow, just outside New Glasgow. It was still dark outside, barely six in the morning. The thirty-eight-year-old RCMP constable was off duty for the weekend, so the phone was unplugged. If there was an emergency, Baker reasoned, someone from Stellarton detachment would be over to find him. Baker opened the door; sure enough, it was one of the sergeants.

But this was an emergency beyond anything Baker had ever faced in his twenty years with the Mounties. There had been an explosion at the mine, the sergeant said, and someone had to field the inevitable calls from reporters. The officer who normally handled media relations for the eighteen-member detachment was in Cape Breton, on a diving team scouring Sydney harbour for the gun used in the McDonald's restaurant killings. The task now fell to Baker.

Baker showered quickly and pulled on his uniform. Dealing with the media was a new assignment, but he had watched a lot of reporters in action. A native of the Newfoundland outport of Brown's Arm, Baker had spent several years in Ottawa on VIP detail, driving armoured limousines during state visits. The Queen, Ronald Reagan, Trudeau—Baker had driven them all. He had watched the way his famous passengers handled the onslaught of journalists armed with microphones and television cameras, and he had been impressed with how adeptly they handled the questions fired their way.

The Westray mine loomed in the distance behind the RCMP detachment building on Foord Street, Stellarton's main drag. Once inside, Baker had to decide where to send reporters and camera crews as they showed up in Plymouth. He thought of the fire hall, but he soon learned that miners' families were gathering there. The only other choice was the Plymouth Community Centre, just across the road from the fire hall. He dispatched an officer to commandeer the building for a media centre, then turned to the phone. He answered call after call from reporters, some from as far away as New York and London. But Baker could tell them little; all he knew was that there had been an explosion, and miners were trapped underground. He had no idea how many men were in the mine, or how serious the explosion had been. After almost an hour, Baker put down the phone and headed to the mine site to get some hard information.

After a briefing from Gerald Phillips, who was coordinating the rescue effort, Baker convened a press conference at the community centre at 10:15 a.m. Perched on a hill facing the mine site, the centre was a three-storey former schoolhouse with a newer meeting room grafted on one side. It was normally the domain of Boy Scouts and dart players, the venue for teen dances and potluck suppers. The entire building was badly in need of a new coat of white paint. Only about a half-dozen reporters were on hand, mostly from local papers and radio stations. Baker set up a stacking table at one end of the auditorium, in front of a beige curtain adorned with vertical stripes in zigzag patterns of darker brown. It was to became a familiar backdrop in the days ahead.

Details were still sketchy five hours after the blast. There had been "an explosion of significant proportion," Baker announced, and as many as twenty-three men were believed trapped a kilometre and a half below the pithead, the entrance to the mine. Until then, radio reports had estimated there were seventeen missing miners. The explosion, Baker confirmed, had knocked out electricity underground, and cut off ventilation. A rescue team had already gone down into the mine, he added; more teams were en route from Cape Breton and the salt mine in Pugwash, about an hour away.

* * *

THE REPORT FROM the first rescue team was not good, and Steven Cyr knew it. Wearing breathing apparatus, the team had reached the fifth crosscut, about eight hundred metres into the mine. Damage was extensive all the way down. No. 1 slope, the main access tunnel for men and machinery, was littered with debris. Each crosscut was fitted with a stopping, a bulkhead made of steel-reinforced concrete, with doors that can be opened or closed to direct the flow of air; the force of the explosion had reduced them to rubble. Steel arches lay flat on the floor; the plastic on the vent tubing had been burned away, leaving only the wire ribbing. Seven-tonne electrical transformers had been tossed around like toys, and had crumpled as if they had been tinfoil. The methane reading at No. 5 crosscut was 2.5-per-cent methane, below the explosive zone. But the carbon dioxide level was dangerously high—800 parts per million. Cyr tried to convince himself there was still hope for the men who had been digging coal hundreds of metres deeper inside the mine.

Cyr had grown up in Westville with some of them. Eugene Johnson, who had been a few years ahead of him in school, was captain of his mine rescue team. He had known Mike MacKay's wife, Beverly, since childhood—she was one of his sister's best friends. And he had worked overtime shifts with a couple of other men on B-crew. It was mid-afternoon before Cyr finally got a chance to see the damage for himself. With his team captain among the missing, Cyr was attached to a Devco mine-rescue team from Cape Breton. This would help the newcomers become familiar with the layout of the Westray mine.

The descent into the mine after the explosion was not unlike what one might conjure in picturing a descent into hell. The atmosphere was a deadly cocktail of noxious gases—methane, carbon monoxide, carbon dioxide. Beams and arches that once held up the roof were twisted, or lay useless on the mine floor. Smashed equipment and mounds of rubble littered the tunnels, forcing rescuers to advance on foot. There was the ever-present risk that the weakened roof would collapse, or that a spark or still-smouldering patch of coal would touch off another blast. More than courage is needed to face such dangers; it takes a combination of determination and toughness, mental as well as physical. Cyr was a newcomer to mine rescue, but he was driven by the same resolve as all the other draegermen; every effort had to be made to find the men, dead or alive. They owed it to the families waiting far above, to their comrades, and to themselves. Being entombed in a mine for all time was every miner's worst fear.

Draegerman—the word comes from Draeger, the German company that manufactured the breathing equipment used by mine rescue crews earlier in the century. The crews at Westray carried oxygen masks connected to a fifteen-kilogram oxygen pack, which was strapped to their backs like an oversized metal knapsack. The pack contained enough air for four hours. There were five men on each crew, and they worked in tandem; one team would forge ahead to check conditions or carry out repairs, while the other one acted as backup. Communication between teams was impossible, so the backup had to be ready to search for the advance team if it failed to return before the air supply ran out. The first task was to re-establish the flow of fresh air. This would allow bare-faced miners to enter—men who could clear debris and do more-extensive repair work, free of the burden of breathing equipment. The deeper into the mine that ventilation was restored, the deeper the draeger teams could operate.

Cyr's team went in about four o'clock Saturday afternoon to erect a temporary plastic seal over one of the crosscuts. He was shocked at the severity of the damage, barely believing it was the same mine he had left just twenty hours earlier. In some places, the roof was still creeping, threatening future cave-ins. Even veteran draegermen from Cape Breton were stunned by the extent of the damage. Cyr quickly lost track of time; he was running on adrenaline and nerves.

* * *

IT WAS TWO O'CLOCK on Saturday afternoon before Bert Martin got the phone call. A woman asked if Glenn Martin had worked the night before. Bert, a grandfatherly, grey-haired man with a ruddy face, confirmed he had. The woman thanked him and hung up. There was no explanation, no mention of an accident or explosion, not even an acknowledgment that the caller was from Westray. By then, Bert and his wife, Jean, were certain Glenn was among those trapped underground; Chris and his brothers had already come over to break the grim news. Glenn had been scheduled to work, and Bert knew he was too conscientious to skip a shift.

Bert just shook his head. How could Westray, or any other mining company, be that disorganized? At the McBean colliery, where he had worked, every miner was issued a brass identification tag; it was hung on a board each time a man went into the mine. Westray, as it turned out, had a tag board—but it was never used. And Bert knew something else from his mining days. A draegerman for eleven years, he had taken part in rescue operations at two mine disasters in Springhill, Nova Scotia, including a 1956 coal-dust blast that killed thirty-nine miners. Bert was familiar with the kind of damage an explosion caused when it ripped through the cramped workings of a mine. He felt certain there was no hope of a miracle, no hope of finding Glenn or any of the others alive. But he kept his thoughts to himself.

* * *

IVAN BAKER could hardly believe his eyes when he returned to the community centre at about three o'clock for a second press briefing. The auditorium had been transformed into a jumble of electrical cables snaking their way to cameras and microphones, monitors and tape decks. Dozens of reporters were already stationed at the community centre, and scores more were on the way. By then, the Westray explosion was the lead item on radio newscasts across the country; CBC Newsworld and local television stations were installing satellite dishes outside the building so they could go live with updates on the rescue effort. Reporters for newspapers with Sunday editions were already tapping stories into portable computers. The American networks, including CNN in Atlanta, were beaming the story to the world; the explosion had thrust the rural hamlet of Plymouth, Nova Scotia, into the international spotlight.

This time the spotlight was on Curragh executives Colin Benner and Marvin Pelley, who had just flown in from Toronto. Baker introduced the two men, and the company made its first official announcement since the explosion. Benner, dressed in a dark blazer and a plaid shirt left open at the collar, took a seat at the table, which was now covered with microphones and tape recorders. Twenty-six men were trapped underground, he revealed, three more than previously believed. Rescue teams had already descended nine hundred metres into the main tunnel, more than halfway to the mine's work areas. Eleven rescue crews were on site, working to re-establish power and ventilation. They hoped to reach the trapped men in as little as twelve hours, and Benner promised to be back by seven o'clock with an update.

Information was still scarce, but the terse briefing marked the debut of Curragh's point man for the Westray crisis. Benner, a forty-seven-year-old career mining executive with just a touch of grey at the temples, was the perfect spokesman for a corporation under fire in the age of instant TV coverage. He had the grooming and good looks of a movie star or a model on the cover of *GQ*. And when he faced the cameras, Benner was the picture of grace under pressure. He exuded calmness, confidence, and just the right degree of toughness as he deftly fielded reporters' questions. For millions of Canadians keeping vigil in front of their television sets, Benner came to symbolize the Westray drama.

But Benner was more than just another pretty face. As Curragh's president of operations, he was responsible for activities at all the company's mines, Westray included. The explosion was stark proof that his fledgling Mine Planning Task Force, with its mandate to make the Westray mine safe, had been too little, too late. Now that the lives of twenty-six Westray miners hung in the balance, it made

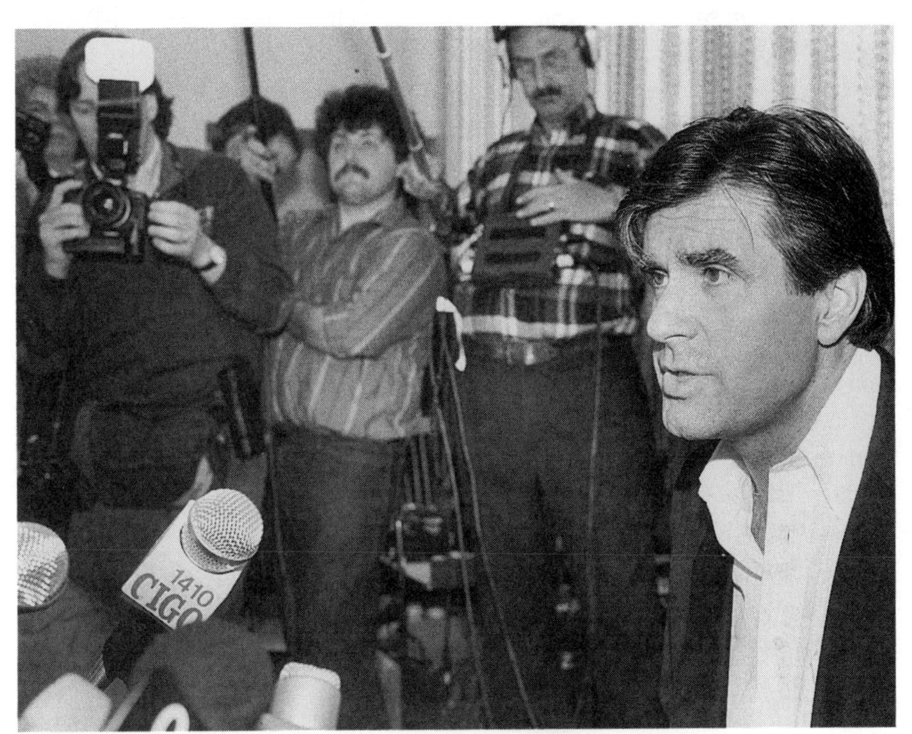

For millions of Canadian keeping vigil in front of their television sets, Colin Benner, the Curragh spokesman with the movie-star looks, came to symbolize the drama unfolding at Westray. [Photo by Len Wagg, courtesy of the *Chronicle-Herald* and the *Mail-Star*]

sense that he should be on the hot seat. The fact that he was a natural on TV was a bonus for Curragh, as the company coped with a public relations nightmare.

Benner's stellar rise in the mining industry was the stuff of Horatio Alger. The son of a Northern Ontario prospector, Benner went into the mines as a labourer in his late teens. Seven years later, he went back to school to get an engineering degree. He joined Denison Mines in the mid-1970s and ended up in charge of that company's giant uranium operation at Elliot Lake, Ontario. In mid-1990, he joined Curragh and was reunited with Denison's most prominent alumnus, Clifford Frame. He had a reputation as a no-nonsense executive, willing to tackle tough problems head-on. "Colin is a straight shooter," one Curragh insider noted admiringly. "He doesn't bullshit people."

And Benner knew how to handle the media. Before jumping to Curragh, he served a stint as manager of Denison's potash mine in Sussex, New Brunswick. The Denison-Potacan mine had a safety record that was spotty at best—seven miners died underground, three of them in cave-ins, in five years. In early 1990, when CBC Television in Halifax produced a documentary about safety concerns at Potacan, Benner went on camera to defend the company. "Safety is first at Denison-Potacan Canada," Benner assured interviewer Gerry Whalen. "Our motto is: 'Safety and order equals security. And security and order equals safety.' " Two years later, with twenty-six men trapped after a massive underground explosion, only the company names had changed. Benner was again in charge of damage control.

* * *

ROBERT HOEGG OF STELLARTON and a handful of other veterans of Pictou County's collieries gravitated to the community centre to monitor the rescue effort. And they were more than willing to talk about the disasters and accidents that had punctuated the area's mining history. The death toll was appalling: as many as 650 men had been killed since mining began in the early nineteenth century, 246 of them in less than a dozen explosions. As the experienced miners well knew, the deadliest and most explosion-prone seam was the Foord. The county's worst disaster was a 1918 explosion at the Allan shaft, which mined a section of the Foord seam adjacent to the Westray workings; that blast killed 88 miners.

Hoegg, a former mine manager and draegerman in his seventies, offered his blunt assessment of the situation to a newspaper reporter. "I can't see any hope, myself," he said, putting into words what most people were trying to banish from their minds. "The Foord seam," Hoegg said, "has never been anything more than a graveyard, in my opinion." There was already speculation that methane was the culprit, but Hoegg suggested that methane alone was not to blame. The force of the explosion, evidenced by the heavy damage to the portal covering, suggested another notorious enemy of the miner—coal dust. "Coal dust is just like gunpowder," Hoegg explained. "It would go through the workings just like a bullet going out of a rifle." He talked about stone dust, how it prevents coal dust from igniting, and how it can localize any explosion that does occur. If coal dust had been a factor in the Westray explosion, it was clear that safety had been seriously compromised at the mine.

Elmer MacKay was singing a different tune. The local MP, Westray's longtime friend in Ottawa, held an impromptu press conference Saturday on the front steps of the community centre to extend his sympathies to the families of the missing men. But reporters quickly raised the safety issue. MacKay said he had no first-hand knowledge of safety concerns, other than "what I've read from time to time." He had been underground at Westray in the past and felt it was safer than other collieries he had toured. "There appeared to be so much more air, the tunnels appeared to be so much larger, the method of getting the coal out so much more modern, that I felt reassured."

* * *

WITHIN HOURS OF THE EXPLOSION, an army of volunteers was in the field, attending to the needs of those caught up in the disaster. The Ladies Auxiliary of the Stellarton Fire Department took kitchen detail, serving hot meals to families at the Plymouth Fire Hall and rescue workers at the mine site. A day-care centre was set up at Stellarton's Sharon St. John United Church to look after the children of rescue workers, volunteers, and miners, including four youngsters who had fathers trapped in the mine. Donations of food—everything from soft drinks to canned goods—poured in from local businesses and restaurants. Jim's Taxi in Stellarton offered free transportation. Even the growing contingent of reporters was looked after. Sandwiches and pizza appeared out of nowhere, and people on the Plymouth exchange gave up extra lines so that dozens of phones for the media could be installed at the community centre.

The four major towns in the heart of Pictou County—Trenton, New Glasgow, Westville, and Stellarton—are clumped together around old colliery sites and ageing factories that speak of a bustling, more-prosperous past. The combined population is less than 25,000. One town flows into the other, making civic boundaries invisible to outsiders. There are long-standing rivalries between the communities, played out on hockey rinks and softball diamonds; there are stubborn differences settled with fists outside Friday-night dances. But jealousies vanished in the face of a mining disaster. Anyone over forty-five could vividly remember the last time miners paid the ultimate price for mining coal; a 1952 explosion at the MacGregor mine in Stellarton killed nineteen men. And countless families could name a brother, father, grandfather, or great-grandfather who perished in the mines.

The Westray explosion, the first disaster to strike the area in more than a generation, touched off a spontaneous outpouring of sympathy and support within the community. Volunteers displayed almost superhuman strength and character, working through the night without complaint; some people took vacation days so they could help out. When the day care's appeal for help was broadcast on the radio, more than one hundred people called to offer their time. Roddy Smith, a counsellor for troubled teens from Trenton, was asked why he put in twenty-hour shifts at the day care. "When somebody hurts in Pictou County," he explained to a reporter, "everybody hurts."

Normal routines came to a standstill as people rolled up their sleeves to help out, or spent the weekend glued to radios and television sets, waiting for the latest news from the mine. The annual Johnny Miles Marathon, named for a Nova Scotia coal miner who won the Boston Marathon in the 1920s, was cancelled; the race had been scheduled for Sunday, and the route included the cordoned-off road past the mine. Bingos, auctions, a tree-planting day for Plymouth's Cubs and Scouts, and other events were put off for another day. The local AM radio station, CKEC in New Glasgow, abandoned its lively mix of country and top-40 tunes in favour of sombre instrumental music, reminiscent of the way the deaths of presidents were commemorated in the Soviet Union.

* * *

MINE OFFICIALS CONVENED another press conference about seven o'clock Saturday night, as promised. Gerald Phillips, who was coordinating the rescue effort, joined Colin Benner to give a first-hand report on progress underground. Phillips, his white mining helmet grimy with coal dust, said the trapped men were believed to have been working in three groups—one group in the southwest section, the others split into two work parties in the North Mains. Draeger crews were re-establishing ventilation as they advanced, he noted, but the missing miners each had a self-rescuer, which could "exchange bad air for pure air" for up to four hours.

Asked about the cause of the explosion, Phillips said that would be determined after the rescue operation was completed. But he noted that the mine had won the John T. Ryan award, the mining industry's highest honour for safety. Westray, he assured reporters, was "as safe a mine as there is."

Only a handful of reporters were still hanging around the community centre when Benner returned at midnight; most had wandered off to motels for the night. He started by reading a two-page press release describing Westray's computerized methane-detection system. Gas readings in the hours before the explosion were well within government guidelines. "The indications are that there may have been an instantaneous build-up of methane gas in the mine which led to an explosion." Curragh, he pledged, would be conducting "a full, independent investigation" of the blast "in due course," in conjunction with government agencies responsible for mine safety. When a reporter asked why the gas build-up had not been detected, Benner speculated that it had happened too quickly for the methane monitors to send a signal to the surface.

Then Benner updated the rescue operation. Teams had advanced to No. 10 crosscut, one of two entrances to the southwest section. The roof was in good condition, and methane was recorded at a comfortable 1.25 per cent in the area. Benner estimated that rescue crews could reach the southwest section's production faces within three hours. Outside, as a cold spring rain pelted down on beleaguered Pictou County, the Westray rescue effort entered its second day.

Benner was back at daybreak. By then, CBC Newsworld was broadcasting each Westray press conference live to a waiting world. Rescuers were within 350 metres of the coalface in the southwest section, he said, but blocked tunnels had slowed their progress. He estimated it would be another two to three hours before they entered the area. More than forty draegermen were on site, including fresh teams that had just arrived from a mine in Bathurst, in northern New Brunswick. The mood was still optimistic. Benner even discussed procedures for bringing survivors to the surface, where they would receive medical attention. "I wouldn't

Rescue workers at the mine's shattered portal. Scores of draegermen descended on Westray from mines in Nova Scotia and New Brunswick, risking their own lives in the desperate search for survivors. [Photo by Len Wagg, courtesy of the *Chronicle-Herald* and the *Mail-Star*]

want to raise any hopes in anyone," he cautioned, "nor would I want to suppress anyone's hopes."

* * *

FOR THOSE GATHERED across the road at the Plymouth Fire Hall, hope was all that remained. Westray officials dropped in occasionally to update the rescue effort; Gerald Phillips, Colin Benner, or Eldon MacDonald, Westray's accountant, handled the briefings. The long wait for news from underground was like slow torture. Many cringed each time someone from Westray entered the room, fearing they were about to announce that everyone was dead. Phillips prolonged the agony each time he walked in, by picking up a cup of coffee before heading to the front of the room to make an announcement.

At any given time, about one hundred people were crammed into the upstairs room at the fire hall. Tables were set up, each with a vase of carnations and a box of tissues as a centrepiece. Wives and parents of miners from Alberta, Ontario, Newfoundland, and other provinces were flown in at Westray's expense. Volunteers provided a steady supply of food, coffee, and soft drinks, but some chose to go out for meals as a much-needed break from the tension. Most relatives lived in

Relatives and friends of the trapped miners gathered at the Plymouth Fire Hall, which overlooked the mine site, to wait and pray for a miracle. [Photo by Len Wagg, courtesy of the *Chronicle-Herald* and the *Mail-Star*]

the area, or had motel rooms; for those who preferred to stay close to the scene, rows of cots were set up on the concrete floor of the lower-floor garage.

The family of each missing miner was assigned a counsellor. These counsellors, drawn from the Aberdeen Hospital's mental health centre and other local agencies, worked shifts so that someone would be available around the clock. Clergymen and Salvation Army workers were also on hand to offer solace, and at least one doctor was always there. Donna Johnson stayed at the fire hall, keeping vigil with her mother, two brothers, and Eugene's mother. Eugene's brother Robert, a fellow Westray miner, was somewhere in the mine, working on a bare-faced crew in support of the draegermen. The counsellors meant well, but Johnson could not deal with the intrusion into her private grief. At one point, she retreated to the parking lot and sat in her mother's car so she could be alone.

Joyce Fraser's one hope was that Robbie would somehow get out alive to see his unborn child. She spent most of Saturday at the fire hall, but the stress was taking its toll, and her family tried for hours to persuade her to wait at home, fearing she might lose the baby. Westray officials did little to help when they came to the hall early in the evening and read the names of the twenty-six miners who were underground. Robbie was not on the list. Fraser stood up and insisted that her husband had been in the mine at the time of the explosion. "Are you sure he went to work?" asked MacDonald, the accountant. An hour later, a company official apologized for the mistake. But the number of men trapped was still twenty-six. One of the miners named earlier as among the missing, it turned out, had been standing outside the fire hall at the time.

* * *

IN THE HECTIC HOURS after the explosion, a public relations strategy seemed to emerge. Curragh and Westray put on the appearance of openness through Colin Benner's televised press conferences, then worked behind the scenes to drive a wedge between the media and the families of the trapped men. Many of those anxiously waiting in the fire hall for word from the pithead had been told horror stories about safety conditions in the mine. But company officials repeatedly warned them not to talk to journalists or listen to the news on radio and TV. Reporters were blowing the whole incident out of proportion, they said; the media were vultures, trying to use the families' anguish to make money. Satellite dishes set up outside the community centre were microphones, relatives were told, capable of eavesdropping on conversations inside the fire hall. The smear campaign gave family and friends of the trapped men a convenient target for their anger and frustration; with a horde of reporters camped on the doorstep, Westray and Curragh were able to keep a lid on safety problems at the mine.

People inside the fire hall already felt like hostages to the media. The community centre was little more than fifty metres away; a paved road and a row of young pine trees was all that stood between them and scores of journalists and camera operators. Each time a family arrived or left, or someone stepped outside for a smoke or a breath of fresh air, every move was video-taped or photographed. Initially, at least, there was little else to record on film; the company refused to allow cameras near the mine site. The ban, Curragh's Marvin Pelley said, was imposed "out of courtesy and respect" for the families of the missing men. But Ivan Baker managed to negotiate a better arrangement for the media; RCMP officers escorted one or two camera crews at a time to a vantage point overlooking the mine property, where they could record draeger crews entering and leaving the shattered portal.

A couple of incidents made it easier for the company to keep the media at bay. The RCMP, at the families' request, ordered journalists to stay away from the fire hall. At least one camera crew went over anyway, and was promptly chased away by an irate family member. A story circulated through the fire hall that reporters were being pushy and insensitive, asking some relatives how it felt to have a loved one trapped in the mine. The rumour, true or not, soured most on the idea of speaking to the media. Bruce Stephen, the brother-in-law of missing miner Angus MacNeil, was one of the few to wander over to the community centre. "I think you always have hope," he said, surrounded by a group of reporters. "But although we do have hope, it's pretty dim."

* * *

A DRAEGER CREW inching its way into the southwest section discovered the bodies at mid-morning on Sunday, May 10. Eleven men had been working in the area at the time of the blast; all had died within a few seconds of the methane-generated fire that touched off the explosion. The faces and hands of all the victims had a pinkish tinge, a sure sign of poisoning from the carbon monoxide produced when the methane ignited. Most had singed hair and clothing. One man, his face severely burned, had apparently been caught directly in the path of the blast.

At least some of the miners had known that an explosion was imminent; a couple had time to grab self-rescuers from their belts before they succumbed to the fumes and fell to the floor. The man operating the continuous miner had managed to sprint some distance from the coalface before the blast, suggesting that it had been triggered by a spark given off as his machine gnawed its way through the seam. At least one man had no warning of the catastrophe; he was found sitting upright on the seat of a vehicle, his hands still firmly clutching the steering mechanism.

* * *

THE TERSE ANNOUNCEMENT sent a shock wave through the fire hall. "Something we feared has come true," Gerald Phillips told the families huddled around their tables. "Eleven bodies were found in the southwest section." A woman began screaming, drowning out his words. One of the Mounties who had showed up at the hall with Phillips said the bodies were being taken to the New Glasgow arena, where relatives could make a positive identification. But the names of the men were known, and their families would be told one by one, table by table. Isabel Gillis was the first to find out she was a widow. When they told her Myles was dead, she kept her hands clenched to the bottom of her chair, refusing to budge. She was in a daze. As long as she stayed in the fire hall, she felt, Myles was still alive; the moment she walked out the door, he would be dead. Donna Johnson collapsed and could barely breathe. Her brother carried her to a motor home parked behind the hall, where the women were being taken, one after the other, to recover from the shock. Alex Ryan, Eugene's buddy and fellow miner, followed to make

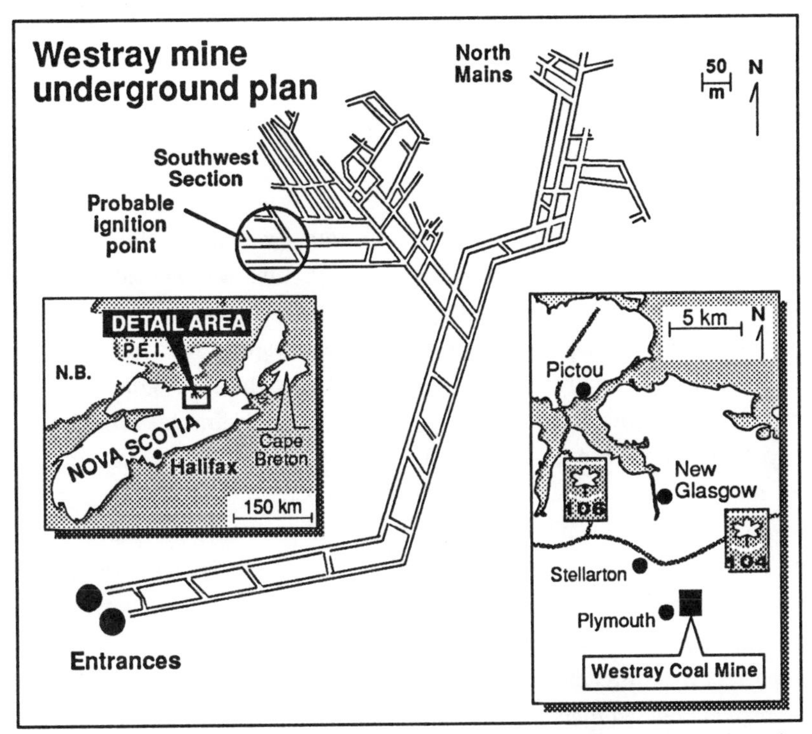

Eleven bodies were discovered on May 10 in the southwest section, the area where the explosion was believed to have been touched off. Fifteen other men were working in the North Mains when the mine exploded. [Graphic by Tina Leighton and Peggy McCalla, courtesy of the *Chronicle-Herald* and the *Mail-Star*]

sure she was all right. "I just want to go home," she sobbed. Eugene had been her whole life. Now that he was gone, she felt like she wanted to die, too.

* * *

"LADIES AND GENTLEMEN, this is a report that we all hoped we would not have to give." It was a few minutes past one o'clock on Sunday afternoon, thirty-two hours after the explosion. Colin Benner looked shaken and downcast as he sat before the cameras to deliver the bad news. Relatives were already filing out of the adjacent fire hall in tears, clinging to one another for support. "Rescue teams have reported that they located the bodies of eleven men in the southwest area of the Westray coal mine. Based on the preliminary reports of the mine rescue crews, it would appear that these men died instantly as a result of an explosion. The waiting families have been advised of the situation."

He paused for a moment as reporters scrambled to telephones to get the news out. His voice was halting, his normally confident tone subdued in defeat. "I can only add that our deepest sympathy goes out to the families at this tragic, tragic time. I know you'll join us in offering our prayers to these families, their friends, and the fellow employees of these men. At the present time we are a community in grief, and we ask that the privacy of our employees and their families be respected."

* * *

JOYCE FRASER, under doctor's orders to stay home, watched Colin Benner's announcement on television. None of the eleven victims was named, but she was sure one of them was Robbie. All she wanted was to be alone. She went outside and sat on the back step, away from the relatives waiting with her at the apartment. The sun had finally broken through, bathing the dead grass and leafless trees in its warm glow. Her five-year-old, Robbie, was riding his bike in the back yard with David Johnson, Eugene's son. She watched Donna Johnson and her people return from the fire hall, but she resisted the urge to walk next door and ask about Robbie. A few minutes later, the agonizing wait was over. Fraser's mother and father-in-law, who had been awaiting word at the fire hall, drove up. Robbie's body, they said, had been found. Fraser, six months pregnant, was now a widow with three children under the age of ten.

Reporters were already showing up outside, looking for quotes. Fraser's brother and nephew fended them off, but someone had to tell the kids. That was a mother's job. Fraser had kept her children clear of the TV for two days, but they knew that people were trying to save their father and the other trapped miners. Robbie had trouble understanding why they could not get them out of the mine. Next door, Johnson took her two sons, Michael and David, into a bedroom to tell them in private. "We won't be able to go fishing any more," one of the boys protested. "Dad knew all the secret spots."

* * *

DONALD CAMERON, the man who had invested a massive chunk of his political capital in Westray, drove up to the community centre at half-past four to face the press and the nation. The Nova Scotia premier had been at his Pictou County home when Westray manager Gerald Phillips called early Saturday with the news that the mine had exploded. At first, he thought he was having a bad dream. Cameron had spent a few hours at the mine site Saturday night, and Phillips and other company officials had phoned to keep him abreast of rescue operations throughout the weekend. By Sunday afternoon, the bad dream had turned into a nightmare. Eleven men had died and, as each minute passed, it seemed more likely that another fifteen would also be found dead. The premier's attire was laid-back—blue wool sweater, black slacks—but his mood was sullen. As he took the seat normally reserved for Colin Benner, Cameron's face bore the sombre look of a man watching his good intentions slowly turn to disaster. Kevin Cox, a reporter for the *Globe and Mail,* took one look at Cameron and scribbled the word *chastened* in his notebook.

"I, like all Nova Scotians, and especially the people in Pictou County, am deeply saddened at the events of the last two days," the premier said. "I hoped and prayed like all others that somehow the men would be rescued. . . . It is hoped the mine rescue teams will be able to reach the other men, and they can save them." A reporter asked if there would be an inquiry. "The inquiry will be automatic. . . . We will go out and hire some outside expertise to assist in this inquiry. But the first priority must be to get to the remaining men." Cameron was asked about safety. "It's not for me to pass judgment on the safety of the mine," he protested. "There's been a lot of things said about it. And we're not going to help anyone today by rehashing the political statements, and trying to separate them from fact." That was Cameron's standard response when faced with any criticism of Westray; he always dismissed it as politically motivated, based on lies and distortion. But miners were dead, and Cameron conceded that questions about safety had to be addressed. "Clearly, we have a major disaster," he added. "Something was wrong."

Cameron was asked if he regretted having worked so hard to bring the mine into production. The question was inevitable, and he was ready for it. "Well," he began, "that's crossed my mind a thousand times in the last two days. Our government gave this business a $12-million loan at full rate of interest; we've helped other companies far more than that; we've helped other coal mines far more than that." He was referring to the heavily subsidized Cape Breton Development Corporation mines, which had long been a whipping boy of Westray's supporters. "But, boy, I want to tell ya," he added, "if I could have, in any way, foreseen what has happened here in the last two days, I'm sure we all would have made different judgments." Cameron's voice trailed off for a moment as he collected his thoughts. "I just feel so badly for the families, the young children, and the wives and the parents. It's very difficult to accept and express."

Jim Nunn, a local CBC-TV news anchor with a knack for getting under the skin of the ruling Tories, seized on Cameron's reference to "political statements." Cameron backed off, seeking refuge behind the families of the dead miners. "There were so many lies told about this, the Pictou mine, it's hard to separate [them] from the facts," he said. "The reality is that we have a major disaster today, but I'm not sure that we're going to help those people that are grieving, and all other people that feel so badly about this, getting into a political argument about things that were said in the past." As for the impact the disaster might have on his own political fortunes, he said, "that's for the people to decide."

Cameron expressed his sympathies for the families of the dead men, and held out hope for relatives of the fifteen miners still missing. "You never want to raise false hopes, but the fact that the tunnels are blocked could be an advantage; that could block some of the effects of the explosion." Cameron was then asked to look further into the future, to a day when Westray might once again be allowed to mine coal. That decision was not up to him alone, he said, but he did offer a prediction. "There's no way the mine's going to open soon again," he said. "Maybe never."

* * *

IT TOOK MOST OF THE AFTERNOON for rescue crews, still labouring under breathing equipment, to remove the bodies. The dead had to be put in bodybags, carried to the main slope on stretchers, then driven to the surface on tractors. There were too many bodies for the Aberdeen Hospital to handle, so they were taken by ambulance to a makeshift morgue set up at an arena in New Glasgow. The dead were identified before they were brought to the surface; Westray draegermen could recognize their fallen comrades without looking at the name sewn on each man's coveralls. But the RCMP required a positive identification. Gerald Phillips sent Lenny Bonner and two other Westray employees to the rink to carry out the gruesome task.

Bonner looked on as medical attendants pulled back the sheets covering the faces of the dead men, one after another. Robert Doyle, just twenty-two, a Plymouth boy and volunteer firefighter, who lived on the family farm bordering the mine site; Larry Bell, another Pictou County native, who played in a local band; Robbie Fraser and Eugene Johnson from Westville; Myles Gillis—he and Bonner had worked together at the Gays River mine; Trevor Jahn and Ferris Dewan, best friends in their mid-thirties, both experienced coal miners from Alberta; Harry

McCallum and Eric McIsaac, native Nova Scotians, both married with two children; Romeo Short from Triton, Newfoundland; and Adonis Dollimont, another Newfoundlander, who was married and had a teenage daughter. Bonner had counted many of them as friends; he had been talking to some of them only two days earlier, as they came on shift. Now they were dead. It would be months before Bonner could sleep, or even close his eyes, without seeing their faces.

* * *

THE ANNOUNCEMENT THAT ELEVEN MEN were dead came as no surprise to Carl Guptill. He knew there was no hope from the moment Roy Feltmate's wife, Bernadette, called early Saturday morning to say there had been an accident at the mine. It was the call that had haunted him in his sleep. Guptill put down the phone and woke his girlfriend. "They're all dead," he told her.

Now that bodies had been found and hope was waning for the remaining men, Guptill felt it was time to fulfil his promise to Mike MacKay and the others who had gathered in Feltmate's kitchen the month before. He had monitored the rescue effort on TV with mounting anger, watching Gerald Phillips and Colin Benner brag about Westray's safety record and gas-monitoring systems. It was time to set the record straight: on Sunday afternoon, Guptill sat in a reclining chair in his Stellarton apartment and talked on camera with a CBC-TV reporter. He looked nervous as he peered out from under a black ball-cap emblazoned with a mine rescue logo. "In my experience at Westray," he said, speaking softly and deliberately, "I seen very little evidence of any safety at all. It wasn't just the methane problem—that was our biggest concern—but there was really nothing safe about the place. Not the attitude of management, or anything."

Guptill's comments were aired Sunday night on the CBC National News. By then, other former Westray miners were coming forward to complain that the mine had been an accident waiting to happen. The most vocal of these men were from Grande Cache, a coal-mining town in the foothills of the Rockies, four hundred kilometres northwest of Edmonton. Dozens of experienced Alberta miners had gone east to take jobs at Westray; Ken Evans, Chester Taje, and many others quit because they feared for their safety.

"There's a litany of stuff that goes on there," Evans told the *Edmonton Journal*. "It's not fit for anybody to work in, nobody, no way." Evans worked at Westray for two years, returning to the Smoky River colliery in Grande Cache just two weeks before the explosion. The last straw for Evans came in early April, when miners were repairing the roof after a cave-in. A Westray supervisor ordered them to use an acetylene torch to cut steel beams, and the flame ignited methane seeping out of the mine floor. Nova Scotia's mining laws require that stone dust and firefighting equipment be on hand when a cutting torch is used underground, but these precautions were not taken. Taje, who had twenty years experience underground, repeatedly warned his supervisors about safety problems while he was working at Westray in late 1991. "From my experience in Alberta and British Columbia," he told a reporter, "the operation just wasn't run in a safe and efficient manner."

A union representative who had been trying to organize Westray's two hundred employees just before the explosion also spoke out. Andy Gillis of the United Steelworkers of America said miners complained to him about high methane levels, cave-ins, poor training, and lax safety standards. "Some of them knew an accident was imminent–they expressed that during the organizing drive."

When he went back on camera late Sunday afternoon, Colin Benner wanted to talk about the rescue mission, not safety. As Marvin Pelley, Curragh's president for corporate development, used a pointer to locate areas on a huge map of the underground tunnels, Benner explained what lay ahead. Six draeger crews were in the mine, and the priority had been removing bodies from the southwest section. Now that the grim task was completed, the section would be sealed at the main slope so that more fresh air could be directed towards the North Mains. Some methane was seeping out of the section, he admitted, but it was not enough to raise the spectre of another explosion. Rescue crews were already making forays deeper into the mine in search of the remaining miners. "I would not want to speculate on the chances of finding the men alive," he said, "but I have to say to you that it's clear with the experience we've had to date that we're quite concerned, obviously."

Reporters began asking about safety. "Was there an indication of coal dust?" one asked. For two days, Benner had eagerly spouted details about gas levels, methane-monitoring systems, and the logistics of mine rescue; but this was a subject he preferred to sidestep. "Ah, we could speculate a number of things that could have occurred," he said evasively, "but I'd prefer not to do that at this time." Another reporter asked for the company's response to the rising chorus of complaints about safety in the mine. Benner deflected the question, saying his press briefings were intended to provide updates on the rescue effort. He did not want to "mix issues," as he put it. "Later on, when time permits, I'm sure you'll get opinion from us on that subject." Then Benner asked to be excused. "We have other business matters to attend to," he said curtly, as he and Pelley walked out of the community centre.

Across the road at the fire hall, family and friends of fifteen missing men were settling in for another gruelling night of waiting, hoping, and praying. Among those still unaccounted for were Roy Feltmate, Mike MacKay, and Glenn Martin. It was May 10, Mother's Day. It was also the day before Glenn Martin's thirty-sixth birthday.

DAYS OF DESPAIR

MONDAY DAWNED SUNNY AND MILD in Pictou County. But the Westray explosion cast a pall over everyone and everything; in homes, in offices, in stores and restaurants, it was the sole topic of conversation. Flags snapping in the stiff wind were at half-mast. Black ribbons adorned store windows in Stellarton. A bouquet of flowers, once intended as a Mother's Day gift, lay at the foot of the town's memorial to those killed in earlier mining disasters. Radio stations were listing the names of the eleven men removed from the mine, but most people had already found out via the grapevine who had been found and who was still missing. It was a day alive with the promise of spring; it was also a day for planning funerals and mourning the dead.

Far away from the sunshine and spring breezes, draegermen continued to pick their way towards the remaining miners. By mid-morning, Colin Benner was back on television with a report on the progress of the rescue effort. Crews had scouted the entrances to the North Mains overnight, locating two possible entries and recording low readings of methane and carbon monoxide. Bare-faced miners were installing large, inflatable bladders to temporarily seal off the southwest section. That would direct more fresh air towards the North Mains, enabling other bare-faced crews to press even farther into the mine. But the obstacles were formidable—mounds of caved-in rock, flattened roofing beams and, in one area, a metre of water.

"The advance now is going to be much slower than we've experienced over the last forty-eight hours," Benner warned reporters when asked to estimate how long it would take to reach the trapped men. "It could be eight to twelve hours; it could be several days." He was equally guarded about the prospects of finding survivors. "By nature I'm an optimist, and we have to remain optimistic," he said, "but I think we have to remain cautiously optimistic, and we are seriously concerned, judging [from] the conditions the crews have reported to date."

* * *

ONE HUNDRED FORTY KILOMETRES to the southwest, in Halifax, the legislative chamber at historic Province House fell silent for one minute on the afternoon of May 11, in memory of the eleven miners confirmed dead. The 1992 session of the House of Assembly had been a stormy one; opposition parties, invigorated by the Conservative government's single-seat majority under a neophyte premier, Donald Cameron, had been out for blood. But on this day, politicians of all stripes were united as they hung their heads in recognition of the weekend tragedy at Westray.

"We must find out what caused this disaster that has brought so much grief and heartache to so many people," Cameron told the legislature. Officials of the Department of Labour were already on the scene and investigating the explosion, he said, but the magnitude of the disaster demanded that "a broader and more encompassing inquiry also be held." To that end, as Cameron had promised the previous day in Plymouth, the government would set up an independent public inquiry to determine what had gone wrong at Westray. The probe, which would be headed by a Nova Scotia Supreme Court judge, to be named later, would have wide investigative powers under the Public Inquiries Act and the Coal Mines Regulation Act.

Bill Gillis, interim leader of the opposition Liberals, and New Democratic Party leader Alexa McDonough joined Cameron in expressing sympathy for the bereaved families and praise for the draegermen risking their lives to search for survivors. But McDonough, whose party had been critical of the government's poor record on labour issues, served notice that the gloves would come off in the days ahead, inquiry or no inquiry. "The fact of the matter is that this is a disaster that was not an act of God," she said. "We have a responsibility to say that we must learn from this situation, and in the future, hopefully, it will be possible to prevent similar tragedies, similar losses."

McDonough's words were still grating on Cameron as he faced a scrum of reporters before driving back to Plymouth to monitor the rescue effort. "There will be lots of time for politicians to point fingers and lay blame," he said. "Surely, out of respect for the families, they could set aside that for a bit." He promised to set "no limits" on the inquiry, which would be free to scrutinize everything from safety concerns to the political decisions that led to the mine's development. He did not feel Westray officials had misled him about safety standards but, he pointed out, "I'm not a mine engineer, so you take the advice of people." Asked if he had let his constituents down by not asking more questions about safety at the mine, Cameron became testy and turned the tables on his interrogators. If the government assisted a company where there was a loss of life, he said, the government should not automatically be held responsible. "Do you really believe that any government, members of a political party, or any company would do things knowing that people's lives would be at risk?"

The other politician under fire over Westray was Leroy Legere, the minister of labour. Complaints by former miners about safety at Westray, he contended, were news to him and to his department's officials. "From the reports I've had, actually the methane levels were quite low at Westray," he said. "We were assured that the mine was following all the safety procedures as best they could, and like any mine, there's always a danger and that's where we're at." He speculated that the mine would not be allowed to reopen until the Labour Department completed its investigation, and predicted that the investigation, headed by the province's chief mine inspector, Claude White, could take as little as a month. The message from Cameron and Legere was clear–there had been an unfortunate accident, and the provincial government was not about to shoulder the blame.

In Ottawa, opposition politicians also showed restraint in deference to the families of those still trapped in the mine. For its part, the government showed that, like its provincial counterpart, it was not about to take the rap for the Westray explosion. Mines and Energy Minister Jake Epp defended the Mulroney government's generous assistance to the project, assuring reporters that his department had studied the methane risk before Curragh was given an $85-million loan guarantee in 1990. "The conclusion of the report was that the mine was technically feasible, but we were all aware of the Foord seam having high methane deposits." The federal cabinet had been confident that Curragh could manage the risk, under the watchful eyes of provincial inspectors. "With new technology, there would have to be a learning curve in which both new technology and mine operation would have to be very closely monitored by the Nova Scotia government," Epp said.

But one of the federal government's own agencies had warned, as far back as 1987, that the Westray project posed grave financial and safety risks. Devco, the Crown corporation operating Cape Breton's mines, sent a confidential memorandum to the federal and provincial governments, advising against funding a new mine on the Nova Scotia mainland. The memo dealt mainly with the economics of the project, claiming Westray would cut into Devco's market for thermal coal and lead to the layoff of one hundred Cape Breton miners. There was a cursory reference to the complex geology of the Pictou coalfield, which is crisscrossed and fractured by numerous faults. "As a result of such difficulties, its seams have given off large volumes of gas and [have] proven extremely liable to spontaneous combustion," Devco warned. The Devco memo, made public back in 1988, played a prominent role in news coverage as the search for survivors entered its third day.

* * *

As MONDAY WORE ON, the Westray disaster threatened to turn into a public relations fiasco for Curragh. The company's shares took a beating on the Toronto Stock Exchange, tumbling 70 cents to close the day at $3.70. Investors and brokers were edgy about the effect of the explosion on the finances of Westray's parent company. Curragh's head office in Toronto called in Reid Management, a Toronto-based communications firm, which recommended decisive action: the flurry of news reports alleging unsafe conditions had to be tackled head-on, and quickly. That advice was faxed to the mine site in Plymouth, and at five o'clock Colin Benner launched a counteroffensive.

"We are aware of rumours circulating to the effect safety procedures were not strictly followed at Westray," Benner told a televised press briefing, his voice betraying a hint of anger. "We consider these rumours to be an affront to our people." Allegations being bandied about in the media were "absolutely false" and "most defeating in this time of sorrow and anguish," he contended. The company was taking steps to plug the leaks; Marvin Pelley circulated a memo on Monday to all Westray employees, warning them not to release any information or documents without approval. But Benner made no mention of the gag order, choosing instead to deliver a pointed lecture to the scores of reporters standing in front of him. "Some people are assuming that human error is the only possible cause of such a tragedy," he said. "There are miners in Pictou County who will tell you that this is just not so. Mother Nature cannot always be predicted or controlled."

* * *

THE FARTHER DRAEGERMEN went into the mine, the more devastation and debris they found. The explosion had blackened the walls and roof of the main tunnel. In areas where ventilation had been restored, the air was thick with the smell of soot and crushed rock. Concrete stoppings looked as if they had been disintegrated with a phaser out of "Star Trek." Steel arches had been knocked to the floor like dominoes. The hum of electrical transformers and the whirring of fans had been silenced; the mine was cloaked in an eerie, deathlike quiet.

By Monday, Shaun Comish had made three trips underground. About seventy-five draegermen on fifteen teams were taking turns exploring the shattered mine, searching for a route into the North Mains. One of Comish's first tasks was to help bring out one of the eleven bodies found in the southwest section. And each trip underground brought new hardships and new dangers. Rescuers were forced to wear hipwaders in one area, as they picked their way over submerged debris. Every movement—carrying a stretcher, digging through fallen rock by hand, cutting through wire roofing-mesh, or walking up the steep main slope—required extra energy, because of the heavy oxygen packs. The often-unbearable heat fogged up their face masks and left their coveralls drenched with sweat. One man imagined his hair was on fire. Another threw up into his face mask and was forced to eat the vomit; removing the mask and its oxygen supply in the methane-filled mine would have meant instant death.

On the surface, in the comfort of the Plymouth Community Centre, Colin Benner tried to give some idea of what the rescuers were going through. "You can picture it [as] something like climbing a steep hill in the dark, wearing scuba gear," he told a televised briefing. "I cannot emphasize enough the personal bravery and commitment of these mine rescue crews, who have found within themselves extraordinary reserves of physical and mental endurance." To show their appreciation for the rescue teams, relatives of the trapped miners signed a poster that was tacked to a wall of the draegermen's staging area at the mine site. It gave many of them the courage and strength to go back into the mine.

The danger of another explosion or a cave-in was ever-present. Methane levels were sometimes perilously close to the explosive range. Draegermen often walked through areas where arches and timbers had been blown down, leaving nothing to hold up the roof. Some felt the change in pressure that signalled a rockfall somewhere in the mine. Steven Cyr's team was deep in the North Mains, crawling over mounds of fallen rock, when they heard the roof groaning behind them. They stopped in their tracks, knowing there was no other way out if the passage caved in. "C'mon, let's go—never mind that," the team leader said, coolly. "We'll see about that when we get back." Cyr and the others pressed on.

But the constant tension and exhausting work took its toll. Draegermen were sent to local motels for meals and rest between trips underground, but most found it impossible to sleep—especially after retrieving bodies. It was the first time many draegermen had handled a corpse. The scene in the southwest section reminded Comish of film footage from Bosnia, except the bodies were those of his friends and co-workers, not victims of some faraway war. Other draegermen simply lost their nerve. Several crews were forced to return to the fresh-air base or the surface after a team member panicked or cracked under the stress. Stephen Thorne, a reporter with the Canadian Press, managed to get a few draegermen to describe their harrowing ordeal when they returned to their motel. "It was like a horror movie," one man said in a shaky voice. "Worse. You couldn't plan something like that." Another young miner was too overcome with emotion to speak. "My God," he sobbed, "the stories . . . the stories."

* * *

In towns and cities far away from Plymouth, Nova Scotia, tens of thousands of people were keeping vigil via television. TV sets in Halifax taverns, normally dedicated to sporting events, were tuned to the all-news channels. In Sydney, detectives out looking for suspects in the McDonald's restaurant murders sought the latest news about Westray each time they returned to the police station. For the crew of the oil rig *Rowan Gorilla III,* drilling some 250 kilometres off the Nova Scotia coast, the events at Westray served as a reminder of the dangers of their own industry. Bruce MacKinnon, the award-winning editorial cartoonist for the *Chronicle-Herald* and the *Mail-Star* newspapers in Halifax, summed up the sombre mood across the province in a compelling image: the map of Nova Scotia, fractured by a miner's pick driven deep into the heart of Pictou County.

The Westray disaster made headlines and newscasts around the world. The *New York Times* sent a reporter to Plymouth. People from as far away as Japan and Australia called to express sympathy for the families. From California to Cairo, Pictou County natives were able to monitor the rescue effort on the TV news. In Shanghai, China, a former coal miner named Jonathan Fon picked up an English-language broadcast about the disaster on his shortwave radio.

In Western Canada, the drama unfolding at Westray was the main topic of conversation on the streets of Grande Cache, Alberta. The Smoky River mine employed 500 people out of a population of 3,600. A succession of experienced miners from Grande Cache had been lured east to take jobs at Westray; many had

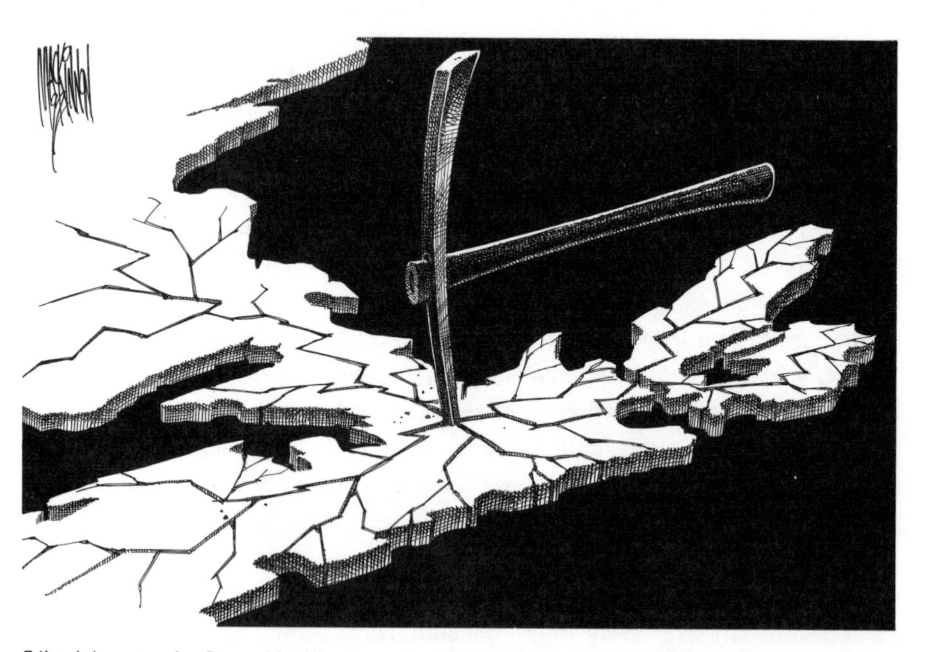

Editorial cartoonist Bruce MacKinnon captured the emotional impact of the disaster with the poignant image of a miner's pick stuck through the heart of Pictou County. [Courtesy of the *Chronicle-Herald* and the *Mail-Star*]

returned, bitter and full of stories of unsafe mining practices. Others had soldiered on, and now they were among the dead and the missing.

There were men like Trevor Jahn, a thirty-six-year-old who left Grande Cache in the summer of 1991 to take a job at Westray. A native of the mining town of Coleman, Alberta, "T. J." had been digging coal since he was seventeen. Westray used Joy continuous miners, the same machines used at the Smoky River mine, and Jahn was hired to show workers how to operate them. With his girlfriend, Bonnie Atkings, her son Jesse, and two cats in tow, he moved to Pictou County. But within a month, Jahn was on the phone to his mother back in Alberta, complaining that the mine was dusty and the roof was falling in. He toughed it out for another nine months, saving up money for the move back out west.

In April, Jahn told his family that he was leaving Westray in June, after he and Bonnie got married. The couple was going to have a baby in the fall. But Jahn never got married, and did not live to see his child. He was among the first eleven miners confirmed dead. "He always said mining was in his blood, and that you've got to do what you've got to do," Jahn's sister, Faye Gibos, told a reporter after the Westray mine exploded. Jahn was buried in Coleman. A few of his old mining buddies gathered at his graveside, cracked open a bottle of rye, and passed it around. Then they poured the rest on Jahn's coffin. "Trevor worked hard and he played hard," one of them explained. "He'd have wanted it that way."

* * *

IN THE EARLY HOURS of Tuesday morning, word came from the mine site that the rescue effort had reached "a critical stage" and that there could be "specific information about our missing men" within hours. But these hopes were dashed at dawn, when Benner announced that the search for survivors would have to be shelved for twelve hours. A draeger crew had been forced to retreat after discovering a patch of coal still smouldering from the explosion. The fire was easily extinguished, but a large rock-fall was blocking their path, and methane readings were higher and oxygen levels lower than previously recorded. The twelve-hour delay, he said, would give crews time to establish a fresh-air base deeper inside the mine, in preparation for a thorough search of the North Mains. That would put rescuers within two hundred metres of the two areas being mined when the explosion occurred.

Benner had stopped at the fire hall to talk to the families before making the announcement. Since there would be no more briefings until evening, many people filed out to their cars and headed to homes or motels to try to get some sleep. For relatives and friends desperate for news from underground—good or bad—the wait seemed endless; the delay only added to the tension and the frustration. But there was still the faint hope that someone had survived. "I think that there has to be hope at this point in time, or these people cannot continue this vigil," said Bruce Stephen, who was waiting to find out whether his brother-in-law, Angus MacNeil, was dead or alive. "Hope is certainly fading, and any light there might be is certainly dim. But that light has not gone out."

Volunteers still had meals to cook, children to care for, and solace to offer. About fifty Salvation Army workers, some from as far away as Halifax, went to the fire hall in shifts to comfort families. More than one hundred police officers from the RCMP detachment and neighbouring towns handled traffic control and roadblocks, some giving up days off to help. As the days dragged on, volunteers and emergency workers became exhausted and burned out; the biggest problem was persuading them to go home for a few hours of sleep.

Reporters, holed up tantalizingly close to the scene, began chafing under the restrictions imposed on the media. Wilkie Taylor, the veteran Pictou County bureau chief for the *Chronicle-Herald,* recalled how draegermen, relatives, and reporters mingled freely at the pithead when he covered the famous Springhill mine disasters of the late 1950s. Despite the high-tech tools that had revolutionized journalism since then, reporters covering the Westray disaster faced a wall of police and barricades separating them from the story. Some reporters questioned the right of the RCMP to block off the public road leading to the mine site, a step justified as necessary to ensure access for emergency vehicles. That, and the RCMP's decision to post a plainclothes officer at the community centre, led some reporters to grumble that the police were too cozy with the company. One columnist sarcastically christened the force "the Royal Curragh Mounted Police."

The news coverage of the disaster unleashed a flood of sympathy for the families of Westray's stricken miners. Donations began pouring in from people around the world who had been touched by the tragedy. Norma Ruddick, whose husband, Maurice, survived nine days trapped in the Springhill mine in 1958, travelled to Plymouth to offer moral support. "I told them to never give up hope, even if one person comes out alive," she said. "I hope [that] before I go home, they have their loved ones back." Inside the fire hall, the wood-panelled walls were plastered with hundreds of brightly coloured cards and letters from schoolchildren, offering encouragement and support. "We are still hoping for you," one child wrote on a card crafted from construction paper and decorated with hearts and a yellow sun. "May God be with them all. We feel like we know them." A pupil at a New Glasgow elementary school pasted a tissue-paper flower to the front of her card. Inside was the cheery message: "Hope this flower will get you through your worries!"

* * *

Clifford Frame, the mastermind of the Westray mine, arrived at the community centre just before sunset on Tuesday, May 12, to make his first public statement since the explosion. Frame and his wife, Catherine, had just arrived in Tokyo when Colin Benner tracked them down with word that, half a world away, the Westray mine had exploded. Within hours, the Frames embarked on the long flight back to Canada. They arrived in Pictou County about midnight Sunday, worn out from lack of sleep. Frame spent Monday at the mine site and was briefed on the rescue effort, while Catherine dropped in at the fire hall with kind words for the families.

Frame, glassy-eyed and subdued, read from a prepared text. "I'm not a Bay Street or Vancouver promoter," he began. "I am a mining engineer, and I have been in the business over forty years." He knew what the families of the fifteen missing men were going through, he said, because he had lost his younger brother in a mining accident. "My sympathies go out entirely, and we've concentrated all our efforts on the people that were left underground and on their wives and their families. . . . Those families are typical mining families. They're strong people. The kind of people I love."

Frame's effort to hobnob with the common folk—and downplay his image as a multimillionaire fresh from the corporate boardroom—relied on more than just words. He sat before the cameras and microphones in a business suit, but wore neither a necktie nor his expensive cuff links. The ubiquitous Monte Cristo cigar was nowhere to be seen. "We hope the time left before final rescue is not too long," Frame continued, praising the rescue effort as "absolutely first-class." There would be financial compensation for families stricken by the disaster, he said, but details would be worked out after the missing men were accounted for.

Questioned about safety at the mine, Frame said he was perfectly well aware of the Foord seam's high methane content and troubled history. "I told you I had forty years in mining. I don't know any mine that doesn't have methane in it," he snapped. "There's going to be an inquiry, and we're going to bring our own experts in to examine the mine workings. So in all probability we will be unable to answer many of those questions while the investigation goes on. . . . The important thing right now is to get to the remaining men down there and get them to the surface."

Benner, seated at his boss's side, announced that a new fresh-air base had finally been established in the mine. Draegermen were poised to make a final push towards the two areas of the North Mains where crews had been digging coal at the time of the explosion. Tests indicated the air in those areas was breathable, raising the possibility that cave-ins had shielded parts of the North Mains from both the force of the explosion and the poisonous gases it created. But Benner remained noncommittal about the chances of finding anyone alive. "We're still treating it as [a search for] missing men," he said.

* * *

WESTRAY TRAINING OFFICER Bill MacCulloch, pinch-hitting for Benner, updated the rescue effort at one o'clock Wednesday morning. Rock-falls clogging tunnels in the North Mains were "significantly more extensive" than officials had first thought, he told a group of bleary-eyed reporters. Draeger crews had tried three separate routes, but each one was impassable. A fresh crew had been dispatched to explore one remaining route, but MacCulloch dashed hopes of an imminent breakthrough. "The extent of these blockages has been very disheartening to all concerned."

* * *

IN TOWNS AND VILLAGES surrounding Plymouth, church bells pealed Wednesday as funerals were held for five of the eleven men who had been brought out of the mine. In Eureka, nestled in the rolling hills a few kilometres to the south, hundreds paid their last respects to Larry Bell, a lifelong resident. Wesley United Church was packed, and a speaker was set up outside for the overflow crowd on the church lawn. "Why this happened, we cannot say," Rev. Charles MacPherson told mourners gathered in Stellarton for the funeral of Plymouth native Robert Doyle, "any

"Why this happened we cannot say, any more than we can answer why a tornado devastates a city." Mourners embrace at the graveside of Plymouth's native son Robert Doyle, who was only twenty-two. [Photo by Len Wagg, courtesy of the *Chronicle-Herald* and the *Mail-Star*]

more than we can answer why a tornado devastates a city, or why an airplane is plucked from the sky." In Westville, the mournful cry of the bagpipes led the procession at Robbie Fraser's funeral.

Eugene Johnson's funeral was held at St. Paul United in Westville, a brown brick church just around the corner from his house. About six hundred people turned out to say farewell, including MP Elmer MacKay. Beside the open coffin was a floral arrangement in the shape of a guitar. Eugene's song to the Westville miners was read aloud. The last stanza was difficult for everyone to hear.

After we finish and our day's
work is done,
We shower up and it's home on
the run.
Can't wait to get there for our
darlings to see,
And our children waiting there
for me.

Donna Johnson, head bowed and arms around son David, follows Eugene Johnson's casket out of a Westville church. Johnson was among the first eleven victims brought to the surface. [Photo by Len Wagg, courtesy of the *Chronicle-Herald* and the *Mail-Star*]

Donna Johnson and her seven year-old, David, wept as the words were read. Her other son, Michael, could not face going to the church; he would pay his respects later, at the cemetery. After the service, Donna, her head bowed and one arm firmly around David's shoulders, followed Eugene's casket down the front steps of the church. Newspaper photographers camped across the street captured the image, a stark reminder of the human cost of the unfolding tragedy.

* * *

As Pictou County paused to bury its dead, draegermen continued their dangerous work far below. Rescuers had dug their way over a massive rock-fall and installed a ladder to get down the other side. The tunnels beyond the cave-in looked clear, and fresh crews were sent down to continue the search. But as the day progressed, there was nothing but bad news from underground. Just past noon, Colin Benner announced the discovery of the bodies of three miners in the North Mains. Six hours later, another body was found, bringing the death toll to fifteen.

Two of the men were still at the rock bolter they had been operating, he said. The blast had driven the other two against a wall. The unidentified bodies were being left in the mine until all areas of the North Mains had been searched. "It's discouraging; I'd be less than honest if I didn't say that," Benner said. "But we're not going to give up." The search continued overnight, and another body was discovered early Thursday morning, crushed and pinned under a mass of rock. At the request of the families, draeger crews turned their efforts to recovering the bodies already located.

* * *

STEVEN CYR AND SHAUN COMISH were among those tapped to bring out the bodies. It was a risky mission. The North Mains had been devastated when the explosion blasted through the dead-end tunnels, directing the lethal force straight to the working areas. Two tractors had been slammed together, leaving a jumbled mass of metal. The only way to tell that there was more than one vehicle was to count the wheels; there were eight. The tunnels were part maze, part obstacle course, clogged with fallen rock and debris. One opening was barely big enough for a man to crawl through. In another area, ropes were tied to bodybags so they could be hoisted up one side of a mound of debris and lowered down the other side.

Comish's team put two of the men into bodybags. The bodies were black, as if burned by a blowtorch. They looked like something out of *Predator* or some other gory action movie, he thought. Comish had been working in the same area twelve hours before the explosion, but he tried not to think about that. He knew there was no hope of finding anyone alive, but he tried not to think about that, either. He just went through the motions, did what he had to do, struggled to keep his emotions in check. His eyes, shielded by his face mask, filled with tears.

Cyr joined a draeger team from New Brunswick for the trip underground, his seventh in six days. They retrieved the body of Jim Munroe, a miner in his late thirties from the Trenton area, the father of two children. Grief and frustration greeted them at the portal. Some of the draegermen were in tears. One rescuer was kicking the side of an ambulance; he was that distraught. Cyr assumed he was related to one of the men just recovered. As the bodies were put in ambulances, Cyr's team removed their mining helmets as a show of respect.

Then, Clifford Frame showed up, shaking hands all around. "Thank you for a job well done," he told Cyr. "You did your best." It finally started to sink in. There would be no more forays underground, no more bodies recovered. The rescue effort had been called off. Fifteen bodies had been brought to the surface. Another had been located, but it was covered in debris and could not be removed. The bodies of the other ten men were somewhere in the North Mains, buried under debris or trapped behind the mounds of fallen rock that choked the tunnels. One of those still underground was Mike MacKay. Cyr knew he had failed Mike's widow—failed to bring her husband back, if only so that his body could be buried. He wondered how he could ever face Beverly MacKay again.

* * *

ONE HUNDRED TWENTY-NINE HOURS after it began, the vigil was over. Colin Benner arrived at the fire hall about 1:00 p.m. Thursday, May 14, to announce that the rescue had been suspended. The flicker of hope that had sustained families and friends for more than five days was gone, replaced with the grim reality of death. Groups of men and women, weeping and clinging to each other for support, streamed out of the fire hall and into their cars. Chris Martin had been in and out of the fire hall since the explosion, getting updates on the rescue. But on Thursday afternoon, badly in need of a break, he took his sons fishing. His missed the announcement, but Benner only made official what Martin had known in his heart for days—Glenn was gone. So were the twenty-five others who went into the mine with him on the night of May 8. More than twenty women, most of them in their late twenties or early thirties, were widows; more than forty children were fatherless.

Benner, his face ashen with fatigue, arrived at the community centre at 2:10. Reporters, who had witnessed the anguished scene that unfolded outside the fire hall, knew why he had come. "Ladies and gentlemen," he began, reading from the prepared speech in his quivering hands, "we have some final news that we have no choice but to deliver. Within the last hour we have been provided with further particulars with respect to the mine rescue operations that have led us to conclude that there is no reasonable possibility that any of our men who were underground at the time of the explosion have survived." The words were like knives, striking at the hearts of reporters in the community centre and people across the country watching the live TV coverage. The Westray mine would yield no survivors.

"Conditions in the areas presently being explored are sufficiently unstable to require a full review of mine rescue operations," Benner continued. No more draeger teams would be sent underground until the review, expected to take twenty-four hours, was complete. "We profoundly regret having to decide that there is no further hope for the lives of our men. It remains our hope that we will have an opportunity to re-enter the mine in an effort to recover the casualties.

"I must add that nobody could ever have tried harder, or more valiantly, than all the draegermen and bare-faced crews, and Gerald Phillips and all at central control. On behalf of everyone, we grimly share the sorrow [of] Pictonians, and Nova Scotians, Canadians, and others who have been on vigil all over the world. We extend our deepest condolences to the brave family members who demonstrated extraordinary faith, and hope for their loved ones, to the very end." Benner's voice was nearly breaking. "There is nothing more I can say at this time." A few reporters shouted questions, but Benner silently brushed past them and left the room.

* * *

IVAN BAKER'S LAST DUTY on Thursday was to release the names of the remaining victims of the Westray explosion. It had been a long week; the constable had been putting in nineteen-hour days since Saturday, catching only a few hours of sleep here and there. But life went on. At one point, Baker had to scurry down to the courthouse in New Glasgow to testify at a bail hearing in an assault case. But his personal inconveniences paled when he saw the strain etched on the faces of the people in the fire hall, waiting and praying for a miracle. Now he had to go on live TV at the community centre and make it official; there would be no miracle at Westray.

Speaking in a precise, clipped voice honed by years of testifying in court, Baker identified the four men brought to the surface earlier in the day: Remi Drolet, married with a two-year-old boy, an experienced miner from Elliot Lake, Ontario, whose younger brother had also died in a mine; George James Munroe, the last man brought to the surface; Peter Vickers from Antigonish, who left behind a wife and daughter; and Larry James, a thirty-four-year-old who had come all the way from Wales to work and die in a Pictou County mine.

Then, Baker named the eleven still in the mine: Glenn Martin; Roy Feltmate; John Halloran; Mike MacKay; John Bates, Westray's oldest victim, at fifty-six; Angus MacNeil from Strathlorne in the Cape Breton Highlands, married with three children; Randy House from Daniel's Harbour, Newfoundland; Bennie Benoit, a Glace Bay native survived by his wife, three daughters, and a young grandson; Wayne Conway, a father of three from Truro, Nova Scotia; Danny Poplar, a father of two from Joggins, another Nova Scotia community built on coal mining; and Stephen Lilly, a forty-year-old with three sons who came to Nova Scotia from his native Yorkshire in 1987.

Baker felt the moment was too poignant to let pass without some words of comfort for the families of the victims. But he wanted to keep it short, to keep his own emotions in check. On behalf of the RCMP, he extended condolences to those plunged into mourning. "I hope I can offer you some small comfort in the fact that these men will not be forgotten." Then he asked to be excused while another officer repeated the announcement in French. Baker worked his way through the ring of reporters and ducked into a side room; he needed a few minutes alone to pull himself together. Then he drove to the Stellarton detachment to tie up some loose ends.

* * *

LATE THURSDAY, Dr. Roland Perry, Nova Scotia's chief medical examiner, announced the results of autopsies conducted on the bodies of those pulled from the mine on the weekend. The men suffered varying degrees of burns on their skin and clothing, he said, but all had died almost instantly from carbon monoxide poisoning. "The miners would have been rendered unconscious very rapidly," Perry assured their relatives, "and they would not have suffered." Carbon monoxide levels in the miners' blood were extremely high—between 65 per cent and 80 per cent. By comparison, he said, a police officer breathing in exhaust fumes while directing traffic would have about a 10-per-cent concentration of the gas in his bloodstream.

As Perry spoke, a car pulled into the community centre parking lot. The man in the passenger seat was Robert Johnson, and he wanted to talk to a reporter about his brother. As Pam Sword of the *Chronicle-Herald* took notes, Johnson talked about Eugene—how he liked to play country songs on the guitar, how much he enjoyed hunting and fishing. He recalled seeking Eugene's advice on applying for a job at Westray, and that his older brother told him he had to make up his own mind. And he remembered the last time he saw his brother, when he dropped by Eugene's place for a chat, just two days before the explosion. "I feel like I lost a brother, a friend, and a fellow coal miner," Johnson explained, struggling to put his feelings into words. "It's a sad loss."

That loss was felt far beyond Pictou County's borders. Prime Minister Brian Mulroney called the Westray explosion "a great tragedy and an enormous loss" for all Canadians. "Our heart goes out to the families, the children, and to all involved—and also our congratulations, to the courageous draegermen who are there trying to save lives." All week, letters of support and donations for the families of the victims flooded in from around the world. The United Steelworkers of America, which had been trying to organize Westray's workers, gave $10,000. Michelin Canada, a subsidiary of the French tire-maker, and Pictou County's largest employer, donated $20,000. A thirteen-hour benefit concert in Halifax, featuring the folksinger Valdy and Cape Breton's renowned miners' chorus, the Men of the Deeps, raised more than $30,000. The City of Ottawa and mining communities scattered across Canada and the United States launched fundraising drives. As money poured in, the Plymouth Mine Disaster Fund was created to meet the needs of the families and educate the children left fatherless. Branches of the major chartered banks across Canada agreed to accept donations. The fund would eventually reach almost $2 million.

In the days that followed, Westray's victims were remembered and mourned across Canada. A memorial service was held in Hinton, Alberta, for John Bates, who had worked for years at the nearby Cardinal River coal mine. In Prince George, British Columbia, more than sixty people turned out for a church service organized by a woman who had known one of victims. The disaster hit home in Cumberland, a coal-mining community located just north of Nanaimo on Vancouver Island. The area's collieries, like those in Pictou County, were notoriously gassy; methane explosions and other mishaps killed almost three hundred men before mining ceased in the 1960s. The annual miners' memorial day was dedicated to the memory of those killed at Westray; as a special tribute, a rose was placed on the grave of each of Cumberland's fallen miners.

The tragedy touched people even farther afield. In Ystradgynlais, a Welsh village near Swansea, friends recalled how Larry James had moved to Canada to find work after the government of Margaret Thatcher closed down the local collieries. "Our hearts go out to the families of those dead miners," said Tom Jones, a village councillor and himself a former coal miner. "If you're from a mining area, you know it could happen to you." And Westray was still on the mind of U.S. President George Bush when Brian Mulroney visited Washington on May 20. "I just wanted to tell you and Mila how strongly we all feel about that tragedy, and we ask that you send our love and our blessings, and ask God's blessings on those families," Bush told Mulroney at a dinner at the Canadian embassy. "Our thoughts and indeed our prayers ... are with you over the tragic incident in Nova Scotia."

* * *

ON MAY 15, the day after the rescue operation was called off, Premier Donald Cameron announced that Mr. Justice Peter Richard of the Nova Scotia Supreme Court would head the public inquiry into the disaster. Richard, a sixty-year-old appointed to the bench by the Liberals in 1978, was given wide powers to summon witnesses and hire mining experts to determine the cause of the explosion. But the inquiry's terms of reference were not limited to the events leading up to the explosion. Richard was asked to determine whether the mine had been properly established and properly run, given the troubled geology of the Pictou coalfield. He was to root out any "defect" or "neglect" that had contributed to the explosion, and decide whether it could have been prevented. He was told to ensure that Westray had complied with all mining and safety laws and regulations. And finally, he was given carte blanche to investigate "all other matters" he considered relevant. That opened the door to an airing of the political deals and decisions leading to the mine's development.

It was a broad mandate, and Cameron acknowledged that he, as one of the mine's biggest promoters, would be under the microscope. "It is essential that Nova Scotians know all the answers to all the questions surrounding and leading to this tragic event," he told the legislature. "Nothing, and no person with any light to shed on this tragedy, will escape the scrutiny of this inquiry." Opposition party leaders welcomed the announcement, but promised that the government would be facing tough questions about Westray when the legislature resumed sitting the following week—long before the inquiry opened public hearings.

In the meantime, the government refused to release reports of inspections carried out at Westray before the explosion. That information was being compiled for the inquiry, a Labour Department spokesman said, and the government wished to avoid "any situation occurring where the information is released prematurely." Researchers for the Liberal party were more forthcoming, supplying reporters with inspection reports obtained under the Freedom of Information Act. The reports only covered the period of May to September 1991, when the mine was just entering production. But they showed that inspectors had been concerned from the start about coal dust, methane levels, cave-ins, and underground storage of fuel. The company was also told to remove a contractor's bulldozer that was not approved for underground use.

Leroy Legere, the labour minister, downplayed the reports. "At all times, the inspectors were satisfied that the safety procedures were being followed," he insisted. The last inspection of the mine was in late April, he said; "things were being done as they should have, and if they weren't, they were addressed." Legere had either been poorly briefed or was misleading reporters; truth was, Westray had been ordered to clean up coal dust after that April visit, and no follow-up inspection had been done. The minister went on to predict that the government's actions would stand up to the inquiry's scrutiny. "I don't think we're going to find that anything slipped through their fingers. They were on the ball," he said of the inspectors. "I've checked and have found no place where the company was given any special exemptions on the safety aspect."

Despite Legere's confidence, there was mounting evidence that the Westray mine had been an accident waiting to happen. Miners who had held their tongues during the rescue operation went public, complaining of high methane levels, lack of training, and use of unauthorized equipment in the mine. Complaints to management and government inspectors, they said, were ignored. Ten miners held a news conference in Stellarton on May 18 to thank draeger teams, police officers, volunteers, and everyone else who pitched in during the rescue effort—even the schoolchildren who sent cards and letters. Although the miners were reluctant to talk about safety, Lenny Bonner revealed he had seen methane readings of between 3.5 per cent and 3.75 per cent on the shift before the explosion. By law, mining should have ceased and the area should have been evacuated. "Did you report those?" a reporter asked of the readings. "It was known," Bonner replied. "Several corners were cut at times," admitted Randy Facette, an experienced coal miner from Alberta. "We were working to rectify [the problems]. Let the inquiry handle that." The inquiry, another miner promised, would be an "eye-opener."

Westray officials refused to discuss the allegations. "Specific matters of safety procedures are the subject of a provincial inquiry," the company said in a written statement. "We will leave those questions to be fully dealt with in the course of our investigation and the inquiry." Legere, who still claimed Labour Department inspectors knew nothing of the problems, criticized the miners for coming forward after the explosion. "No-one in a safety area . . . can abdicate their responsibility," the minister scolded. "It bothers me that, after the fact, people are coming out and saying all of these things were not being done."

* * *

UNTIL MAY 9, Clifford Frame would have sold all or part of the Westray mine to anyone with the money; now he was stuck with a wrecked mine that hung around his neck like an albatross. The disaster was a heavy blow to a company already losing millions of dollars. The loan-guarantee agreement with the federal government left the company on the hook for up to $25 million if the project was abandoned. Brokers were openly speculating about Curragh's ability to weather the storm, and the uncertainty drove down stock prices. Curragh shares took a beating on the Toronto Stock Exchange in the days following the blast, dipping as low as $2.80. By the end of the week, the stock recovered, but at $3.45, it was still down about a dollar from trading before the explosion.

More bad financial news had come hours after the rescue effort was suspended. With Westray idle, Curragh needed coal from the Wimpey strip mine in Stellarton to fulfil its contract with the Nova Scotia Power Corporation. But on May 14, Nova Scotia Environment Minister Terry Donahoe said Curragh's application to dig 200,000 tonnes per year at Wimpey was on hold until the company showed how it would deal with dust, noise, pollution of nearby streams, and other environmental concerns.

To restore confidence in his battered mining empire, Clifford Frame presided over a press conference at the Plymouth Community Centre on the afternoon of May 15 to discuss Westray's future. Peering over his reading glasses as he delivered a five-page speech, he announced that the one-day review of the rescue operation had been completed. "Since there is no possibility of survivors," he said, "mine rescue operations will not be reactivated." Curragh would cooperate with the public inquiry, he said, and Westray would be reopened only if such a decision was "consistent" with the inquiry's findings.

Frame insisted he had to hold out some hope for the approximately 160 surviving Westray employees, who were likely to lose their jobs in the wake of the explosion. Curragh remained committed to repairing the underground mine and would continue to seek a permit to strip-mine coal at the Wimpey site. Then, in a clumsy bid for support from the bereaved families, Frame held out the possibility that bodies still in the mine could be recovered. "Should we be permitted to rehabilitate the mine, I make my personal commitment to the families that we will make every effort to recover the eleven men." And Frame promised money. "Any coal production from the open-pit mine at Stellarton and, if appropriate, from underground, to a certain sum, will be contributed to a fund through a levy," he said. "This fund will be established for the benefit of the families and children left behind." No dollar figure was put on "a certain sum," and he refused to answer questions from reporters.

Frame was already working to raise Westray from the ashes. Earlier in the week, as draeger crews continued their desperate search, he had dropped in to see Stellarton Mayor Clarence Porter, a vocal opponent of mining at the Wimpey pit. Now that the explosion had disrupted Westray's coal supply, Frame hoped Porter would reconsider his stance. The rescue effort was still in high gear; Porter was appalled that Frame's priority was restoring Westray's cash flow. But Frame was not the only one thinking ahead. Despite the disturbing allegations of unsafe practices, some miners were already talking about going underground if Westray reopened. Others were eager for jobs at the strip mine. With unemployment in northeastern Nova Scotia hovering at about 18 per cent that spring, a lot of people were hungry for work. Any work. Even if it meant working for Westray Coal and Clifford Frame.

It was also business as usual for Gerald Phillips. On May 16, the Saturday following the blast, Steven Cyr donned his draeger gear for another trip underground. Phillips wanted a team to check a previously unexplored section of the No. 2 main slope, the tunnel equipped with the conveyor belt to bring coal to the surface. The team was to check for bodies, but there was no reason to think anyone had been in that area when the mine exploded. Cyr suspected the real purpose of the mission was to check for damage. Back on the surface, the team told Phillips that the tunnel was in good shape. Phillips was buoyed by the news, and talked about driving new tunnels off the main slope. Cyr could barely control his rage; eleven bodies were still in the mine, and, as far as he could tell, all Phillips was thinking about was getting the mine back into production.

* * *

Prime Minister Brian Mulroney and Nova Scotia Premier Donald Cameron, two of the politicians who had helped make Westray a reality, were grim-faced at the memorial service for the twenty-six miners. [Photos by Len Wagg, courtesy of the *Chronicle-Herald* and the *Mail-Star*]

ONE BY ONE, the names of Westray's twenty-six victims echoed through the cavernous stadium in New Glasgow on May 19. As each name was read, a child or parent or sibling stepped forward to place a rose at the foot of a wooden cross, shrouded in white. Miners spared by the explosion laid a wreath. The Men of the Deeps sang the "Miners' Memorial Hymn." There was a standing ovation for the draegermen and bare-faced miners who had searched in vain for survivors.

A week earlier, the stadium had been pressed into service as a morgue; on this, Nova Scotia's official day of mourning, the rink was draped in black for a memorial service organized by local churches. Three thousand people were crowded into the stadium, the most allowed by fire regulations; another three thousand stood outside in the afternoon sunshine, listening to the service on loudspeakers. Among the mourners were the political heavyweights who had helped revive Pictou County's coal industry: Prime Minister Brian Mulroney was on hand with his wife, Mila; a grim-faced Donald Cameron was accompanied by his wife, Rosemary.

Randy Facette, dressed in coveralls and clutching his mining helmet, read a passage from the Bible in memory of his co-workers. "Goodbye, our brothers," he said. Clifford Frame took the podium, and also quoted scripture. "Make sure that you do not grieve about them, like the other people who have no hope," he read. "We believe that Jesus died and rose again, and that it will be the same for those who have died in Jesus. God will take them with him." Governor General Ray Hnatyshyn brought condolences from the Queen. "Never doubt that your sorrow becomes part of the history of Canada," he told the families of those killed. "In the dark moments of the past ten days, we kept watch with you. We were awed by your courage and devotion throughout this ordeal." In his homily, Rev. Angus MacLeod of St. Gregory's Church in New Glasgow asked that questions and recriminations about Westray be left to the public inquiry. "In respect for our dead and their families, let us put aside unnecessary controversy," he pleaded. "It is not a time to try to score points for any side."

But even as MacLeod spoke, opposition parties in Ottawa and Halifax were sharpening their knives for the political battles ahead. After members of Parliament observed a moment of silence for the Westray dead, opposition critics served notice that the time had come for answers about the public largesse heaped upon Westray's owners. Liberal House Leader David Dingwall, a Cape Breton MP and staunch foe of Westray, said the federal government would be called on to defend its "preferential treatment" for the project. "There are many serious questions," New Democrat Rod Murphy told the House of Commons. "They must be asked. They must be answered."

Back in Nova Scotia, Isabel Gillis had questions of her own. The day of the memorial service, she granted an interview to CBC Television News so she could thank the draegermen who had searched for Myles and the others. But she also accused Westray and Curragh of raising false hopes that at least some of the men had survived the explosion. "The things that were told to us–to me, it wasn't the truth," she said, recalling the gruelling vigil at the fire hall. "They were coming and giving us hope and more hope, but really they knew from the very beginning that there was just no hope."

Then Gillis recalled that, not long before the explosion, her husband had made her promise to seek a full investigation if he died in the mine. "At the time I thought that Westray would take care of the men, so I didn't really worry about it," she explained in a quiet voice. "I was wrong." She was not alone. As grieving turned to anger, as promises faded into delays, as allegations became prosecutions, it would become painfully clear that virtually everyone had been wrong about Westray.

THE ROOTS OF DISASTER

BLACK GOLD

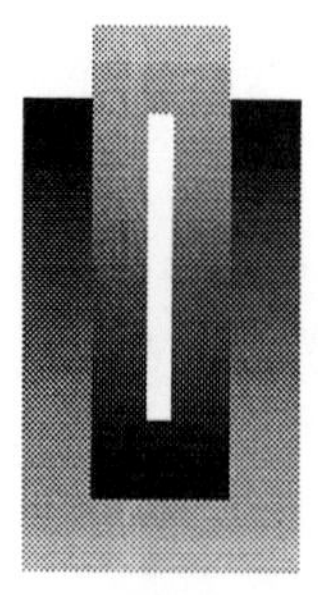

F THERE IS A MORAL in the tale it is that human beings should not have to earn a living in dark holes thousands of feet from sunshine and fresh air, holes wherein too many men have died. The history of coal mining in Pictou County differs from coal mining history elsewhere only in the names of people and places. It is a sorry history.

With these words, James M. Cameron, the prolific chronicler of Pictou County's past, launched *The Pictonian Colliers,* the definitive history of the area's coal-mining industry. It is indeed a sorry history, because geological forces have split and twisted the dozens of coal seams buried beneath Pictou County's rugged hills, challenging the skills of generations of mining engineers. A sorry history, because the money to be made tapping the thick seams of high-quality coal has lured a succession of entrepreneurs and investors to their financial ruin, squandering millions of dollars of public money.

But most of all, it is a sorry history because the price of mining Pictou County coal has been measured in lives, not just dollars and cents. From 1866 to the early 1970s, 576 men and boys died in Pictou's collieries. Of that number, 246 perished in methane and coal-dust explosions. The remaining 330 fell victim, one or two at a time, to other underground hazards. Many were crushed in cave-ins; others were suffocated by gases, run over by coal cars, or mangled after becoming caught in machinery. Those numbers do not tell the whole story; for the first fifty-seven years after commercial production began in 1809, no records were kept. Cameron estimates that accidents and explosions during the early years claimed as many as seventy-five more lives, bringing the number killed in a century and a half of mining Pictou County coal—before the opening of the Westray mine—to roughly 650.

The death toll in Pictou's mines is comparable with the number of county residents—647—who died fighting in World War I and World War II. In Pictou County, monuments to the war dead barely outnumber those dedicated to the memory of those killed closer to home, beneath their native soil. Stellarton and Westville, towns built around collieries, each have memorials bearing the names of scores of men killed in the worst of the mine disasters. Westville's monument commemorates an 1873 explosion that killed 60 men in the Drummond mine, Canada's first major mining disaster. A miner, carved in red sandstone, stands stiffly at attention atop the memorial, clutching a lantern in one hand and a lunchpail in the other. For a century, he has been a silent witness to the rise and fall of Pictou County's coal industry. And at his feet, chiselled into the stone along with the names of the dead, is a stark warning to those who choose to dig the coal that lies below: "Be ye also ready."

In the days after the Westray explosion, James Cameron was sought out by journalists who wanted his opinion on the latest disaster to befall Pictou County. "I'm not surprised, as much as I regret it," he told one reporter. "There has been a long history of explosion and fire in the seam of coal that the Westray people are operating." Curragh Resources and Westray Coal joined a long list of mining companies that had tried to tame the volatile Pictou coalfield, and failed. "With the benefit of hindsight, it was a gamble," Cameron noted. "But there are always entrepreneurs willing to take a gamble."

* * *

Burned-out buildings at the pithead of the Drummond colliery in Westville after the 1873 explosion that killed sixty miners. It was the first major mine disaster in the Pictou coalfield. [Courtesy of Dalhousie University Archives]

THE PICTOU COALFIELD, covering an area about 120 square kilometres in the heart of Pictou County, is the most heavily mined field in Canada. That wealth of black gold, plundered by man in less than two centuries, is the product of millions of years of geological upheaval and physical change. Nova Scotia's coal deposits—confined for the most part to Pictou County, Cape Breton Island, and the Springhill area—originated in the lush swamps and lagoons of the Late Carboniferous period, three hundred million years ago. Coal is made up of the remnants of ancient plants and trees, transformed into carbon and turned into rock by the forces of time and nature.

The Micmac people were aware of the presence of coal long before the first Scottish settlers arrived in Pictou County aboard the *Hector* in 1773. Their legends told of places where the earth caught fire, where the rivers burned and thundered. An area of what later became Stellarton was known to the Micmac as Fire Hill after a coal outcrop caught fire and burned, day and night. They also told stories of flames shooting out from the banks of what is now called the East River, not far from the future site of Plymouth; methane, leaking from coal seams exposed by the river, had caught fire. By one account, the name Pictou comes from the Micmac word *bucto* (fire).

The first white settler credited with exploiting Pictou County coal was James MacGregor, a Presbyterian minister. In the winter of 1798, he heated his home near the East River with coal dug from an outcrop on his property. A coal fire in pioneer Nova Scotia was enough of a novelty to amaze candidates who visited the minister during an election campaign. Local blacksmiths soon followed MacGregor's lead, using coal exposed on the surface to fuel their furnaces. In 1809, John MacKay opened the county's first mine on his property, tapping what was then known simply as the Big Seam, a band of coal up to twelve metres thick, later called the Foord seam. MacKay prospered during the War of 1812, expanding his operation to meet a fuel shortage. Twelve shiploads of coal from the pit were sent to Halifax in 1815, but the war's end left MacKay saddled with debt. In 1817, bankrupt and imprisoned by his creditors, MacKay begged the colonial government for money so he could get out of jail. MacKay was the first Pictou County coal-mine operator to dip his hand into the public purse; he was by no means the last.

A decade later, another man's debt—and the need to avoid embarrassment for the Royal Family—opened the door for the first major mining venture in the Pictou coalfield. When the Duke of York needed money to pay off his debts, King George IV, his brother, came to the rescue. In a blatant abuse of royal prerogative, the King granted the duke exclusive right to all the minerals of Nova Scotia, which was then thought to be rich in copper. The duke promptly transferred the lucrative sixty-year lease to a London jewellery firm, in order to settle his accounts. The jewellers had already formed a company, the General Mining Association, with an eye to opening mines in South America. In return for conveying the lease, the duke extracted a promise that the GMA would develop mines in Nova Scotia.

The GMA soon discovered that Nova Scotia's mineral wealth lay not in copper, but in the abundant coal of Pictou and Cape Breton. In 1827, coal miners and machinery were sent to Pictou County to establish a full-scale mining operation. By the fall of that year, a steam engine was hoisting coal from a seventy-metre shaft dug at Stellarton, then known as Albion Mines. The mine employed 330 men by 1840, and was producing almost ten thousand tonnes of coal per year, mostly for export to the United States. Coal mining was the driving force behind Pictou County's own version of the Industrial Revolution: foundries, coke ovens, brickworks, and steam-driven power plants began to push aside the farms of the original settlers. Pictou County coal fuelled the *Royal William* when it became the first ship to cross the Atlantic under steam power in 1833. The industry also brought another technological marvel of the age, the railway. In 1839, three of the first locomotives to operate in British North America began puffing their way along ten kilometres of track, carrying Albion Mines coal to the shores of Pictou harbour to be loaded onto ships.

The push for self-rule in Nova Scotia spelled the end of the GMA monopoly in 1857. The company continued to operate the Albion colliery until 1872, but by then new operators had appeared on the scene. The capital and expertise came from outside Nova Scotia, following a pattern that started with the GMA and was repeated in the 1980s by Curragh Resources and Clifford Frame. The Intercolonial Coal Company, created by a consortium of Montreal investors, opened the Drummond mine in 1868, in the process establishing the town of Westville. New York money bankrolled the Acadia Coal Company, which opened mines in the Stellarton area in the mid-1860s. In 1867, a British entrepreneur opened the Foord pit to tap the rich seam of the same name; the ruins of the sandstone tower that housed the mine's pumphouse still stand at the northern edge of Stellarton. To the east, the Vale Coal, Iron, and Manufacturing Company started mining in the Thorburn area in the early 1870s. Vale was owned by a Montreal financier, Hugh Allan, a confidant of Prime Minister Sir John A. Macdonald and a promoter of the Canadian Pacific Railway.

By 1875, Pictou County's five collieries produced 250,000 tonnes of coal per year and employed almost 1,600 men and boys, nearly 1,000 of them underground. The coal was bituminous, too high in impurities for making steel, but fine for heating and producing steam. Pictou County coal warmed homes in Montreal and Halifax, powered steamships and locomotives, and fired the boilers of factories. Oil extracted from the "stellar" coal found in one seam—so called because its high petroleum content made it burn very brightly—was used to light Boston's streets, and later lent its name to the town, Stellarton. Despite the flurry of investment and mine development, the output of Pictou's collieries was eclipsed by that of Cape Breton, where twenty mines were in operation by the 1870s. The reasons had more to do with geology than business savvy. Pictou's coalfield was fractured by numerous faults that made coal seams difficult to follow, increasing the cost of extracting coal. To make matters worse, Pictou's mines were prone to underground fires and gas explosions that could spell death for a colliery, and the men who worked it.

* * *

THE COMPLEX STRUCTURE of the Pictou coalfield has long fascinated mining experts. "The number and thickness of the coal seams . . . the fiery nature of both coal and the shale beds, combined with the variable and faulted nature of the strata, all occurring in a small superficial area," one geologist noted early in this century, "mark out the Pictou coalfield as one of the most interesting carboniferous deposits known." Interesting, and dangerous; in layman's terms, the thick, broken coal seams produce too much methane and are unusually susceptible to fire. In the late 1930s, the former chief engineer of the United States Bureau of Mines declared mining conditions in the Pictou area to be "the most difficult found in any coalfield" in Canada, the U.S., or Europe.

The Pictou field is less than twenty kilometres long and about six kilometres wide; major faults split it into three distinct sections. At the extreme east is Thorburn; the middle area is known as Albion or Stellarton; and to the west is Westville. Each area has an array of coal seams of varying size, depth, and quality—and these bear little resemblance to seams only a few kilometres away. Within each section, seams are broken by smaller faults, shifting the coal upward, downward, or sideways—at the whim of nature. All of this makes life extremely difficult for engineers and mine operators.

And then, there is the methane hazard—no secret to the early miners. The flaming waters of Micmac legend were apparently put to good use by the first settlers; tradition holds that women dug holes in the banks of the East River, near outcrops, then set fire to the escaping gas so they could boil water to wash clothes. Richard Smith, the first manager of the Pictou operations for the GMA, observed in 1835 that methane seeped from the coal with a pressure similar to that of a steam boiler, and with a sound he likened to "a hundred thousand snakes hissing at each other." When miners first tapped the seam in the Albion mine, he claimed, the gas "roared" as they struck it with their picks, often producing a loud crack "like the report of a pistol."

The miners called methane "black damp" or "fire damp," terms that perfectly captured the essence of the invisible, explosive gas. Methane is a byproduct of decomposition, sometimes called "marsh gas" because it is produced in swampy or marshy areas. The gas is found in all coal deposits, sealed in the rock during the slow transformation of prehistoric plant material into coal. It sometimes collects in fissures and pockets within the coal, stored under pressure until released by a miner's pick or the cutting head of a modern mining machine. Methane has been a particular hazard in the Pictou coalfield, in part because greater amounts of coal are exposed during the mining of thick seams like the Foord. Numerous faults have split the field's seams, leaving areas of broken coal where methane can collect. Although non-poisonous, the gas can be lethal to miners if it reaches concentrations so high that it displaces oxygen. At lower concentrations–between 5 per cent and 15 per cent of the air in a mine–methane is explosive if ignited by a spark or flame.

Compounding the gas menace is the threat of spontaneous combustion, particularly of coal in the Stellarton area. The phenomenon is not fully understood; it has long been known that piles of coal stored on the surface will heat up from within and sometimes catch fire. The presence of oxygen, combined with the pressure of piled-up coal, produces heat, sometimes enough to result in flame. Underground, the hazard is greatest in mined-out areas, where coal left in pillars is crushed under the weight of the covering layers of rock. James Cameron has documented forty-eight major underground fires in the Pictou coalfield since mining began, and there have been countless smaller outbreaks. Methane also collects in such areas, providing all the elements for a sudden–and deadly–explosion.

* * *

THERE IS NO RECORD of a mine explosion or fire in the Pictou area before the General Mining Association arrived on the scene in 1827, probably because the early pits were relatively shallow. In 1832, a fire killed more than a dozen horses in the company's Store pit, which tapped the Foord seam. From then on, methane, spontaneous combustion, or a combination of the two exacted a heavy price in human lives. Explosions in 1836 and 1838 killed six Store pit miners. Two men died in an 1858 blast in the Cage pit, and three others lost their lives when the Bye pit exploded in 1861; both of these were GMA mines sunk into the Foord seam. In addition, GMA operations were plagued by a number of non-fatal explosions and fires, permanently closing some mines and playing havoc with the company's finances. At one point, the GMA threatened to shut down completely unless the government helped by reducing the royalty on coal; the request was granted.

Mining operations expanded rapidly after the GMA monopoly was broken in the late 1850s; and more mines meant more deaths. In 1873, an explosion at the Drummond mine in Westville killed sixty men. Another forty-four died when Stellarton's Foord pit blew up in 1880. Five years later, a blast at the McBean colliery in Thorburn claimed another thirteen lives. Miners not killed outright in the explosion were often overcome by carbon monoxide, which was carried to other areas of the mine through the ventilation system. Carbon monoxide, dubbed "after damp" by the miners because it flooded the mine after an explosion, could kill men far removed from the explosion. Some explosion victims were found sitting upright or in the midst of eating lunch—chilling evidence of how quickly and silently the gas did its deadly work.

In the early years, those who survived an explosion had to fend for themselves. Some, badly burned and mangled, managed to crawl to the surface. Volunteers were often dispatched to search for survivors or fight fires, but many perished in secondary explosions or were overcome by carbon monoxide. After the turn of the century, the Acadia Coal Company purchased breathing apparatus and organized Pictou's first draeger team—the Acadia Rescue Corps. The equipment was ungainly—the early helmets resembled those worn by deep-sea divers—but allowed rescuers to work in spite of the noxious gases produced by fires and explosions. In the disaster-prone Pictou coalfield, the rescue corps was soon put to the test. Draegermen toiled for two weeks to put out a major fire at the Albion mine in 1913, saving the colliery. From that point on, draeger crews shouldered the perilous duty of scouring exploded mines for survivors and bodies. In the mid-1930s, Pictou County draegermen became international heroes after they helped rescue

The Stellarton-based Acadia Coal Company organized a rescue corps just after the turn of the century. Despite the ungainly breathing apparatus they wore, the early draegermen were adept at fighting mine fires and searching for explosion victims. [Courtesy of Dalhousie University Archives]

two men who had been trapped for eleven days in an abandoned gold mine in Moose River, seventy kilometres northeast of Halifax.

The 1880 explosion ended mining on the Foord seam, but only temporarily; vast profits awaited anyone who could successfully mine the world's thickest deposit of coal. In 1905, Acadia Coal sunk the five-hundred-metre Allan shaft on the outskirts of Stellarton, a stone's throw from the abandoned Foord pit. The mine cost one million dollars to develop, an astronomical amount at the time, but never fulfilled the hopes of its original New York–based owners, or the Belgians who took over in 1907. Like the Westray mine, the Allan shaft never reached its output targets. The seam's thickness and the high volume of methane made it costly to ventilate and susceptible to explosion. A Pittsburgh-based manufacturer of safety equipment once described the Allan shaft as the most dangerous mine in North America, if not the world.

The Foord seam continued to take its toll in blood. The Allan shaft exploded in 1914; the mine was idle, but the blast claimed the lives of two mine officials who went underground to investigate. That, however, was merely a prelude to the worst disaster to strike the Pictou coalfield. On January 23, 1918, another explosion ripped though the Allan shaft, killing eighty-eight of the ninety-seven men working below. Despite the carnage, the mine was back in production before the year was out. Four Allan miners died in a 1924 explosion, and another seven in a 1935 blast. Explosions rocked the mine in 1929, 1931, 1941, and 1950, but no-one was killed. During its forty-six years of operation, the Allan shaft was also beset by fires that forced the company to seal up tunnels and abandon workings long before the coal was exhausted. An underground fire closed the mine for good in 1951.

Death and disaster have been the twin curses of all coal-mining areas. Some 425 workers died in the Springhill mines of northern Nova Scotia over a ninety-year span ending in 1960; more than one-third of these deaths occurred in major explosions in 1891 and 1956. In the Cape Breton coalfield, which dwarfs the Springhill and Pictou fields in size and extent of mining operations, more than 1,300 miners have been killed on the job since Confederation. But Cape Breton has suffered only two major explosions in its long mining history—a 1917 blast killed 65 people, and a 1979 explosion claimed 12 lives. A 1914 explosion at a colliery in Hillcrest, Alberta, near the British Columbia border, ranks as Canada's worst mining disaster: 189 men died, leaving 130 women without husbands and 400 children without fathers. "Practically the whole male population of this town was wiped out," said a newspaper account at the time.

Despite the appalling record of death in Canada's collieries, however, nowhere has the potential for explosion and sudden death been greater and more persistent than in Pictou County's mines.

* * *

IT IS SURPRISING that the Pictou coalfield's history of fires and explosions is not worse, given the primitive equipment and procedures of the early mines. Despite the ever-present methane, open-flame lights were the norm in Pictou County's mines until the middle of the nineteenth century. The first miners worked by candlelight or whale-oil lanterns. An enclosed mining lamp—the aptly named "safety lamp"—was developed in Britain in the early 1800s, but it did not appear in Pictou collieries until the 1850s and 1860s. Even then, the use of open-flame lamps persisted in Thorburn's less-gassy mines until after the turn of the century. The oil-burning safety lamp was surrounded by metal gauze that prevented the flame from igniting gas, and the lamp doubled as a gas detector: if the flame turned blue or went out with a puff, methane was nearing explosive levels; if the flame paled and went out, carbon monoxide was present. Electric lights, attached to a miner's cloth cap or helmet and powered by batteries, came into vogue during the 1920s. But they did not totally displace safety lamps; mine officials were using them to detect gas until the 1960s.

In the early days, some miners would clear methane out of their working areas by flapping their arms or their coats. But small fans, cranked by hand, were the favoured method. As mines got bigger and deeper, ventilation systems were developed. At first, large underground furnaces were used to create an updraft, sucking air out of the mine and drawing in fresh air. Strangely, although the furnaces introduced yet another source of open fire, no Pictou County mine explosions were attributed to them. By the late 1800s, most mines used large fans, installed on the surface to pump in fresh air for miners and horses, and flush out gas.

Improved ventilation meant deeper shafts and slopes could be sunk, after the more-accessible coal seams were exhausted. But a new hazard arose—coal dust. Most of Pictou's early mines were shallow, cool, and damp enough to keep dust in check. But the deeper the mine, the warmer and drier the surrounding rock. A strong flow of fresh air from the surface compounded the problem, extracting moisture as it flushed out methane. The dry coal dust was not explosive on its own, and would not burn on contact with the flame in a miner's lamp. But it soon became tragically apparent that dust could fuel a deadly secondary explosion if a spark or flame ignited methane. The destructive force of the 1873 blast that tore through Westville's Drummond mine was attributed to coal dust. But mining engineers and geologists were sceptical about the hazard. Edwin Gilpin, a Nova Scotia mine inspector, apparently went out on a limb in 1894 when he asserted that a number of the province's mine explosions "have had their area and their destructiveness to life materially increased by coal dust." That view gradually gained acceptance. Writing in 1917, an official of the federal Department of Mines noted that Gilpin had reached his conclusions at a time "when the explosive properties of coal dust were not so freely admitted as is the case today."

Recognizing the coal-dust problem was one thing; dealing with it was another. The simplest way to keep dust out of the mine air was to water down the roof, walls, and roadways. But in the 1920s, it was discovered that limestone, ground into a fine powder—called stone dust or rock dust—could prevent explosions. Limestone made coal dust inert, containing an explosion or fire by choking off the fuel supply. In 1926, the Acadia Coal Company began stone dusting in the explosion-prone Allan shaft. The practice soon became a routine, part of a rookie miner's rite of passage; when Gordon Clark went underground at the Allan as a teenager in the late 1930s, he stone-dusted full time before beginning his training as a miner.

Stone dust repeatedly proved its worth as a life-saving measure. In 1935, an explosive charge, used to break up coal, ignited methane in the Allan mine, killing 7 men working nearby; stone dust was credited with confining the explosion to a small area of the mine, sparing the lives of another 184 men working underground. When the Allan mine again exploded in 1950, all 72 men working underground survived because stone dust localized the blast. Nineteen men died in Pictou County's last mine disaster before Westray, a 1952 explosion at Stellarton's MacGregor mine, but the death toll could have been higher. Stone dust choked out the explosion's flame before it reached 9 other men who were underground. Thirty-nine miners died when one of Springhill's collieries exploded in 1956, but the dust shielded another 88 miners from harm.

* * *

FOR THE FIRST SEVENTY-FIVE YEARS that coal was mined in Nova Scotia, safety laws were non-existent. "Every mine owner worked pretty well as he saw fit," concluded an inquiry into the 1873 Drummond explosion. Miners were left to their own devices, working in dangerous conditions in a daily gamble to earn their pay. The province's first laws governing coal mining were enacted in the 1850s, but they only dealt with leasing arrangements. It took the deaths of sixty miners at Drummond in 1873 to force the government to consider the welfare of the men in the pits. After that tragedy, the Coal Mines Regulation Act was expanded to include some safety measures. Pictou County coal miners were among the first workers in the province to organize, and unions stepped up the pressure for improved safety laws. But the impetus for change invariably came only after a major disaster. In the wake of the Foord pit explosion of 1880, the province passed a series of safety-related measures: certification of mine officials and underground workers; regular checks of mines by government inspectors and miners' committees; mandatory testing for gas; a ban on smoking and matches underground. Pictou County's explosion-prone seams became a deadly proving ground—Nova Scotia's safety laws, Pictou historian James Cameron claims, were a model for collieries around the world.

Since disasters played a crucial role in making mines safer, there had to be an effective means of investigating those disasters. Until World War I, coroner's juries heard evidence and attempted to determine the cause of explosions, as was the case in all accidental deaths. But learning from past mistakes proved difficult; more often than not, the best witnesses to what had happened—the men in the area of the mine where the explosion originated—were dead. Most juries were left with second-hand testimony and conjecture, making it difficult to isolate the cause of the disaster. But juries that included miners were most effective. For example, a miner served on the jury that investigated the 1880 Foord pit explosion; the inquiry recommended a number of safety measures that eventually became law.

In the wake of the 1918 Allan shaft explosion, the coroner's jury was abandoned as an investigative tool. In its place, the Coal Mines Regulation Act provided for the appointment of a special examiner to probe mine fatalities. The special examiner was given wide powers to determine what caused the incident and whether safety regulations had been followed, and, if appropriate, to recommend changes in mining methods or laws. But rooting out the causes of explosions or other mining mishaps remained an inexact science. Judge George Patterson, named special examiner to probe the 1924 Allan blast, which killed four men, was unable to pinpoint what set off the explosion. The judge relied solely on the testimony of survivors and the opinions of Allan mine officials; if independent mining experts were available to review and interpret the evidence, Judge Patterson chose not to retain their services.

Even when faced with clear evidence of negligence or wrongdoing, Nova Scotians have a tradition of chalking up coal-mine disasters to fate, bad luck, or the inherent dangers of working underground. A coroner's jury discovered that the 1873 Drummond explosion was ignited when a miner used blasting powder in an area known to be gassy and dangerous. One man's negligence had caused the deaths of sixty of his comrades, but the jurors simply expressed "regret" that powder was used. Time had little impact on the tendency to forgive and forget. A royal commission set up to probe the 1956 explosion at Springhill's No. 4 mine concluded that a power cable had been severed by runaway coal cars, touching off methane and coal dust. Government inspectors and officials of the Cumberland Railway and Coal Company had known "for some time" that the cable was in a dangerous location, the commission found. The mine's owners were chided for breaching the Coal Mines Regulation Act, including a failure to stop mining when methane reached dangerous levels. Nevertheless, the commission attributed the explosion to "an unfortunate combination of circumstances for which no blame can be attached to any individual."

The assumption that mine disasters were acts of God rather than acts of man reached its pinnacle in 1958 with the infamous "bump" at Springhill's No. 2 mine. In what became known as the "Springhill miracle," eighteen men were rescued after being trapped almost four kilometres underground, some for as long as nine days. The media hoopla and euphoria unleashed by the discovery of survivors overshadowed the grim fact that seventy-five men perished. Also overlooked was any serious attempt to find out what caused the disaster. Bumps are minor earthquakes, triggered when layers of rock overlying deep mines like those in Springhill collapse, filling in the mined-out areas. Ian McKay, a historian at Dalhousie University, has concluded that the bump was a man-made disaster, a direct result of the way the mine's owners chose to mine coal. At the time, miners and their union leaders warned that a major bump was inevitable unless the mining procedure was changed, but their protests were ignored. After the disaster, the press of the day dismissed the bump as an unfortunate—and unforeseeable—tragedy. In 1959, a royal commission absolved the mine's operators from blame, while offering no explanation of what caused the disaster. More research into the phenomenon of bumps was recommended—a meaningless suggestion, since Springhill's collieries never reopened.

Despite the reluctance to point fingers, inquiries into disasters have been an important vehicle for making Nova Scotia's coal mines safer. The 1979 explosion at the Cape Breton Development Corporation's No. 26 mine in Glace Bay—the last major Nova Scotia coal-mine disaster before the Westray blast—was a case in point. Sparks, given off as a coal-cutting machine struck sandstone, ignited methane and coal dust; stone dust limited the extent of the explosion, but twelve miners died from burns and carbon-monoxide poisoning. An inquiry chaired by mining engineer Roy Elfstrom found that the Canada Coal Mines Safety Regulations—the law governing Cape Breton's federally owned collieries—did not reflect modern mining methods. Elfstrom outlined a series of amendments designed to clarify and update the regulations, and the federal government responded by enacting fifteen of his recommendations. But Devco was the only operator to benefit from the lessons of No. 26 colliery. The Coal Mines Regulation Act, which set the standard for the few remaining Nova Scotia coal mines that were not Devco-operated, was not updated to reflect Elfstrom's findings.

* * *

THEY STARE OUT from old photographs, their faces and baggy clothes blackened by coal dust, their grim expressions the result of years of toil in a dark, subterranean world. Coal mining has always been a dirty and brutal job, fraught with the prospect of injury or death. For the men who attacked the coal armed only with picks and blasting powder, the pay was a pittance. Until World War I, miners were lucky if they earned three dollars a day; they were expected to buy their own blasting powder and pay to have their picks sharpened. Their families lived in the shabby, company-owned row houses that squatted near the collieries, far from the fine homes of mine officials.

Wages and working conditions have improved over the years, the product of the clout of unions rather than the benevolence of mine owners. And mechanization has transformed mining; horses and hand tools have given way to heavy machinery for cutting and hauling coal. But coal mining remains as much a way of life as a way of earning a living. "We people who became miners were bred from mining parents," says Robert Hoegg, who started at the coalface in his teens and rose through the ranks to become underground manager at the McBean mine. "It was only the natural thing that you followed your father into the coal pits."

Coal mining has not only been a man's job; boys as young as ten worked underground during the nineteenth century, sometimes making up as much as 20 per cent of the workforce. Children drove horse-drawn coal cars or helped operate the ventilation system, working twelve hours a day or more for a fraction of a miner's pay. It was an apprenticeship for the day they would be old enough, and skilled enough, to work at the coalface. Like their elders, they faced the danger of injury or death in cave-ins or explosions. After 1923, the age limit for boys working in the mine was raised from twelve to sixteen. It was 1951 before the government made it illegal for anyone under eighteen to work in the mines.

For the families of miners who went to work and never came back, life was bleak. Explosions turned scores of women into widows, leaving them penniless and with small children to raise. Most families deprived of the breadwinner faced a grim future unless they had relatives willing to help, or sons old enough to replace their slain fathers and brothers. There was no compensation for miners killed or injured on the job until 1889, when the government began paying seventy dollars, plus two dollars a week for two years, as a death benefit. Around the turn of the century, relief societies were set up to help stricken families, and disasters usually drew donations from businesses and the public. But financial assistance could not ease the heartbreak of the many widows left without even a body to bury. Most victims of the Foord pit explosion remained entombed in the mine for decades, until miners in the adjacent Allan shaft broke into the old workings. In 1926, the skeletons of nine men were found at their workplaces, where they had died forty-six years before. The remains were brought to the surface for burial, and middle-aged Dominic Nearing, three months old at the time of the disaster, was able to attend the funeral of his father, Foord pit miner Joseph Nearing.

For those who answered the call of the coal, the danger and the dirt and the heavy work added up to a job they loved—and sometimes loved to hate. "The only job worse than coal mining," offers Bob Gould, a retired Cape Breton coal miner, "is trench warfare." The analogy is apt; veteran miners are like war veterans, survivors of the long struggle to wrest coal from the earth. The stories they tell are stories of fallen comrades and a constant dance with death; their limbs are scarred and bent from wounds suffered in their underground battlefield. And the camaraderie among coal miners is like that of men tested and hardened in battle, men who have looked death in the face and soldiered on. Jim Linthorne, who worked at the McBean mine in Thorburn until it closed in 1972, watched helplessly as his father was torn to pieces by a coal-cutting machine. That horrifying experience did not keep Linthorne from going back underground, however. "You don't dwell on it," he said, shrugging off the danger. "We just done our work."

For others, one close call was enough. "You're always a little apprehensive and nervous. You'll make a mistake if you're not," says Dave Melanson of New Glasgow. "If you have good men and a good mine, it's a pretty safe place." Melanson had been mining coal for twenty-five years when the Allan shaft exploded in 1950. He escaped injury, but some of his co-workers were so badly burned that they lost their ears. Melanson lost his nerve, and quit the next day.

Bert Martin, who left the Pictou County mines unscathed thirty years ago, only to lose a son at Westray, has fond memories of working in the McBean colliery. "I didn't begrudge a minute of it. There's something about it that gets in your system," he says. "You're not really your own boss, but you think you are. When they give you a place to work, you kind of run it the way you want to. It gives you a little bit of pride." But for Martin and many others, the real incentive for digging coal was the scarcity of other work. When Scott Maritimes opened its Pictou County operation in 1967, Martin was among the first to snap up a safer job at the pulp mill.

* * *

NOVA SCOTIA'S COAL INDUSTRY reached its peak in 1913, when more than seven million tonnes of coal was brought to the surface. Cape Breton's mines dominated the industry, supplying almost 80 per cent of that year's output. Pictou County's production peaked at about one million tonnes early in the century, then went into a gradual, but steady, decline. Nova Scotia coal became a tough sell in the lucrative marketplace of Central Canada; American collieries were closer, and could supply coal at a lower price than their east-coast competitors. The coal industry became increasingly dependent on government handouts to ease shipping costs and keep the mines running. The 1920s and 1930s were lean years in Nova Scotia's coalfields, marked by bitter strikes as mine owners cut wages to compensate for slumping sales.

Most Pictou County miners were at the mercy of a single employer, the Acadia Coal Company. A merger in 1886 left Acadia in control of most mines in Stellarton and Thorburn; the only other major player in the Pictou area, the Intercolonial Coal Company, continued to run the Drummond mine. Acadia, in turn, was swallowed up in 1919 by a larger corporate empire. The Dominion Steel and Coal Company, or Dosco, as the new conglomerate came to be known, also operated Cape Breton's collieries and steel industry. Acadia became a subsidiary—a poor cousin to Cape Breton's ailing coal industry—and slipped into receivership in 1933, the worst year of the Depression, after subjecting its miners to layoffs and wage cuts. A royal commission set up by the province later determined that Acadia's profits had been skimmed off to pay dividends to shareholders, rather than finance repairs and new equipment. The provincial government wrote off some $300,000 in unpaid coal royalties and taxes, and Acadia's operations limped on. In 1957, Dosco was taken over by the British conglomerate Hawker Siddeley, completing a circle that began with the General Mining Association more than a century earlier; much of the Pictou coalfield was once again controlled from the boardrooms of London.

After World War II, Pictou County's mines began to close, one by one, leaving hundreds of miners and surface employees out of work. An old enemy, fire, was one of the culprits; the collieries were also becoming too old, too deep, and too expensive to operate. A fire closed the Allan shaft in 1951, throwing 400 people out of work. The MacGregor, hit by a fatal explosion in 1952, was closed by fire in 1957. In 1953, the Intercolonial ceased operation after almost a century of mining; its Drummond mine in Westville continued to operate on a smaller scale, its workforce slashed to 100 from 380. Stellarton's Albion mine, another relic from the nineteenth century, closed in 1955 because the coal deposit was exhausted.

By the 1960s, markets for coal were fast disappearing, hastening its decline. Fuel oil was cheap, and in most people's minds, coal belonged to another age. Railways switched from steam engines to diesel locomotives; coal furnaces were fitted with oil burners to heat homes and businesses. The Greenwood Coal Company, a small, locally owned operation, closed its Thorburn pit in 1966. Then, Hawker Siddeley announced it was shutting down its mines in Cape Breton and Pictou County, a move that could have sounded the death-knell for the province's coal industry. But the federal government stepped in, creating a Crown corporation, the Cape Breton Development Corporation, to run the mines. To cushion the blow, Devco would phase out the island's collieries over a fifteen-year period and create new jobs for thousands of displaced miners. The McBean colliery in Thorburn, Hawker Siddeley's only remaining Pictou County operation, was given a four-year reprieve. After eating up more than one million dollars in government subsidies, the McBean closed for good in 1972, throwing 175 men out of work.

As the years passed, slopes and shafts were boarded up or filled in; the crumbling buildings clustered around pitheads were demolished. The only vestiges of Pictou County's coal-mining past were the black hills of waste coal, the row upon row of company houses, and the memorials to those who went underground and never came back. The government attracted new industries to the county to give work—and hope—to people whose lives had once revolved around the shifts at the mines. The Scott Maritimes pulp mill and a Michelin tire plant created hundreds of new jobs. But many others were forced to pack up and join the exodus of Maritimers "down the road" in search of jobs in Central and Western Canada.

The Drummond mine was all that remained of a once-great industry. In the mid-1970s, about fifty men worked at the Westville mine, extracting coal from pillars left behind during earlier mining operations; the mine produced only about 25,000 tonnes per year. For Eugene Johnson and Alex Ryan and a new generation of Pictou County coal miners working at Drummond, the future was dim. When James Cameron published *The Pictonian Colliers* in 1973, he proclaimed that the local coal industry was "breathing its last—unless an unforeseen and unanticipated miracle occurs."

The miracle arrived in an unlikely form—the Organization of Petroleum Exporting Countries and the energy crises of the 1970s. Imported oil was prone to price hikes and embargoes at OPEC's whim; almost overnight, Nova Scotia coal became an attractive alternative to fuel the province's electrical generating stations. The last large reserve of Pictou County coal was a section of the Foord seam located east of Stellarton, beside the abandoned workings of the explosion-plagued Allan shaft. Mining the Foord seam, as James Cameron observed, has always been a gamble. But by the 1980s, thirty years after the Allan mine hoisted coal for the last time, entrepreneurs and politicians were once again willing to take the risk.

KING COAL'S REVIVAL

HE REBIRTH OF NOVA SCOTIA'S COAL INDUSTRY was triggered by a major shift in government policy that seemed to have nothing to do with coal mining. In 1973, the Liberal government of Premier Gerald Regan took control of all the province's electrical utilities, creating a Crown agency called the Nova Scotia Power Corporation. The move was supposed to standardize service and save money, but there was an unwelcome side effect–power rates became a political issue. The crunch came in 1973, when an OPEC embargo boosted oil prices to unprecedented levels. NSPC, which generated the bulk of its electricity by burning imported crude, was forced to pass on the higher cost to its customers. Electrical rates jumped almost 50 per cent in 1977, giving Nova Scotians the dubious distinction of paying the highest rates in Canada. After the Progressive Conservatives crushed the Regan Liberals at the polls in 1978, power rates were widely viewed as the main cause of the government's downfall.

The lesson was not lost on the new premier, John Buchanan. An affable lawyer from the working-class Halifax suburb of Spryfield, Buchanan was renowned for his ability to remember names and faces; long before "networking" became a buzz word, Buchanan was busy making friends with just about everyone. His managerial abilities and business acumen might have been suspect–he came to power, it was later revealed, saddled with close to one million dollars in personal debts–but no-one questioned his political savvy. Buchanan's Tories campaigned in the 1978 election on a promise to subsidize power rates until alternative sources of energy could be tapped. Harnessing the powerful tides of the Bay of Fundy was one possibility, but the logical solution was to fall back on the province's long-neglected reserves of coal. The supply was secure, the price stable, and a revitalized coal industry could create thousands of jobs for Nova Scotians.

Under the Tories, NSPC received a new mandate. Millions of dollars were spent building new coal-burning generating stations in Cape Breton and converting oil-fired plants to coal. In 1980, just over 20 per cent of the province's electricity was generated from coal; within six years, coal accounted for 70 per cent of Nova Sco-

The Conservative government of Premier John Buchanan made Nova Scotia coal the cornerstone of its energy policy. By the mid-1980s, coal accounted for 70 per cent of the provincial electrical utility's output.

tia Power's output. The Cape Breton Development Corporation, created to phase out the island's collieries by 1982, changed its course and opened new mines to feed NSPC's growing demand for thermal coal. By the mid-1980s, the utility was Devco's major customer, providing a market for more than two million tonnes of Cape Breton coal per year.

The switch from oil to coal was expensive, but the Buchanan government was determined to hold the line on power rates—and avoid political suicide. When NSPC secured a 30-percent rate hike in 1983, the government used subsidies to slash the actual increase to consumers. Then, NSPC announced it would seek no new rate increases until at least 1986, helping the Tories sail unscathed through the 1984 provincial election. Critics complained that the Crown-owned utility was being run in the best interests of the ruling Tories, not in the best interests of the Nova Scotia public. There was no doubt that Buchanan, ever the astute politician, had put the utility in the hands of people he could trust. In 1980, he appointed a Conservative fundraiser, Joseph Macdonald, as NSPC chairman; Macdonald was a senior partner in the Halifax law firm McInnes Cooper Robertson and was one of Buchanan's closest advisers. Three years later, Louis Comeau, a former Conservative MP, was named president. Under the Buchanan government, NSPC had become, in the words of one Halifax economist, "a political toy."

* * *

NOVA SCOTIA'S COAL INDUSTRY was undergoing a renaissance, but most of the action was on Cape Breton Island. In Pictou County, only the Drummond mine remained in production; there were also some small-scale projects to reclaim coal from the mounds of waste at old colliery sites. With coal back in demand, however, mining companies began to give Pictou County a second look. British-owned George

Wimpey Canada did some preliminary work on a strip mine in Stellarton during the late 1970s—the so-called Wimpey pit—but the project was abandoned.

In 1976, the federal government commissioned a report on the prospects for renewed underground mining in the Pictou coalfield. Prepared by the Canada Centre for Mineral and Energy Technology, a division of Energy Mines and Resources Canada known by the acronym CANMET, the study concluded there was enough coal for a "medium-size mine" with an annual output of up to a half-million tonnes. But the report identified a number of drawbacks, such as the depth of the seams, the high-ash content of the coal, "old workings with gas and water problems," and the lack of a local workforce trained in modern mining methods. F. Grant, the CANMET scientist who wrote the report, doubted that a private operator could be found, and this would leave development of the mine up to the government. "The problem is to mine these coal reserves efficiently and safely to the benefit of the mine, consumer, community, and province," he concluded.

By the early 1980s, despite CANMET's gloomy outlook, a couple of privately owned Canadian mining firms were willing to try. Suncor teamed up with Brinco Mining in the summer of 1981 to explore untapped coal seams in the Westville area. Test holes were drilled, but the seams were deep, reserves were small, and the coal retrieved from core samples was too high in ash for thermal use. Brinco, which had invested a half-million dollars in the exploration program, cut its losses and dropped out of the partnership in 1982. Despite the setback, Suncor decided to go it alone. "We thought it was a bit premature to can the whole thing, and we went back to the drawing board," a Suncor official would later recall.

The company also had deeper pockets than Brinco. The Canadian subsidiary of the Pennsylvania petroleum giant Sun Company, Suncor was one-quarter owned by the Ontario government and boasted revenues in the $1.5-billion range. The company was one of Canada's largest producers of oil and gas, with extensive holdings in Western Canada that included a swath of northern Alberta's tar sands. In the early 1980s, Suncor was in the midst of drilling programs in the Arctic and off the coast of Nova Scotia; in 1984 alone, the firm spent more than $100 million on exploration and development. Suncor's modest mineral exploration program, handled out of the company's Calgary offices, included a 7,000-hectare chunk of the Pictou coalfield.

After the Westville deposit failed to pan out, Suncor's engineers turned their attention to the adjoining Stellarton block. At the company's request, the Nova Scotia government extended the exploration lease to include sections of the Foord and other seams. A three-year program of drilling and seismic work was launched in 1983, centred on an area east of Stellarton. As drilling rigs popped up in the woods and fields, Suncor officials tried to temper expectations that coal mining was about to return to Pictou County in a big way. "The very earliest that we could have a mine, assuming everything falls into place and on a very optimistic scenario, is five to seven years," a company spokesman cautioned in a newspaper interview.

Within a year, it seemed things were indeed falling into place for what Suncor was calling the Acadia Coal Project. The drilling program identified a mineable reserve of coal in the Foord seam, and Suncor began laying the groundwork for development. "The project at last appears to be bearing fruit," a Suncor official told Nova Scotia Mines and Energy Minister Joel Matheson in early 1984. In May, John Shillabeer, the Suncor engineer in charge of the project, flew to Halifax to brief the minister and to discuss a possible coal contract with officials of the Nova Scotia Power Corporation, which operated a generating station in Trenton. NSPC had operated a succession of power plants at Trenton; the latest incarnation was designated Trenton 5. A provincial election was in the works, and Suncor asked the government to exercise restraint. "Exploration and mine development is a high-risk business, and we would ask you and your colleagues to continue exercising discretion," Ray Moss, manager of Suncor's coal and minerals department, told Matheson after the meeting. "We must not risk inadvertently raising people's expectations prematurely."

There was a major obstacle to an underground mine that had nothing to do with markets or geology; Suncor wanted the rights to all surface coal within its exploration leases. Since 1979, all Nova Scotia coal lying less than 130 metres deep had been in the hands of Novaco, a provincial Crown corporation that operated a strip mine in Cape Breton. As early as 1982, Suncor had floated the possibility of gaining access to surface coal, including the Wimpey pit in Stellarton. It made economic sense; mining surface coal first could help finance a more-expensive underground mine, and would prevent another company from using Wimpey coal to steal Suncor's markets. Suncor raised the issue at the May 1984 meeting with Matheson, who agreed to consider transferring the rights.

There was another, smaller player on the scene—Antigonish-based Pioneer Coal Company—but Pioneer was not competing for Wimpey coal, because it had a strip mine of its own. Pioneer took over the Drummond mine in Westville in early 1984. That summer, a cave-in blocked the main slope; before the debris could be cleared, a fire broke out underground, closing the mine for good. That enabled Pioneer to open a strip mine at Drummond. Pioneer, like Suncor, pinned its hopes on a long-term contract to sell coal to NSPC's Trenton 5 power plant. In a sense, Pioneer enjoyed an advantage over Suncor: because leases on the Drummond site predated the creation of Novaco, they included rights to surface coal.

Strip mining was a dirty, noisy business, and in Pictou County, where towns had sprung up haphazardly around long-abandoned collieries, it was inevitable that an open-pit mine would be on somebody's doorstep. Back in 1981, Westville residents had solidly rejected a strip mine proposal. Underground mining was a palatable alternative, a point Suncor stressed when it made its pitch to the Nova Scotia government. "Underground mining is part of the local culture," Moss told Matheson. "A proposal for a significant underground operation which also contains some incidental strip mining is more likely to succeed than one which is for strip mining alone."

Suncor's request put the government in a quandary. Failure to grant the surface rights could scuttle the entire Pictou County coal project. "It may be a major deciding factor when the Suncor board [of directors] is making their 'go–no go' decision on a mine in Nova Scotia," Ed Bain, the province's manager of coal development, warned in March 1985. Cabinet had the power to take the surface rights back from Novaco and transfer them to a private company. But open-pit mining was such a touchy political issue that Matheson chose a more-subtle approach. The minister asked Novaco to surrender Pictou County surface rights to the government, which would retain the right to approve or reject any strip-mining proposals. Although Suncor would not get the surface coal outright, the company agreed to the compromise. "We are anxious to facilitate the development of the coal resource in the area in the interests of the people of Nova Scotia," Matheson assured Novaco chairman Malcolm Turner.

Novaco, for its part, was wary of the government's willingness to dance to Suncor's tune. "If surface mining was to be started in Pictou County, I can hear as loud an outcry as has been our sad experience in Cape Breton," Turner reminded Matheson in a letter in April 1985. While not wanting to appear "high-handed and difficult," Novaco needed more information about Suncor's plans before it would consider handing over the rights. "My real concern would be that they would do their strip mining and then, having taken their profits, decide deep mining was not feasible," Turner bluntly told the minister. "One cannot help feeling it would be analogous to signing a blank cheque, and this I will not do." But the final decision would rest with Novaco's political bosses.

* * *

SUNCOR FORGED AHEAD with the Acadia project, despite the uncertainty over surface-coal rights. By 1985, the company had spent four million dollars on exploration, optioned land for surface facilities, and retained consultants to prepare feasibility studies. Suncor opened an office in New Glasgow that summer and began assembling a team to move the project towards the development phase. Newspaper advertisements appeared, seeking a mine engineer, a geologist, a marketing specialist, and an engineer with experience in coal processing. But Suncor remained determined to keep the project low-key; the cryptic ads said the jobs were connected to a "possible coal-mining venture in Nova Scotia." When the Halifax *Chronicle-Herald* tracked down John Shillabeer in August 1985, Suncor's chief mining engineer confirmed that another two million dollars would be spent on the project in the coming year. "We've found some coal that we believe is of a quantity that's saleable," Shillabeer said from his Calgary office, cautioning that the odds against developing a underground mine were still five to one.

With the project now in high gear, Suncor decided it was time to let the Nova Scotia government in on its plans. At the end of September, Shillabeer and Ryan Moore, Suncor's government liaison, travelled to Halifax to brief provincial representatives, including the three members of the legislature from Pictou County. At the closed-door meeting, Moore outlined an ambitious plan to open a $95-million colliery on the Foord seam at Plymouth, just east of Stellarton. The mine would employ about 250 people and produce about 600,000 tonnes of coal per year for twenty years. At a royalty rate of 27.5 cents per tonne, that would put some $165,000 a year into the provincial treasury. Although the mine was "currently only at the study stage," Moore said, production could begin as early as 1989. But there was one catch: for the mine to be viable, the Nova Scotia Power Corporation would have to take at least 500,000 tonnes per year.

Suncor finally went public in December. Shillabeer announced that Suncor had targeted the Trenton power plant as the main market for its proposed mine. Trenton 5 burned 400,000 tonnes of Cape Breton coal annually, much to the chagrin of Pictou County residents who were out of work. But a Suncor mine, located almost next door to Trenton, was expected to be able to ship coal to the plant at competitive rates. And Shillabeer touted one other advantage of Pictou County coal; it was lower in sulphur than Cape Breton's coal, and less harmful to the environment. "This low-sulphur coal, which reduces the acid rain problem, would be particularly attractive to the province, and might justify a doubling of the Trenton plant," Shillabeer claimed.

For the Cape Breton Development Corporation, the spectre of competition on its home turf could not have come at a worse time. No. 26 colliery in Glace Bay, scene of a fatal explosion in 1979, closed permanently after a fire in 1984. The mine had been Devco's only source of metallurgical coal for the export market, and the closure left the corporation largely dependent on selling thermal coal to NSPC. It had been a lucrative market; Devco won an impressive price increase—from about $45 per tonne to a rate of about $65—when it negotiated a five-year contract with the utility in 1984. But losing Trenton 5, which accounted for one-fifth of Devco's annual sales to NSPC, would be a severe blow.

The opening salvo in the emerging coal war between Cape Breton and the Nova Scotia mainland was fired in October 1985, just as Suncor was putting the finishing touches on its plan. Vision, a Cape Breton group that promoted regional development, questioned the logic of developing a Pictou County mine at a time when Devco needed to open new collieries to recapture its export markets. "To develop the capacity to be internationally competitive, secure export markets and to become profitable," Vision argued in an article published in the *Cape Breton Post,* "Devco requires a stable domestic market."

That view was not unanimous; Jake Campbell, head of the United Mine Workers of America in Cape Breton, was of the opinion that there were "enough markets to go around if Devco goes after them." And Dr. Dieter Birk, director of a coal-testing laboratory in Sydney, fired off his own opinion piece to the *Post,* lambasting Vision's assumption that competition was bad for business. "The fear expressed by some Cape Bretoners that a Pictou County project might hurt Glace Bay is typical of the parochialism that often interferes with any Maritime initiative," he scoffed.

Cape Breton's concerns received short shrift in Pictou County. Mainland miners were being treated like "lepers," said the head of the union representing Pioneer Coal workers, most of whom were out of work. "Imagine if the situation were reversed and NSPC tried to ship a shovelful of mainland coal to burn in a Cape Breton power plant," said Peter Johnston. "All we ask for is a little bit of fairness." Town councils in Trenton, Pictou, New Glasgow, Westville, and Stellarton passed resolutions in January 1986 calling on the provincial government to build a second generating station in Trenton. The coal needed to fuel both Trenton 5 and a possible Trenton 6 would come from an expanded Pioneer coal operation and the proposed Suncor mine. "I think Pictou County needs the jobs just as much as Cape Breton," said Pictou councillor Kate Kennedy. "The bottom line," added Bill MacCulloch, executive director of the Pictou County Research and Development Commission, "is we need employment in Pictou County and we need it now." MacCulloch could have been speaking for himself; years later, when Westray went into production, he snagged a job running the mine's training program.

* * *

BY THE FALL OF 1985, even before the company went public, Suncor's newly assembled team of engineers was digging into mounds of data generated by the company's four-year exploration program. The mine engineer was hired in November 1985—Gerald James Phillips, a thirty-five-year-old with impressive coal-mining credentials. Born in England, Phillips was in the midst of a meteoric career that had taken him from the coalface to the boardroom. At fifteen, he left school to dig coal for the National Coal Board, the Crown corporation that operated Britain's collieries. Phillips rose steadily through the ranks, earning a promotion to underground manager while still in his early twenties. In the summer of 1975, he immigrated to Canada to work at a mine in Coleman, Alberta, as a foreman. Over the next five years, Phillips bounced from mine to mine, always moving to positions of greater power and responsibility. By 1979, he was underground manager at McIntyre Mines in Grande Cache, Alberta, where he supervised three collieries producing more than one million tonnes annually. Within a year he was put in charge of the $80-million Cardinal River Coals mine being developed in Hinton, another Foothills town west of Edmonton.

Phillips oversaw all aspects of getting the Cardinal River mine up and running, everything from feasibility studies and budget projections, to negotiations with unions and government regulators. "I have had extensive experience in dealing with government officials, such as safety, engineering, environmental, forestry, and ministry-level officials," he wrote in his résumé. Phillips was versatile; he had held management-level posts at collieries on both sides of the Atlantic, and was familiar with a variety of mining methods. He also claimed to have expertise in controlling costs and achieving what he called "high-speed development in coal"—music to any mine owner's ears. To round out his résumé, Phillips noted that he had been responsible for mine safety in the past, and claimed to have dealt with Alberta government officials "regarding the development of new mine safety regulations."

But Phillips's career on the fast track had not been without its bumps. In February 1980, a few weeks after he left McIntyre Mines to join the Cardinal River project, four miners died in a cave-in at McIntyre's Reiff Terrace mine in Grande Cache. The men, part of a crew cutting coal from a pillar, were crushed under a metre and a half of rock; two other men, shielded by machinery, escaped injury. It was Alberta's worst mining disaster in almost forty years, prompting the government to set up a commission of inquiry headed by Gerald Stephenson, a Calgary-based mining consultant.

In his report, Stephenson ruled that the crew had extracted coal from the pillar despite warnings from miners on an earlier shift that a collapse was possible. But he concluded that McIntyre's management had followed "a faulty mining sequence," weakening the roof in the area where the cave-in occurred. "During 1978 and 1979 the central area [of coal] was completely and prematurely developed into relatively small pillars. Conditions deteriorated rapidly," he explained. "Given the previous experience of mining this seam by room and pillar at Grande Cache, the company should have foreseen the problems which would be caused by cutting these areas into pillars of low strength, long before they could be depillared." During 1979, Stephenson added, reports by mine inspectors and the miners' safety committee revealed "serious problems in maintaining an adequate standard of [roof] support."

Despite these findings, Stephenson did not single out McIntyre officials for censure. The thrust of his report, made public in 1981, was that steps should be taken to prevent similar accidents in the future. "Everyone involved in decision making at underground mines in Alberta, and in the rest of the country, would be well advised to study the sequence of events at Reiff Terrace which, starting in 1976, led finally to the accident on February 28, 1980," Stephenson said. Although Phillips was underground manager at Reiff Terrace for all of 1979—the period when roof conditions in the mine were at their worst—he was not called as a witness at the inquiry's hearings. And his name did not surface in Stephenson's final report. This was surprising, given his key role in the mine's chain of command; Phillips himself later listed "safety, coal production, and planning of underground operations" as among his responsibilities at McIntyre.

But officials of the miners' union at Reiff Terrace at least settled some old scores when they testified before the Stephenson inquiry in the fall of 1980. And one of the targets of their wrath was Phillips, who had little time for unions and their complaints. Alex Gallacher of the United Steelworkers of America told the inquiry that the mine's safety committee was constantly at odds with Phillips over repairs or unsafe conditions. "We would note something in the mine, go see him, and we would get [into] some argument with him as to how it should be repaired or what can be done to prevent something happening again." On one occasion, Phillips mistakenly accused Gallacher of shutting down a roof bolter because the ventilation system in the area was not operating. "Mr. Phillips came down, started shouting, and he said something about [how] we could have a good relationship here if I would mind my own business."

Reports by Alberta government inspectors, entered as evidence at the inquiry, also provided damning evidence of safety standards at Reiff Terrace. In the year leading up to the cave-in—Phillips's tenure as underground manager—inspectors repeatedly complained about poor roof and road conditions and inadequate ventilation in the mine. After a miner was crushed to death in a May 1979 cave-in, inspectors had criticized Reiff Terrace's management for not having a "clearly explained" procedure for removing coal pillars. They demanded that stone dust be spread to prevent the build-up of explosive coal dust. "Systematic rock dusting must be an integral part of the mining operation," inspector John Greenwell pointed out in October 1979. A month later, the same inspector again noted that "systematic rock dusting of conveyor lines must be done."

Gallacher, an experienced miner from Scotland, told the inquiry he had difficulty convincing management—and his fellow miners—that stone dusting was important. When he raised the issue through the union's inspection committee, mine officials told him additional stone-dusting was not required. He also suggested that McIntyre install dust barriers—a series of shelves covered with loose stone dust. In an explosion, the shelves collapse, releasing the dust into the air and snuffing out the flames. It was a safety measure Gallacher had seen used in Britain, where he, like Phillips, had worked for the National Coal Board. But Phillips scoffed at the idea, telling Gallacher the barriers were not needed at Reiff Terrace. "We tried to stress the importance to Mr. Phillips of the rock dust, and I am sure he understood the importance," Gallacher testified. "I believe he came from the NCB, so what more can I say about it?" Phillips, it was clear, had his own views on the value of stone dusting as a safety measure.

The controversy over his brief tenure at McIntyre was long forgotten by the time he signed on with Suncor. Phillips had the expertise Suncor needed for its fledgling project in Pictou County. He was young and ambitious, with the energy and determination needed to get the new mine into production. And if the past was any guide, he would fit in well in Pictou County. He was a family man; by the early 1980s, he and his wife, Catherine, were raising five children. He was active in the community, serving on the school board in Alberta. And his passion for golf and darts—two sports that provide as much opportunity for socializing as they do exercise—ensured that he would be a good ambassador.

As chief mine engineer, Phillips was responsible for planning everything from the proper procedures for blasting access tunnels, to the design of the ventilation system. He was also expected to do what he did best—bring government officials onside. In November 1985, within weeks of being hired, Phillips outlined the Pictou project to Patrick Phalen, the director of mining engineering for the Nova Scotia Department of Mines and Energy. Phalen was obviously impressed; he invited Phillips to comment on proposed revisions to the Coal Mines Regulation Act. In order to bring the newcomer up to speed on Nova Scotia's safety laws, he sent Phillips a copy of the new Occupational Health and Safety Act, which had just become law.

Phillips lost no time drawing up safety procedures for the proposed mine. In early 1986 he forwarded a "Manager's Ventilation Plan" to Walter Fell, the province's chief inspector of mines, for review. The twenty-two-page document described the steps Suncor would take to prevent underground fires, such as ensuring that "no oil, grease, canvas, or other flammable material shall be stored in the mine except in a fireproof receptacle." There were also detailed procedures for dealing with dangerous levels of methane; the only men to remain in an affected area were to be those sent to flush out the gas. "Normal operation of the mine can be resumed only after [the] degassing operation has been successfully completed and the working places are examined by a mine official," Phillips wrote.

Tucked at the back of the ventilation plan was a three-page document for dealing with another underground hazard—coal dust. Phillips adopted the standard prescribed in the Coal Mines Regulation Act: stone dust was to be spread to reduce the proportion of combustible coal to no more than 35 per cent of the dust in the mine. If there was methane present, the percentage of stone dust was to be increased by 1 percentage point for each .10 per cent of gas in the air, a provision that, once again, paralleled the wording of the act. The stone-dusting rules were to apply to all areas of the mine, to within ten metres of the working face. Phillips also envisioned a program of regular testing; dust samples were to be collected monthly from "one or more representative places in each mining area," and more frequently in sections of the mine where visual inspection suggested there was not enough stone dust. Phillips's commitment to tackling the coal-dust problem looked good on paper; whether his attitude about stone dusting had changed since his days in Grande Cache remained to be seen.

* * *

THE CONSTITUENCY OF PICTOU EAST cuts a swath across the eastern end of Pictou County, from the beaches and cottages of the Northumberland Strait to the overgrown farms that dot the county's interior. The electoral boundaries rub shoulders with the towns of Trenton, New Glasgow, and Stellarton, but Pictou East is predominantly a rural riding. This is the home of people with surnames like MacDonald, Cameron, MacKenzie, Fraser—descendants of the county's original Scottish settlers. Pictou East was created when the county was split into three ridings in 1949 and, until the mid-1970s, voters sent more Liberals than Conservatives to the legislature in Halifax. But in the 1974 provincial election, a young dairy farmer named Donald William Cameron claimed the seat for the Tories.

Cameron was a local boy made good, a graduate of the science program at McGill University in Montreal who had come home to take over the family farm in Egerton, a speck on the map about ten kilometres east of New Glasgow. He was a clean-cut, clean-living man, a non-smoker who only rarely allowed himself to indulge in a glass of wine. His demeanour was as austere as his lifestyle; he seemed uncomfortable in public, even shy. Cameron and John Buchanan, his premier and party leader, were a study in contrasts. Buchanan was a folksy, grassroots politician, known to climb onstage at political gatherings to belt out songs for the party faithful. That was not Cameron's style; he preferred to toil in the background while others grabbed the spotlight.

The differences between Cameron and Buchanan went deeper than their actions on the hustings. Buchanan's whole approach to politics was rooted in bestowing political jobs and favours. Buchanan, columnist Jeffrey Simpson noted in *Spoils of Power,* his landmark book on political patronage in Canada, was "the classic parish-pump politician, most at ease discussing the fine points of political maneuvering, arranging for the satisfaction of every local desire, administering patronage and injecting partisanship into the most routine matters." Cameron could be as fiercely partisan as Buchanan, capable of attacking opposition politicians and policies with single-minded zeal. But the old-style politics of winner-take-all offended his sense of fairness. And no-one questioned his integrity. After one election, a Tory supporter made a point of reminding Cameron that he had contributed to the Conservative party's campaign fund. "I have an investment in you, boy," the man said. Cameron's defiant reply: "You'll not buy me for $200 or $2 million."

When Buchanan led the Tories to power in 1978, Cameron, just thirty-two, was given responsibility for two portfolios, fisheries and recreation. Back in Pictou East, he was expected to administer the traditional postelection firing of highway foremen, snowplough drivers, and other Department of Transportation workers ensconced by the former Liberal government—and hire loyal Tories in their place. Cameron refused, bucking more than a century of Nova Scotia tradition, and left the employees in place. Despite the outcry in the ranks, Cameron clung to his belief that government hiring should be based on merit, not political stripe. "I don't blood-test someone at my door to see if they have Tory blue blood in their veins," he declared at one party meeting.

After less than two years in the cabinet, Cameron abruptly resigned in the summer of 1980, saying he needed to devote more attention to running his sizable dairy operation. There were rumours of clashes with his cabinet colleagues but, truth was, the farm was going broke, and eventually had to be sold. It was years before Cameron publicly acknowledged the real reason for his departure: the buyer had applied to the provincial government for a loan, which required cabinet approval. Cameron immediately handed in his resignation to avoid a conflict. "To me, it was a matter of principle," he later explained in an address to the legislature. Cameron's high-minded views left him branded as a renegade by some of his hidebound fellow Tories.

Cameron, once more a backbencher, turned his energies to promoting and protecting the interests of his native county. The promise of a new coal mine that would draw investment and jobs to Pictou County quickly captured his imagination. It was only logical that he should go to bat for the project: the proposed mine would be built in the village of Plymouth, on the boundary of his Pictou East riding. In April 1986, during debate in the legislature after the Speech from the Throne, Cameron put in a plug for Suncor and Pictou County coal. The unemployment rate in the Pictou area was nearly 25 per cent, he said, and the government had a golden opportunity to create badly needed jobs. The Nova Scotia Power Corporation's Trenton 5 generating station was in need of upgrading, and municipal leaders had been calling for another plant in the town. At the same time, the Suncor mine promised work for up to 250 people. Burning Pictou County coal in Pictou County generating stations, creating Pictou County jobs—it made perfect sense.

There were other advantages to using Pictou coal to generate electricity, Cameron explained. It was low in sulphur, making it better for the environment—an issue quickly elbowing its way to the top of the public agenda. And Suncor was willing to develop the mine without government aid, a refreshing change from the "hundreds of millions of dollars" in public money poured into developing Devco mines in Cape Breton. "I am not criticizing that," Cameron said of the Devco assistance. "I just want to point out the contrast." Back home in Pictou County, the speech was quoted at length in the New Glasgow *Evening News*, under the headline: "Coal sales could help unemployed."

In hindsight, Cameron's speech was remarkable for a couple of reasons. The mine was still on the drawing boards—but there was Cameron, already taking swipes at Cape Breton and Devco. He even prefaced his comments about Pictou County's jobless rate with the comment that Cape Breton was not the only area where unemployment was high. And Cameron was already talking as if he and the people of Pictou County had a personal stake in the mine's development. "We are not asking for one cent of government money," he declared. "We are asking for the right to sell our coal to our own markets, at our own back door, in our own county."

The announcement about the future of the Trenton power plant came only a few weeks after Cameron's speech. Premier John Buchanan, who was in Pictou County on April 29 to tour the Scott pulp mill with Cameron and other Tory politicians, did the honours after a luncheon at the Abercrombie Golf Club. NSPC was "very close" to announcing the sites of up to four new power plants, Buchanan told reporters, and "we've said consistently that one of those sites will be in Pictou County—in Trenton." Buchanan granted another of Cameron's wishes; the new plant, he said, would burn local coal from the proposed Suncor mine. As a consolation prize, Pioneer was finalizing a deal to sell 100,000 tonnes of coal per year to the Trenton 5 station from an expanded strip mine in Westville.

Behind the scenes, the wheels had been in motion for more than a month. On March 20, NSPC officials had privately told Suncor they were ready to open negotiations on a coal contract. Suncor planned to borrow $68 million of the estimated $90 million needed to bring the mine into production, and an assured market with NSPC was the key to obtaining a bank loan. The government, always willing to help, went to bat for Suncor. John Laffin, the province's deputy minister of mines and energy, pitched the project during a lunch meeting with officials of the Royal Bank of Canada in mid-April. The utility agreed to buy 500,000 tonnes per year for its existing Trenton 5 facility and the proposed power plant. That was the exact number Suncor needed to make the project fly.

As 1986 wore on, the pieces were falling into place for what Suncor was now calling the Pictou County Coal Project. Novaco, despite the reservations of its chairman, relinquished the surface-coal rights on the areas covered by Suncor's exploration leases; those rights reverted to the government for the time being. Suncor was content to apply for an underground mining lease only, and put out tenders on the main access slopes, with construction to begin as early as the fall. The company also applied for exemptions under the Coal Mines Regulation Act during the tunnelling phase, in three areas: use of diesel equipment, blasting methods, and certification of employees. Trenton was not Suncor's only potential customer; the prospect of coal sales to New Brunswick Power and cement plants in the Maritimes prompted Suncor to revise its production figures, raising the projected output to 950,000 tonnes per year by 1991, up from 600,000 tonnes.

The only snag came after Suncor held an open house in New Glasgow in June to unveil an artist's conception of the mine's surface plant. While the "demanding and potentially hazardous" job of mining would require a "core group" of experienced miners, Suncor's John Shillabeer said local people would be taken on as trainees. But the prospect of new jobs and supply contracts for local businesses did not please everyone. Plymouth residents were upset that the mine would be built in their back yards, and that heavy trucks would be rumbling past their doors to deliver the coal to Trenton. Within days, 90 per cent of the residents of East River Road, the proposed truck route, had signed a petition opposing Suncor's plans. Cameron defused the situation by suggesting the government build a new access road to accommodate Suncor's coal trucks. Suncor was "a very, very responsible company," Cameron assured residents, and their concerns would not be ignored.

A few names on a petition was the last thing on the minds of executives at Suncor's head office in Toronto. The price of oil had renewed interest in Pictou County coal; now it threatened to scuttle Suncor's plans. Between January and April 1986, the world price of oil had dropped from $28 U.S. per barrel to as low as $13 U.S. per barrel. Suncor laid off hundreds of workers at its massive oil-sands project in Alberta and slashed its capital budget in an attempt to cushion the blow, but was still headed for a $7-million loss for 1986. As summer turned to fall, the prospect of sinking $90 million into a coal mine in Nova Scotia was rapidly losing its appeal. Suncor's board of directors met in July but postponed a decision because feasibility studies were not complete. The studies were finished within a few weeks, but a September deadline also passed. In November, Suncor announced that the project was on hold until a partner could be found to share the cost of developing Pictou County coal.

* * *

THE DECISION TO SHELVE the project was based solely on Suncor's financial bind. The feasibility studies, delivered in August 1986, extolled the project as technically viable and economically sound. There were two reports. One dealt with the geology of the area of the Foord seam targeted for mining, which had an estimated reserve of 45 million tonnes. The other outlined methods for mining and processing the coal. The geological report had been handled in-house with the aid of a computer system set up at the New Glasgow office; Norwest Resource Consultants of Calgary had been retained to prepare the companion volume. But the entire study was heavily influenced by Suncor personnel; Norwest worked closely with John Shillabeer, Gerald Phillips, and the other members of the New Glasgow–based project team, and relied totally on Suncor's reading of the geology.

The thrust of the Norwest report was that modern mining equipment and methods could meet the challenge of the Foord, providing a combination of techniques was used. Most of the coal would be dug using the longwall method, while the room-and-pillar approach would be used in areas near major faults, where broken rock and coal would make mining difficult. The longwall technique, as its name suggests, uses a machine to cut coal along the entire length of a tunnel driven into the coal seam. Instead of leaving pillars to support the roof, longwall mining uses movable hydraulic jacks to support the roof until mining is completed. Although room and pillar was the preferred method for most of the history of mining in Pictou County, longwalls were used at the McBean and a few other mines. Since the Foord seam was notorious for faults, major breaks in the seam could disrupt production. "Conditions adjacent to major faults in the mine area are poorly understood," Norwest admitted, but employing both mining methods would give Suncor the flexibility and equipment needed to deal with the problem.

Each report dealt with the Foord seam's well-documented potential for spontaneous fires. "Past mining practices were often questionable and ventilation standards low," Suncor's geology report noted; this probably explained the spontaneous fires that bedevilled earlier collieries. Using Foord coal taken from drill cores, Suncor hired an independent laboratory, Hazen Research (International), to gauge its propensity to self-ignite. Haven experiments suggested spontaneous combustion was unlikely, except at temperatures far above those usually experienced in underground mines. While those results were "not conclusive," the Suncor report noted, "they do provide some assurance that Foord coal is not highly susceptible to spontaneous heating and self-ignition."

Norwest did not share that optimistic view. Its report declared that the potential for spontaneous combustion was "relatively high," although the mining procedures being contemplated would "minimize the risks associated with this hazard." Norwest pointed out that the numerous fires in the adjacent Allan shaft had occurred in areas of the mine exposed to air long before mining was completed; this was not the plan for the proposed mine. Norwest criticized Suncor's testing methods and suggested more laboratory analysis, warning that the possibility of spontaneous fire "should not be dismissed too readily." To avoid the risk of fire, Norwest recommended that each working area be tightly sealed as soon as mining was completed.

The Norwest and Suncor studies were unanimous on another safety risk—methane. Despite the Foord seam's deadly history of explosion, both reports concluded the risk was low, as long as enough fresh air was flowing through the mine. "Methane emissions are not expected to be a problem in this mine," Norwest flatly declared. "The Foord seam was not particularly gassy in the Allan mine, and with adequate ventilation, methane dilution and removal will not be a problem." It was a startling conclusion in light of the Allan's reputation as one of the most dangerous mines in the world. Suncor also conducted lab tests to measure the amount of methane given off by Foord coal. The results: "Data from the Foord seam suggested relatively low gas values." The geological report conceded that the samples had been collected near old mine workings, and much of the methane could already have escaped.

Suncor's cavalier attitude toward the methane risk was not surprising; the company had long viewed methane as a commodity rather than as a hazard. Since 1979, an Alberta company, Nova Corporation, had been drilling in the Pictou coalfield with an eye to extracting methane for commercial use. In 1984, Suncor asked the government for the gas rights on its leases; the company had no interest in extracting the gas, but wanted to prevent other firms from sinking wells and disrupting its underground activities. The issue generated an internal debate at the Department of Mines and Energy, where some saw merit in having methane drained off before underground mining began. "There is a considerable health and safety benefit to the practice," argued Patrick Hannon, the manager of mining engineering. Drawing off the gas would improve air quality in the mine, he noted, and reduce the number of shutdowns caused by high methane readings. Suncor, for its part, did not pursue the idea of draining off gas before mining. "I believe the main challenge with respect to methane will be to correctly estimate the methane emission rate and the required ventilation," project manager John Shillabeer noted in a 1985 letter to Hannon.

One mining consultant on the Suncor payroll refused to take the methane risk lightly. In 1986, Derek Steele, an engineer with the consulting firm Dames & Moore in Cincinnati, Ohio, was retained to review plans being developed for the Pictou County mine. Steele, a consultant of international stature, had almost four decades of experience in the coal industry. Over the course of the year, he produced three reports; each concluded that the project was viable, but suggested minor improvements. The biggest challenge, in Steele's opinion, was warding off fires and explosions. Suncor planned to mine the seam to a height of more than five metres, using flexible tubing to flush out methane. Steele questioned whether that measure would be enough to prevent methane from collecting in layers at the roof.

Steele foresaw another danger; bands of iron-bearing rock, commonly known as ironstone or pyrite, were hidden within parts of the seam. The reports prepared by Suncor and Norwest suggested it would be difficult for mining machines to cut through the rock layers. But Steele saw them as a safety hazard, not simply as a technical problem. He was alarmed by the "high potential" for sparks as machinery cut into the ironstone bands. To avoid sparking in the presence of high levels of methane, adequate ventilation at the working face would be essential at all times. Methane would also build up in mined-out areas, Steele cautioned Suncor, and changes in barometric pressure could flush out the gas, creating another explosion hazard.

* * *

AS FAR AS SUNCOR WAS CONCERNED, the methane and the sparks and all the other headaches would soon be someone else's problem. The company was looking for a partner willing to take a majority position in the project, or buy it outright. Suncor officials insisted development could begin as early as the spring of 1987, but one official of Nova Scotia's Department of Mines and Energy dismissed that as "wishful thinking." The director of mining engineering, Patrick Phalen, doubted a start-up was imminent: "A new partner or owner will probably do their own feasibility and design work, which will require at least six months," he predicted in a January 20, 1987, memo to Mines and Energy Minister Joel Matheson.

Two weeks later, the government gave Suncor's sales effort a boost. John Buchanan, in a speech to the Conservative party's annual meeting in Halifax on February 7, announced that the Nova Scotia Power Corporation would forge ahead with a new Trenton power plant. The premier had promised the same thing almost a year earlier, but this time he offered details: the 150-megawatt station would cost $250 million to build and was slated to be in operation by 1991. The plant, christened Trenton 6, would burn 400,000 tonnes of coal annually, Buchanan added, supplied from the proposed Pictou County mine. There was another factor to consider—the environment. Acid rain was fouling rivers and lakes across eastern North America, and Nova Scotia was under mounting pressure to do something about the problem. The province's expanding network of coal-fired power plants was a major producer of the sulphur dioxide emissions that produced acid rain. The government was committed to using low-sulphur Pictou coal at Trenton

6, Matheson told reporters after the premier made his announcement, and the plant would require no special equipment to reduce emissions.

Once again, the Cape Breton Development Corporation viewed the proposed mine as a threat to its hegemony over the Nova Scotia coal market. Devco president Derek Rance came out swinging in the press, accusing the government of jeopardizing coal production and jobs in Cape Breton. In an effort to smooth the waters, Matheson revealed that the Pictou mine had only been promised a contract of 300,000 tonnes per year, leaving Devco to supply the remaining 100,000 tonnes. Rance, however, contended that Devco would eventually lose all its Trenton business. "You can imagine a coal mine that is just a stone's throw away will be in a very competitive position when compared to the transportation costs from Cape Breton." Matheson pointedly reminded Rance that a new NSPC plant was planned for Cape Breton, with state-of-the-art equipment to cleanly burn 400,000 additional tonnes of Devco's high-sulphur coal.

Despite the renewed sniping over which region should enjoy the spoils of coal development, the Trenton 6 announcement had the desired effect. On February 19, Suncor signed an option agreement with Placer Development of Vancouver, a major producer of petroleum that also operated gold and silver mines in Canada and abroad. As Phalen had predicted, the new company wanted time to conduct its own feasibility studies. Suncor gave Placer until July 15 to size up the Pictou property and decide whether to jump into the coal business.

* * *

AS PLACER BEGAN ITS ASSESSMENT, more warning bells were being sounded within Nova Scotia's Mines and Energy department. Robert Naylor, a government geologist who had studied the Pictou coalfield, had serious reservations about Suncor's 1986 geological study. "The report as a whole gives a simplified view of the geology of their exploration areas," he wrote in a March 1987 memo to Patrick Hannon, the department's manager of mining engineering. Much of the "structural complexity" of the coalfield, such as the steep angle of the seams, had been overlooked. "They have also either failed to recognize or include major faults which appear to have been intersected in a number of their drill holes." And, Naylor said, many of the faults located by Suncor did not show up on recent maps of the coalfield. If Placer followed through on its plan to take over the project, he said, the government should look for better use of available geological data, and better documentation of the faults.

Naylor's concerns were taken seriously within the department, and a meeting was set up with Suncor's geologist, Arden Thompson, in mid-April. The discussion was candid, and Thompson agreed with many of Naylor's reservations. But the project's future was now up to Placer, which was drilling four new exploration holes to assess the Foord seam's potential. "Arden feels these holes have to yield positive results or the project will not likely go ahead," Naylor reported to his superiors. Thompson suggested Naylor get together with Hans Bielenstein, the mining consultant preparing a new geological report for Placer. But time was running out; Placer only had a five-month option on the Suncor lease. "Placer finds itself in a somewhat awkward position," Naylor noted. "They have a short time to evaluate the Foord seam. The understanding of the geology of the mine area based on available information is a fairly lengthy process."

Two long meetings with Bielenstein convinced Naylor that the Pictou project was in more-competent hands. He was impressed with Bielenstein, whom he found knowledgeable, willing to listen, and blunt. Bielenstein, a geologist with Associated Mining Consultants Ltd., dismissed Suncor's initial geological report as "incomplete and based on an erroneous approach." Suncor had ignored a recent study prepared by government geologists, a decision Bielenstein said was "difficult to understand, and a mistake." Despite his limited exposure to the project, Bielenstein's "gut instinct" told him the Foord had mining potential. In early May, Naylor told the government's manager of coal development, Ed Bain, to expect "a very professional and in-depth evaluation of the Foord seam" from Placer. Still, he suggested government geologists carefully examine the finished report. "It is unfortunate that [Bielenstein] was not involved with the project earlier," Naylor noted, "as he is now in the unenviable position of trying to evaluate a large amount of information in a short period of time."

Less than two months, to be exact. Work on the feasibility studies proceeded at a frantic pace to meet the mid-July deadline. Placer employees and consultants played an advisory role, leaving much of the legwork and drafting to Suncor's New Glasgow team. Arden Thompson worked closely with Hans Bielenstein on the geological aspects; Gerald Phillips immersed himself in plans for mining methods, choices of appropriate machinery, and drafts of production schedules and cost estimates. The biggest challenge, Phillips confided to Patrick Phalen at Mines and Energy, was keeping the project's price tag low enough to meet Placer's "minimum rate of return" on its investment. The capital cost of developing the mine, estimated to be about $90 million in 1986, had ballooned to $115 million; Placer's bottom line was a project in the $100-million range. That goal, Phillips indicated, was realistic, and he expected the studies to conclude that the project was viable.

The new set of feasibility studies—a stack of five thick volumes—was delivered in July. It was a comprehensive examination of all aspects of the project, building on the groundwork laid in the earlier Suncor-Norwest reports. The study envisioned a mine capable of producing 750,000 tonnes of coal per year over nineteen years, for a total output of about 15 million tonnes. Coal quality was good; it contained less than 1-per-cent sulphur, and the ash content averaged about 24 per cent, which could be refined to marketable levels.

The most significant change was the choice of mining procedures. Norwest's flexible approach, combining longwall and room-and-pillar methods, was rejected in favour of a full room-and-pillar operation. There were obvious cost advantages to employing the same method and equipment throughout the mine; as for flexibility, the Placer study concluded that room-and-pillar mining would provide enough leeway to deal with faulting and the increasing depth of the seam. There were also safety considerations. Room-and-pillar mining allowed mined-out areas to be sealed off, Placer noted, reducing the risk of spontaneous combustion. Norwest's plan to use the longwall methods would have left mined-out sections exposed to the air. Dames & Moore consultant Derek Steele had spotted that flaw when he reviewed the Norwest report in 1986, and his concerns had apparently been heeded.

Otherwise, Placer adopted the prevailing view among Suncor's engineers that Foord seam coal posed no special risk of catching fire, and contained low levels of methane. Placer simply accepted the results of Suncor's laboratory experiments, and conducted no further tests to confirm those findings. But Placer did suggest a new precaution: an electronic monitoring system capable of measuring the levels of methane and carbon monoxide in the mine air. High levels of carbon monoxide would indicate that spontaneous combustion had occurred, minimizing that risk. As for methane, Placer said the ventilation system was designed to handle larger volumes of the gas, if necessary.

One hazard identified by Steele was overlooked in the Placer study. Dames & Moore had warned that ironstone bands within the seam could cause sparks as the coal was cut; maintaining adequate ventilation at the working face was essential, in order to remove methane. Placer's study, however, attached no special significance to the layers of ironstone. "Careful mining practices will segregate most of this material without major operational impact," it concluded. Placer's assessment of the Foord seam's geology, on the other hand, offered more in-depth analysis than Suncor's 1986 effort. No doubt acting in response to the concerns of the Nova Scotia government's geologists, Hans Bielenstein and his team made a point of carefully documenting the location of every known fault in the mining area.

* * *

AS THE DEADLINE for a decision approached, the Nova Scotia government did its part to make the project attractive. At Placer's request, the Nova Scotia Power Corporation more than doubled its offer to buy coal, improving the project's financial outlook. The utility agreed privately to buy 275,000 tonnes of coal for its older Trenton plant, and another 385,000 tonnes for Trenton 6 when it opened in 1991. Devco's worst fear—that NSPC would develop an appetite for Pictou County coal—was becoming a reality. Placer, unlike Suncor, was not shy about seeking direct financial aid from the government. Before it would agree to develop the mine, Placer wanted tax breaks, and a subsidy from the federal government to reduce the interest rate on its bank loan.

Placer also wanted concessions on the thorny issue of surface-coal rights. The issue had already cropped up during negotiations with NSPC on the coal contract. Suncor, acting on Placer's behalf, asked the provincial government in May 1987 to consider granting access to the Wimpey pit on "a contingency basis." Placer wanted to ensure that it could continue to supply coal to NSPC if "emergencies" or "unforeseen circumstances" interrupted production. "Let it be clearly understood that development at Pictou is focused on creating a substantial underground mine," Suncor's John Shillabeer assured the Department of Mines and Energy. "There is no thought of substituting an open pit or strip mine."

It turned out to be a moot point. During the summer of 1987, Placer joined forces with two other major players on the Canadian mining scene—Dome Mines and Campbell Red Lake Mines. The merger created Placer Dome, one of the largest gold producers in the world. Placer's option on the Suncor property was extended one month, but the writing was on the wall. When the directors of the new company gathered in Toronto for the first time, on August 12, they agreed to concentrate their combined efforts on mining gold and other metals. The Suncor purchase was rejected. For the second time in less than a year, Bay Street executives had dashed Pictou County's hopes for a new coal mine.

Both Suncor and NSPC were left in a bind. The utility was finalizing the blueprints for Trenton 6, and needed to know where its coal supply would come from when the plant came on stream in the fall of 1991. Pictou coal was higher in ash than Cape Breton coal, and this had implications for the design of the boilers. And if low-sulphur Pictou County coal was not available, NSPC could be forced to spend tens of millions of dollars on pollution-control equipment to reduce sulphur dioxide emissions. Suncor, for its part, needed a guaranteed market if it stood any chance of attracting another buyer. NSPC agreed to leave the coal contract in place until at least the end of 1987, giving the company some breathing room.

As Suncor put its leases up for grabs, the die was cast. The Nova Scotia government, committed to building a power plant to burn Pictou County coal, had displayed a willingness to accommodate any company willing to open a mine. Surface-coal rights were back in the hands of the provincial government, a possible drawing card for a new operator. The battle lines had been drawn, pitting Cape

Breton against the Nova Scotia mainland in the struggle for coal-mining jobs. Donald Cameron had emerged as one of the mine's most vocal and energetic proponents; Gerald Phillips had emerged as a likely candidate to run it. And mine engineers and consultants, with few exceptions, were downplaying the risks of mining the Foord seam.

Only one element was missing—an entrepreneur with the vision and panache needed to pull it all together. Someone with a history of turning marginal mining projects into winners. Someone with the right political connections. Someone who could get the politicians onside, and maybe even attract government money to sweeten the deal. That someone was Clifford Hugh Frame.

THE FRAME FACTOR

CLIFFORD FRAME WAS A GOOD HOCKEY PLAYER in his youth, maybe even good enough to make it to the National Hockey League. He grew up in Trail, a company town nestled in the British Columbia interior, where his father worked at the Cominco lead and zinc smelter. He played junior hockey for the Trail Smoke Eaters, and at eighteen he got a shot at the big time; the New York Rangers scouted him for their farm team in Winnipeg. But Frame declined the offer. Back in the early fifties, NHL players did not command multimillion-dollar salaries, and Frame's goal was to be a millionaire. Besides, Frame was too ambitious to settle for playing the game; he was more inclined to buy his own team, so he could run it his way.

Frame was born in Russell, Manitoba, on May 28, 1933, in the middle of the Depression. He was only two when his family moved to Trail. By the time he was sixteen, he was playing hockey in the winters, working in the Cominco smelter alongside his father in the summers. It was Frame's first taste of the mining industry, and he liked what he saw; mining, not hockey, became his ticket out of Trail. He married soon after graduating from high school, and enrolled in the mine engineering program at the University of British Columbia in Vancouver. He emerged in 1956 with a bachelor's degree in engineering, ready to begin the long climb to the top. His objective, he would later admit, was simple: "I wanted to be in the position where I was calling the shots in building mines the right way and creating jobs."

After working a year for a B.C. mining firm, Frame took a job with Denison Mines, which was in the process of turning Elliot Lake, Ontario, into a uranium boom town. Frame went underground, first as a shift boss, then as a mine captain responsible for a 150-man crew. He was only twenty-four and, in his words, "a young, energetic buck." Frame left Denison in 1960 to take charge of mine planning for the International Nickel Company in Thompson, Manitoba. He moved from the pits to the front office, expanding the existing operation and working on the development of new mines. But Frame could not escape the harsh realities of work underground. In 1965, his younger brother, Mervyn, died after being crushed between two ore cars at a mine near Trail. He was only twenty-eight. Frame never got over the death; almost two decades later, he would name one of his sons Mervyn in his brother's memory.

In 1966, Frame returned to the Denison fold as assistant manager of the massive Elliot Lake operation in Northern Ontario. Within three years he was back at Inco, heading overseas as vice president of operations for the company's subsidiaries in the South Pacific. His crowning achievement at Inco was the development of a $135-million nickel mine in Indonesia. By the early 1970s, Frame was on the other side of the globe, laying the groundwork for the biggest mine in Europe, a lead-zinc operation in Ireland with a $150-million price tag. While working in Ireland he met his second wife, Catherine. He already had six children from his first marriage.

Frame's talent for getting new mines into production made him a hot commodity in mining circles, and Denison wanted him back. But Frame had a game plan, and he stuck to it. "I didn't want to come to Toronto without lots of mining experience," he later explained to the *Northern Miner Magazine.* "I wasn't a lawyer, I wasn't an accountant, I wasn't a financial man. So I had to be a miner." But in 1975, having proved himself on the international stage, Frame was ready to take the plunge. He returned to Denison a third time, joining head office in Toronto as executive vice-president in charge of exploration and mining operations. He was being groomed to replace John Kostuik when he retired as president, with the prospect of one day succeeding Denison's chairman, Stephen Roman. At the age of forty-two, Cliff Frame, miner and former hockey prospect, had broken into the big leagues.

* * *

DENISON MINES was as brash and as hard-driving as its founder, Stephen Boleslav Roman. Roman emigrated to Canada from his native Czechoslovakia just before the outbreak of World War II, and began speculating in penny mining stocks; by the 1960s, he was in charge of an international mining empire worth billions of dollars. He owed it all to one metal–uranium. Roman got in on the ground floor when a massive deposit of the radioactive ore was discovered at Elliot Lake in 1953. It was the height of the Cold War, and Denison grew fat feeding the United States the raw material it needed to build stockpiles of nuclear bombs and missiles. Later, Denison hammered out lucrative long-term contracts to supply the federal government and Ontario Hydro with uranium for nuclear generating stations.

Roman was the Canadian version of a Citizen Kane–ruthless in his methods, motivated by greed and a lust for power. "Everybody calls me a sonofabitch," he once told an associate. "I might as well act like one." In *The Roman Empire,* an unauthorized biography, journalist Paul McKay described Roman as "tough and tyrannical . . . the consummate predator." Like the fictional Kane, he thought in grandiose terms; Roman spent $25 million to build an onion-domed cathedral on his sprawling Toronto-area estate, fashioned after the church he attended as a boy in Czechoslovakia.

In Roman, Clifford Frame found the perfect role model for an aspiring mining baron. "It's vital to have people to look up to," Frame told the *Northen Miner Magazine* in May 1992. "It's vital to copy and benefit from an older person's experience, assimilating some of their experience and even copying them. Your own personal style and traits will evolve by themselves." Roman, he candidly admitted, was one of those people. "I admired him, and learned an awful lot from him. He was a mentor," Frame gushed years later, when asked about his former boss. "He had a natural aptitude for business and great perception." When Frame was finally running his own company, he would sometimes wonder out loud what Roman would do in a given situation. Frame even emulated Roman's hobbies. Roman kept a herd of purebred Holsteins worth millions of dollars; Frame would eventually buy a farm near Uxbridge, about ninety kilometres north of Toronto, where he tended a couple of hundred Black Angus cattle. Roman was partial to driving Jaguars; Frame ended up owning three.

Roman also introduced Frame to the Conservative party's old boy network. Roman was drawn to politics, but not out of a sense of public duty—political power, and having friends in high places, was good for business. The politician he admired most, until the Watergate scandal broke, was Richard Nixon, who was Denison's lawyer in the United States before becoming the country's president. From time to time, Roman would install a retired politician on his board of directors. (Frame followed that practice once he founded his own company, recruiting former prime minister John Turner as a director.) Roman, a staunch proponent of free enterprise, who saw socialist and Communist conspiracies everywhere, naturally gravitated to the Tories. He even made two runs for a seat in Parliament during the 1970s, losing both times to the Liberal incumbent.

Roman's political dabbling introduced him to a Quebec business executive and rising political force by the name of Brian Mulroney. In the late 1970s, Mulroney, then president of the Iron Ore Company of Canada, organized a fundraising drive for his alma mater, St. Francis Xavier University in Antigonish, Nova Scotia. At Mulroney's request, Roman became the campaign's honourary chairman. Their friendship continued after Mulroney became Tory leader in 1983 and, a year later, prime minister. Roman's companies gave generously to the federal Tories, and Mulroney was once a guest at Roman's fortresslike Caribbean villa. Frame's personal contact with Mulroney during his Denison days was practically nil—the two met once at a cocktail party.

But Frame quickly befriended another high-profile Tory within Roman's circle. Robert Coates was a member of Parliament from Nova Scotia and the Conservative party's national president in the late 1970s. Exactly how and when they met is unclear, but Coates's obvious warmth towards the Curragh chairman shone through even after the Westray explosion. "I've known Mr. Frame for a *very* long period of time," Coates said in June 1992, when asked about his role in bringing Frame and Curragh to Nova Scotia. Coates, like Roman, was a great admirer of former prime minister John Diefenbaker. When the Conservative party spurned the Chief in 1967, Coates was one of the few MPs to stand by his leader to the bitter end. After Diefenbaker died in 1979, his admirers formed a non-profit foundation in his name. Coates became the foundation's first president, while Roman and Mulroney, another Diefenbaker acolyte, served as directors. The John G. Diefenbaker Memorial Foundation's avowed aim was "to promote a better understanding and appreciation of Canadian history." Its main activity, however, was hosting banquets and handing out "Canadian of the Year" awards–small bronze statues of the former prime minister–to outstanding citizens like Roman, Mulroney, and Coates.

There was one thing Frame would never learn at Roman's knee, and that was to put the health and safety of workers ahead of production. Safety standards at Denison's Elliot Lake operation were a disgrace; inadequate ventilation left miners to breathe dangerous levels of radioactive gases and silica dust. "Denison was explicitly warned about the hazards by government mine-safety inspectors, doctors, and union officials as early as 1958," McKay wrote in *The Roman Empire*. "Yet for the next decade, both radiation and dust levels underground at Denison exceeded the existing government standard." In 1974, Elliot Lake miners, who were succumbing to lung cancer at rates far above the norm, staged a wildcat strike over working conditions. The Ontario government belatedly set up a royal commission, which condemned Denison's practices and prompted sweeping new safety standards. Frame was at Elliot Lake while safety practices were at their worst, and later claimed that the "horrendous amount of criticism" convinced him of the importance of mine safety. In an interview with *Report on Business Magazine* shortly after the Westray explosion, Frame was insistent on that point. "I've never stinted on spending money on safety," he declared.

* * *

FRAME SPENT A DECADE in the top echelons of Denison Mines. It was a heady time; Stephen Roman's empire embraced oilfields off Spain and Greece, uranium deposits in Australia, and shipbuilding in Louisiana. At home, Denison had diversified into pulp and paper and cement, and even controlled its own trust company. Denison executives, as befitting their importance, flew to meetings in the comfort of the company's top-of-the-line, $5-million private jet. Frame was in the thick of things, buying new mineral and petroleum properties and negotiating multimillion-dollar financing deals with the banks–one of his first tasks was to oversee a $500-million expansion at Elliot Lake. All the while, he was edging his way up the corporate ladder, joining Denison's board of directors in 1981. A year later, Frame was named president and chief operating officer.

At that point, Roman handed Frame his biggest challenge–development of the billion-dollar Quintette coal project. Among Denison's vast holdings were reserves of metallurgical-grade coal in the Peace River country of northeastern British Columbia. Japan's steel industry was the logical customer, but getting the coal to market would require the construction of expensive road and rail links through mountainous terrain. In the end, the driving force behind development of the coalfield was politics, not economics. British Columbia's Social Credit government, under Premier Bill Bennett, was committed to the project, and that was that. In 1981, the deal was done. The Japanese signed a fifteen-year contract to buy B.C. coal, the lion's share to come from Denison's open-pit mine at Quintette. The B.C. and federal governments agreed to build the roads, rail lines, and Pacific Coast port facilities needed to ship the coal, and to create a new town, Tumbler Ridge, to house thousands of workers and their families. In all, upwards of $1.5 billion in public money was staked on Quintette's success.

Frame took charge of the mega-project, handling the financing and development. A consortium of more than fifty banks anted up the nearly $950 million needed to underwrite the mine. But from the moment production began in 1983, Quintette suffered one setback after another. Production fell short, the coal was of lower quality than expected, and areas of the mine that were supposed to contain coal seams harboured only worthless rock. Denison had scrimped on exploratory drilling before development began, and the company's geologists concluded, belatedly, that the pit had been dug in the wrong place. To compound these problems, the unpredictable marketplace dealt the project a blow. Quintette's viability was based on the premise that coal prices would nearly double, to $120 per tonne, by 1990. But by the mid-1980s, the world price of coal was still about $70 per tonne, and the Japanese, stuck with paying exorbitant rates for Quintette coal, demanded price cuts.

The upshot was that Denison, the banks, the Japanese steelmakers, and Canadian taxpayers took a bath. Denison wrote off its entire $240-million investment in Quintette, while the bankers collectively swallowed a $700-million loss. The B.C. government was forced to continue subsidizing coal shipments out of Quintette, while Denison and the Japanese took their battle over pricing into the courts. To this day, Frame insists that Quintette was "an extremely well done project." Roman biographer Paul McKay, however, begs to differ. "A more senseless waste of money, resources, and human effort would be hard to imagine," he concluded.

Quintette was Frame's project, and the captain went down with his sinking ship. In March 1985, he was fired as Denison president. The autocratic Roman, never shy about blaming his mistakes on others, laid responsibility for the Quintette fiasco squarely on the shoulders of his protégé. "Some of us should have been prepared for these problems, but we were not," Roman told Denison shareholders, leaving no doubt who he had in mind. Frame had seen the writing on the wall. "I felt a sense of relief when I left Denison," he later admitted. "A divorce never happens instantaneously." He lost about $1 million of his own money on the stock market when the Quintette failure sent Denison shares into a nose dive. Within the mining industry, a lot of people felt he was the fall guy, a scapegoat for the collective mistakes of Denison and its government backers. Few thought they had seen the last of Cliff Frame.

* * *

DESPITE HIS SUDDEN FALL from grace at Denison, Clifford Frame was determined to fight his way back to the top. "I didn't want to sit on the beach and pick my nose," as he indelicately put it. This time, however, Frame would do it his way, without having to answer to Stephen Roman or anyone else. In May 1985, just two months after he was fired, Frame formed his own company with two partners: James Hunt, a New York oil analyst and Denison director; and Ralph Sultan, a Harvard University economist and a former classmate of Frame's at the University of British Columbia. The new firm was bankrolled by a group of investors based in Dallas, Texas. Paying homage to his Celtic roots, Frame christened the new company Curragh Resources. *Curragh* is the Gaelic word for *coracle,* a primitive boat made from hides stretched over a wicker frame. But for a gentleman farmer like Frame, the word was synonymous with horse racing. The Curragh is an area of flatlands near Dublin renowned for horse breeding, and Curragh Racecourse is the site of the Irish Derby. The name was apt; the new company was as lean as a thoroughbred. But the odds were against a trip to the winner's circle.

Like the Klondike gold-seekers before him, Frame headed to the Yukon in hopes of striking it rich. The setting for his unlikely comeback was an abandoned lead-zinc mine in Faro, about two hundred kilometres north of the territorial capital, Whitehorse. Cyprus Anvil Mining Corporation brought the open pit into production in 1969, employing close to one-quarter of the Yukon's workforce. But in 1982, Cyprus Anvil's financially troubled parent company, Dome Petroleum, abruptly walked away from the operation, leaving behind giant trucks still loaded with ore. By the time Frame and Curragh Resources bought the property, Faro was almost a ghost town, and the gaping pit was well on its way to becoming a man-made lake.

Within months, Frame had wrung enough concessions out of the federal and territorial governments to put Faro back into production. Ottawa transferred to Curragh a $17-million loan originally negotiated with Cyprus Anvil, and joined the Yukon government in backing a $15-million bank loan for the project. The Yukon government, desperate to get the mine back in production, contributed $12 million to buy an electrical generator and company-owned houses in Faro, on top of an outright grant of $3 million. Others in the mining industry cried foul. "What is the point of providing 100 jobs [at Faro] if you are going to lose 100 jobs elsewhere?" complained one mining executive. It was the same logic Cape Breton coal miners would use a few years later when confronted with government-subsidized competition from Frame in Pictou County—that there were only so many customers to go around.

But the results spoke for themselves, silencing Frame's detractors. A leaner, meaner Faro was back in production by the summer of 1986, doubling the mine's previous output. Even with a reduced workforce of four hundred, Curragh was the Yukon's largest employer. Curragh rode the crest of a wave as overseas markets improved and lead and zinc prices rose in the late 1980s. By 1987, the company was turning a profit, and the government-backed $15-million loan had been repaid.

Frame's triumph at Faro earned him the nod as the *Northern Miner*'s "Mining Man of the Year" for 1987. It was the second time in five years the industry newspaper had singled out Frame for the distinction; he had received the same honour in 1982 for his work on Quintette, before that project became a white elephant. All that embarrassment was in the past, thanks to Faro, and Frame was once again a major force on the Canadian mining scene. "[He] had to prove he was not altogether to blame for the design problems and some of the dubious decisions made at Quintette," the *Northern Miner* noted in December 1987. "And he has done that by turning Faro around a full 180 degrees." The hoopla prompted Prime Minister Brian Mulroney to send along a personal letter of congratulations, in which he praised Frame's "leadership, energy and entrepreneurial spirit."

Frame may have been "Mining Man of the Year," but two Curragh executives had helped him pull off the Faro miracle. One was Marvin Pelley, who had worked under Frame on the ill-fated Quintette project, managing all phases of engineering and construction. Pelley, who retained a trace of his Newfoundland accent, had learned the ins and outs of mining at the Iron Ore Company of Canada in Labrador. Kurt Forgaard, Curragh's president, had extensive experience in the base-metal industry, and he also brought considerable coal expertise to the firm when he was hired in 1986. Forgaard's background included four years as vice president at the McIntyre Mines underground and open-pit operations in Grande Cache, Alberta.

As the Curragh juggernaut gained momentum, Frame added several other key players to his management team. Colin Benner, whose father had helped stake the Elliot Lake uranium find for Stephen Roman, was eventually snatched from the Denison fold. On the legal side, there was George Whyte, a former partner at the blue-chip Toronto law firm McMillan Binch and a director of Maple Leaf Gardens. And to keep the books in order, Frame lured accountant Adrian White away from a top post in international finance at the Bank of Montreal. "There were mixed views on Cliff in those days because of Quintette," White says today. But by 1987, he says, Curragh looked solid enough to have "a sporting chance" of becoming the next success story on the Canadian mining scene. White latched onto Frame's rising star, becoming Curragh's chief financial officer.

For Frame, Faro was a springboard to bigger and better things. In 1987 he joined forces with Giant Resources, an Australian firm that purchased a 46-per-cent stake in Faro. In return, Frame took charge of Giant's Canadian subsidiary, Pamour, which controlled several gold mines. Frame was in an expansionist mood, snapping up a couple of smaller companies through Pamour. He even took a run at Dome Mines, before its amalgamation with Placer Development and Campbell Red Lake Mines—the very merger that left Suncor without a buyer for its Pictou County coal property. And that property, as it turned out, became Frame's consolation prize.

* * *

AS THE SUMMER OF 1987 faded into fall in Nova Scotia, time was running out for Pictou County's long-awaited coal mine. Suncor had until the end of the year to find a buyer, or the Nova Scotia Power Corporation would have to look elsewhere for coal to fuel its new Trenton station. And without a secure market, the Pictou County exploration leases were next to worthless. In six years, the company had spent $6.4 million on its drilling program in Nova Scotia, not to mention the cost of feasibility studies, salaries, expenses, and the like. To break even, Suncor needed to find a company willing to shell out roughly $8 million for a project that only existed on paper. As the call went out for bids, Suncor began pulling up stakes. The New Glasgow office was closed at the end of August, and four of the Pictou project's five employees were laid off. Gerald Phillips, the only one kept on the Suncor payroll, was transferred to the oil sands project in Fort McMurray, Alberta.

A potential buyer was already waiting in the wings. David Rose and Allister MacKenzie, two Cape Bretoners with years of experience in the coal industry, had formed a company called Coalcor Resources; they set their sights on the Pictou project. Rose, a financial consultant in the Sydney area, and MacKenzie, who ran his own engineering firm, had formerly worked for the Cape Breton Development Corporation. Development work on the mine would be contracted to Davy McKee (Stockton), a British engineering firm with worldwide experience in designing and building coal mines. Rose and MacKenzie had tried to buy into the Pictou County property in 1986, but the deal fell through when Suncor began to negotiate with Placer Development. With Placer out of the picture, however, Coalcor geared up to make another offer.

Esso Resources Canada of Calgary, a division of Imperial Oil, also took a serious look. Esso's mining engineers and geologists pored over the feasibility studies, but were far from satisfied with the groundwork laid by Suncor and Placer. In particular, they felt the laboratory experiments were not enough to support Suncor's conclusion that the coal had low methane content. "The Foord seam is a thicker seam and by sheer volume of it being a thicker seam, it has bigger potential for gas release," explained Esso geologist Allister Peach, "We felt that we needed additional work . . . to further understand the coal-gas scenario." Methane was only one concern; Esso's experts also recommended further study of the fault zones within the seam and the potential for roof collapse. The whole process would take at least six months. But Suncor needed a deal by mid-December to meet the NSPC deadline, and there was no time for yet another feasibility study. Esso was out of the running.

By mid-September, ten companies were evaluating the Suncor property. But the search for a new operator was not moving fast enough for the Nova Scotia government. NSPC was out on a limb, committed to spending a quarter of a billion dollars on a new plant in Trenton to burn low-sulphur coal. Despite the enticement of a secure market, the government had watched helplessly from the sidelines as first Suncor and then Placer walked away from the project. John Buchanan, for one, was growing impatient. "It was important that the power corporation do the right thing by the environment," the premier explained later, and Pictou coal would help fulfil Nova Scotia's "moral obligation" to limit acid-rain emissions. At the same time, "it was a good project, one that would create a lot of jobs." The only thing missing was a company willing to take the plunge.

Enter Nova Scotia MP Robert Coates, who was probably one of Buchanan's closest political allies. If Stephen Roman was Clifford Frame's role model, Coates was probably as close as Buchanan came to having a mentor. Coates was three years older than Buchanan, but they started out on the same career path—undergraduate degrees at Mount Allison, a small university in Sackville, New Brunswick, then law at Halifax's Dalhousie University. Coates went straight into federal politics, winning the Cumberland Colchester riding in northern Nova Scotia in 1957. Buchanan practised law in the Halifax area for ten years, until Coates convinced him to run for the Conservatives on the provincial level. When the Tory leadership became vacant in 1971, Coates was among those who urged Buchanan to run, nudging him on the path to the premier's office.

Coates knew just the person to help Buchanan, NSPC, and the people of Pictou County out of their predicament. His old friend Cliff Frame had the smarts and the experience needed to kick-start the stalled Pictou project. The Nova Scotia government, as Frame later put it, "wanted somebody they thought they could work with and not be pushed around by." Frame was the kind of guy who liked doing business with politicians, and he was apt to be generous when party bagmen—particularly Tory bagmen—came looking for money. Between 1987 and 1989, Frame and his companies donated $41,700 to the federal Conservatives, including almost $33,000 in 1988, a federal election year. The Liberals, by comparison, received a paltry $9,500 over the same period.

Frame, with typical aplomb, went straight to the top. He met face to face with Suncor's president, Tom Thompson, to make his pitch. Then, Coates set up a meeting with Buchanan, who was in Ottawa in the latter half of 1987 for a federal-provincial conference. Buchanan and Frame discussed the Pictou project for about an hour over lunch. Buchanan remembers it as nothing more than a get-acquainted session. "At the time, I didn't know who Curragh was; I'd never heard of

Nova Scotia Tory MP Robert Coates, right, shares the podium at a political rally with John Buchanan and Elmer MacKay [rear]. Coates knew just the man to give Pictou County its coal mine—Clifford Frame of Curragh Resources. [Courtesy of the *Chronicle-Herald* and the *Mail-Star*]

them." Frame expressed Curragh's interest in buying the Suncor property, and Buchanan told him to talk to the people at the Department of Mines and Energy. "I was rather pleased that someone was interested in looking at it," Buchanan recalls. "At the time, there wasn't too much interest by too many people."

Almost overnight, government officials back in Nova Scotia were scrambling to make sure Frame got a crack at buying the Suncor project. Officials with the Department of Mines and Energy, who were monitoring the sale, sought assurances that Curragh would be able to submit a bid. Director of Mining Engineering Patrick Phalen, the government's contact with Suncor, reported on September 23 that Curragh's financial advisers, the accounting firm Price Waterhouse, would receive the Placer feasibility studies. "There may have been some delay in getting the information package to the Curragh interests," Phalen noted, but Suncor officials had assured him that "they are treating this inquiry in the same way as the other 10 companies who are interested in the project." That memo ended up on the desk of Joel Matheson, the minister of mines and energy. The name Curragh caught his eye, and he sent back a note seeking confirmation that the firm would be able to make an offer. "Yes, Curragh Resources (Cliff Frame's company) is expected to bid on the project," Richard Potter, an assistant deputy minister, assured Matheson in an October 5 memo. The bracketed description was Potter's; "Cliff" and the Nova Scotia government, it seemed, were on a rather familiar basis.

Curragh submitted a bid before the end of October through one of his companies, Westray Mining Corporation. Frame, once again drawing on his roots, named the firm after one of Orkney Islands off the northern tip of Scotland, the ancestral home of his mother's family. Then, as he had done at Faro, Frame asked the government to share the risk. On November 25, 1987, Westray applied to the federal Department of Regional Industrial Expansion for assistance. Frame asked Ottawa to fully guarantee a $117-million bank loan over fifteen years, plus cover up to 6 per cent of the interest on the loan. The package would cost the federal government $63.9 million during the term of the loan. Coalcor, Curragh's only serious competitor was also seeking government help, but at more modest levels—a $5-million loan from the province, a federal guarantee covering $68 million of an $80-million bank loan, and a 2-per-cent interest-rate subsidy. Total estimated cost to the federal government: $8 million.

The government aid sought by Curragh was far beyond anything previously contemplated for the project. Suncor had been prepared to build the mine without public assistance, as had Esso. Placer would have gone ahead with only an interest-rate subsidy and the promise of tax breaks down the road. But by late 1987, only about eighteen months after Donald Cameron boasted that the mine would be built without "one cent" of government money, the stakes had risen dramatically. For Frame, government assistance was the bottom line. "If I didn't have that encouragement," he told *Maclean's* magazine years later, "I wanted nothing to do with that project."

By November, Curragh and Coalcor had emerged as the leading contenders. Rose and MacKenzie were confident that Curragh, a producer of base metals from a single open pit in the Yukon, was no match for the coal-mining expertise of Davy McKee. "The difference between a rock miner and a coal miner is the difference between a general practitioner and a brain surgeon. A different mentality is required, completely," Rose recalled. "We didn't take [Curragh] seriously because there was no way that anybody in their right mind could have chosen them with no coal experience, or even knowing about coal." Curragh's senior executives, as it turned out, had worked on coal projects in the past, even if Frame's own coal expertise was limited to the Quintette debacle. But Rose was suspicious that Curragh enjoyed a crucial advantage over Coalcor–an inside track with the politicians.

The coal lease was Suncor's to sell, but without a contract to supply NSPC's Trenton stations, there would be no mine. As the bid process unfolded, both Curragh and Coalcor met separately with NSPC to review tonnages, prices, and deadlines. Rose and MacKenzie outlined their business plan, but NSPC officials seemed to be simply going through the motions. Coalcor could offer to supply the coal for free, Rose remembers thinking, and the utility would still not be interested. In desperation, Coalcor tried to take its case directly to the premier. Rose and MacKenzie hounded a couple of Cape Breton Tory MLAs about setting up a meeting with Buchanan, and logged dozens of calls to the premier's office. But the meeting never materialized, and Buchanan later contended he was never told of the calls. While Frame enjoyed direct access to Buchanan through Coates, Coalcor was left out in the cold. "We were totally blocked," Rose charged. "This was a done deal, and we didn't fit into the plan."

Politics aside, Coalcor was the company that lacked credibility as a coal-mine developer, in the eyes of Suncor and the Nova Scotia government. Coalcor offered the British firm's coal-mining experience, but the company still only existed on paper; Curragh, on the other hand, was a going concern, basking in the glow of Faro's success. At one point, Rose phoned Suncor's Calgary office to complain that he was being "danced around," and that the bidding process was skewed against Coalcor's offer. "If you don't want to be danced around," replied Sam Safton, the Suncor official in charge of the sale, "don't play." In mid-November, Rose vented his frustration in a pointed letter to Suncor, complaining of "bias" and NSPC "interference" in the selection of the winning bid. "All we are asking is that the game is played fairly," he wrote. In mid-December, word leaked out to the press that Suncor would sell its leases to Curragh.

That meant Suncor was accepting Curragh's bid of about $7.5 million, leaving Coalcor's higher offer, $8 million, on the table. For Rose, Suncor's willingness to swallow a half-million-dollar loss confirmed that politics had creeped into the bidding process. In mid-January of 1988, about a week before the sale was finalized, Rose and MacKenzie finally convinced someone in government to hear them out. An official of the Department of Development steered them to Pictou East MLA Donald Cameron, who agreed to a meeting at Stellarton's Heather Hotel. For the first fifteen minutes of the meeting, Cameron "made like a junkyard dog," Rose recalls, tearing a strip off the two Cape Bretoners for trying to get their hands on Pictou County coal. Once Cameron settled down, Rose notes, even he seemed frustrated with the government's handling of the sale. But he served notice that Curragh was in, Coalcor was out, and Rose and MacKenzie were wasting their time. "It's hardly likely you fellows will get [the mine] unless there's a change of government," Cameron advised the pair.

Curragh's purchase agreement came with a couple of strings attached. Trenton 6 would have to win regulatory approval, and a sales contract would have to be hammered out with the Nova Scotia Power Corporation. NSPC needed the approval of Nova Scotia's Board of Commissioners of Public Utilities to build the new facility, but faced opposition from two privately owned Nova Scotia power suppliers. At public hearings on the Trenton application, lawyers for Black River Hydro and Atlantic Combustion contended that the province did not need the additional generating capacity. And NSPC, already $1.5 billion in debt, would have to raise power rates "dramatically" to finance a new power plant. Despite the dire warnings, the utilities board gave Trenton 6 the green light on February 23, 1988. The following day, Kurt Forgaard, who doubled as president of Curragh and Westray, announced that construction could begin as early as the fall, assuming a sales agreement was finalized with NSPC. With Curragh at the helm, it was full speed ahead for the Pictou mine.

* * *

CLIFFORD FRAME'S ARRIVAL on the Nova Scotia mining scene did not escape the notice of the Cape Breton Development Corporation. Devco had been sniping at the proposal to open a Pictou County mine for more than year but, by the end of 1987, the threat of competition had become real. In early December, federal officials informed Devco that Ottawa was reviewing Curragh's application for financial help. Devco's management, in an effort to nip the proposal in the bud, promptly fired off a confidential memorandum opposing any form of public assistance. Addressed simply to the "Governments of Canada and Nova Scotia," it ended up on the desks of Premier John Buchanan and Prime Minister Brian Mulroney.

"Devco's board of directors believe they have a duty to respectfully advise both levels of government of Devco's considered opinion that a deep coal mine in Pictou County would not be of benefit to Nova Scotia's economy," the memo declared. In the twelve-page document, Devco presented a compelling case against opening a new mine at a time when Cape Breton's coal industry was struggling to reach self-sufficiency. The Pictou mine would displace coal sold to the Nova Scotia Power Corporation's existing Trenton plant, costing Devco an estimated $12 million a year in revenues. And that would mean cutting 275 jobs at Devco's Lingan colliery, more than offsetting the new jobs promised at a Pictou mine. Adding insult to injury, the federal government, which had poured hundreds of millions of dollars into Devco since 1967, was now being asked to help finance a Devco competitor.

The memo also offered a brief but ominous warning about the consequences of developing a Pictou coal mine. "Potential safety risks would accompany the negative impact on Nova Scotia's economy [that] a deep coal mine in Pictou County would cause," Devco cautioned. "The Pictou coalfield's mining history . . . involves experience with difficulties due to a complicated geological structure with numerous faults. As a result of such difficulties, its seams have given off large volumes of gas and proven extremely liable to spontaneous combustion." It was only one paragraph amid a wide range of economic and financial arguments, but the safety issue had been laid before the political leaders with the power to decide the mine's fate.

When the memo was made public three months later, the safety warning was overlooked in the furore over possible layoffs at Cape Breton's mines. On March 8, 1988, the opposition Liberals launched a two-pronged attack, tabling the memo in the House of Commons and the Nova Scotia legislature. David Dingwall, a Cape Breton MP, accused the Mulroney government of "playing worker against worker, community against community, company against company." While Dingwall scored points in Ottawa, Liberal MLA Guy Brown set a trap for Buchanan in Halifax. During Question Period, the premier assured Brown that the Pictou mine would not cost jobs in Cape Breton. At that point, Brown let the other shoe drop, producing the Devco memo. The document was out of date, Buchanan sputtered, and a new NSPC plant planned for Cape Breton would more than replace lost sales at Trenton. But when reporters called NSPC, they were told the Cape Breton plant would not be in service until 1994, three years after the Pictou mine was supposed to go into production. Minutes after Buchanan was confronted with the discrepancy, a sheepish NSPC spokesman backtracked, and said the premier was right and the utility was wrong. It was a glaring example of how closely NSPC marched to the tune of its political masters. As it turned out, the new power plant at Point Aconi, Cape Breton, did not commence service until the fall of 1993, two years after Westray came on stream.

With the federal and Nova Scotia Tories expected to go to the polls in the fall of 1988, the Liberals were quick to charge that the Westray project was being fuelled by politics. William Gillis, the Liberal finance critic, claimed that NSPC did not need the additional power plant, and the Buchanan government was "trying to buy the voters with their own money." Dingwall told reporters that the mine was designed to "puff up" Revenue Minister Elmer MacKay, the Pictou-area MP, and bolster the Buchanan government in advance of the coming elections. MacKay countered that the "imaginary fears" of Cape Bretoners should not be allowed to block the Curragh proposal. "The Pictou County basin has long been a bastion of coal mining," MacKay pointed out, "and when the mines ran into difficulties in the fifties, the Cape Breton mines were looked after and the Pictou County mines were allowed to close." It was time Pictou County got its fair share of government help.

The day the war of words erupted over the Devco memo, Donald Cameron was back home in Pictou County. So was the federal minister in charge of the Department of Regional Industrial Expansion, Robert de Cotret, who was in Trenton to announce federal aid for the town's troubled railcar factory. DRIE was the department considering Curragh's funding request, and Cameron met with de Cotret to refute Devco's criticisms of the Westray project. But by the end of the day, all hell had broken loose, and Devco's dire warning of layoffs was on all the newscasts. Upon his return to Ottawa on March 9, de Cotret declared that the federal government had "no intention of shifting jobs from Devco to a new project which as yet has to be determined to be financially viable." That same afternoon, Cameron was back in the House of Assembly to denounce the "cheap politics" of the opposition. Devco's warning about layoffs was "rubbish," he said, and the Crown corporation's real fear was of being upstaged by a successful, privately owned mining firm.

The rhetoric obscured deeper forces at work in the Nova Scotia coal industry. Ottawa had never intended to get into the coal business in the first place; Devco was supposed to wean Cape Bretoners from their dependence on coal mining by creating new industries. But the energy crisis had made coal king once again. Devco's management was right to question why Ottawa, which had poured millions into expanding Cape Breton's mines to meet NSPC's growing demand for coal, was now being asked to help establish a competing private-sector operation. But there was merit in Cameron's view that Devco was simply afraid of competition. The Crown corporation had grown dependent on government handouts, and had enjoyed a virtual monopoly on the Nova Scotia coal market for two decades. The Liberals had been in power in Halifax and Ottawa for most of those years, but now the Tories, the proponents of free enterprise, were in control. Privatization was the new political credo, and Devco was a prime candidate for a taste of the real world.

In Pictou County, it was time to circle the wagons. The Cape Bretoners and Devco were not about to scuttle the Pictou area's long-sought mine. Leaders of the county's five municipal governments called a press conference in late March 1988, in a show of solidarity for Westray. "We won't stand for having coal shipped in from Cape Breton when we have it here to burn, and a plant here to burn it in," insisted Stellarton Mayor Ron Marks. New Glasgow's mayor, Jack MacLean, accused the federal government of "dragging its heels" on Curragh's funding request, placing the entire project in jeopardy. Unless construction began within two months, Curragh had told the municipal leaders, the mine would not be in production in time for the start-up of Trenton 6 in the fall of 1991. "Time," Marks warned, "is of the essence."

* * *

TIME WAS ALSO of the essence as Curragh finalized its plans for the Westray project. Curragh had commissioned a new set of feasibility studies back in November 1987, about the time it submitted a bid to Suncor. Kilborn, a Toronto engineering firm, prepared the studies in conjunction with Ohio-based consultants Dames & Moore. The firms were logical choices for the work. Kilborn's president, John B. Mitchell, sat on the Curragh board of directors, while Dames & Moore had done work on the same project for Suncor. The study was ready by early 1988, but the consultants included a disclaimer. "Constrained by shortage of time," the firms said, they had relied on the geological work carried out for Suncor and Placer Development. Plans for mining the Foord seam using room-and-pillar techniques were also developed "under time constraints," based largely on earlier studies.

As a consequence, there were few surprises. Suncor's questionable lab tests were trotted out to downplay the risks of methane and spontaneous combustion. A large ventilation fan on the surface was expected to pump enough air into the mine to handle the "peak methane emissions" indicated in those tests. Placer's plan for an early-warning system of remote sensors to detect methane and carbon monoxide was adopted. Based on Placer's mapping of the locations of faults, Kilborn concluded that Westray's miners could expect to encounter "minor, normal" faulting as they burrowed through the Foord seam. As early as 1986, Dames & Moore had recognized that the ironstone band within the seam could be a source of sparks when coal was mined. Despite the firm's collaboration with Kilborn, however, ironstone was not red-flagged as a safety hazard.

As Curragh's consultants submitted their findings, there were renewed warnings within Nova Scotia's Department of Mines and Energy about the need for more-detailed analysis of the geology of the Foord seam before mining began. Robert Naylor, the government geologist who criticized Suncor's 1986 geological study, and a colleague, William Smith, outlined their concerns in an April 6, 1988, memo to Patrick Phalen, the director of mining engineering. "We feel this is a project which should be entered into cautiously by experienced persons with a high degree of prior knowledge of the geological and engineering parameters involved," they contended. A misreading of the seam's structure would leave room for "significant error" in estimates of the amount of coal to be mined–and its quality. And despite Placer's detailed work, Naylor and Smith were concerned that smaller faults had not been detected. Curragh, they said, should be required to conduct further exploratory drilling to confirm its reading of the geology, working under the close supervision of the department.

* * *

ONE DAY AFTER the Nova Scotia government's geologists put their concerns on record, Donald Cameron rose in the House of Assembly to once again promote the Pictou mine. Cameron heaped scorn on the Liberals, accusing them of supporting the project in Pictou County in an effort to win votes, while telling Cape Bretoners their party opposed it. He made a pledge that drew applause from the government benches: "I am determined, and I believe that we will succeed and we will have that mine developed in Pictou County."

Part of Cameron's April 7 speech was devoted to a brief history of coal mining in Pictou County, based on his reading of *The Pictonian Colliers*. Cameron, no relation to the book's author, James Cameron, even held up a copy and recommended that his fellow legislators take time to read it. The book, he said, showed the disparity between Pictou County, where a once-flourishing coal industry had been allowed to die, and Cape Breton, where coal mines had been rescued with public money. Although fully one-third of *The Pictonian Colliers* is devoted to accounts of fires, explosions, and disasters, Cameron made no mention of the darker side of Pictou County's mining history.

Cameron now says that he was well aware of the death toll from past explosions. He knew from the old-timers around Pictou County that the coalfield was reputed to contain excessive amounts of methane. Because he knew this, he made it his business, long before the mine went into development, to find out whether gas would pose an inordinate risk. He asked officials at the Department of Mines and Energy, and got his hands on the results of the lab tests done for Suncor. In the end, he was satisfied that the methane content of Pictou coal was in the "medium range," manageable with proper ventilation. "It's not the amount of methane, it's [taking] the proper precautions to ventilate it and mine the seam safer," he later noted. Pictou's proposed colliery, he concluded, would be no more hazardous than any other coal mine.

A week after his speech in the House of Assembly, Cameron headed a delegation of four Pictou County municipal leaders that descended on Ottawa to lobby on Westray's behalf. With Revenue Minister Elmer MacKay playing host, the group spent a half-hour with Industry Minister Robert de Cotret, who promised an early decision on the funding request. Then they paid a courtesy call on Brian Mulroney. Cameron and Mulroney needed no introductions; during Mulroney's 1983 by-election campaign to win the Central Nova riding, Cameron helped out, donating his campaign manager to the federal cause.

The delegation was granted a ten-minute audience with the prime minister, allowing time for only a general discussion about Westray. But the prime minister was well aware of their concerns; for weeks, Pictou County municipal leaders had been inundating Mulroney's office with letters, telegrams, and faxes seeking his help. "Without your support," Stellarton Mayor Ronald Marks pleaded in a March 23 letter, "this project appears to be going nowhere." Mulroney's staff wrote back with assurances that the prime minister had noted their concerns before relaying their messages to de Cotret.

The junket brought results. Three weeks later, on May 3, 1988, the Curragh chairman, Clifford Frame, took his own trip to Ottawa to make a case to de Cotret. Robert Coates was there, once again helping open doors for Frame. So was MacKay, Nova Scotia's most powerful presence in the federal cabinet. Curragh was seeking $63.9 million in federal money over fifteen years, and asking Ottawa to take all the risk by fully guaranteeing a $117-million bank loan. But there was a bright side; up to 240 jobs would be created, and the company reckoned that Ottawa stood to recover all but $5.4 million of its $63.9-million investment through taxes.

Before the meeting, however, officials of the Department of Regional Industrial Expansion decided that the contract being negotiated with the Nova Scotia Power Corporation, for about 700,000 tonnes per year, was not enough to make the project viable. DRIE crunched the numbers and calculated that the production levels should be boosted to 900,000 tonnes per year. That was the figure de Cotret floated during the May 3 meeting. Frame was open to the idea of higher production, and came away confident that he could put together a financial plan acceptable to both the federal government and the banks. In a follow-up letter to de Cotret on May 6, he asked for a meeting with DRIE officials within two weeks, "to finalize the details." Curragh had a deadline to meet for supplying Trenton 6, and the project was already two months behind the schedule for having financing in place. All Frame had to do was find someone willing to buy a whole lot more coal, and fast.

* * *

BACK IN NOVA SCOTIA, an eight-year-old political scandal gave the Westray project an unexpected boost. Roland Thornhill, the deputy premier and a senior minister in the Buchanan government, had been the subject of an RCMP investigation in the early 1980s, after he ran into financial trouble. Thornhill reached a settlement with his bankers to write off his debts at twenty-five cents on the dollar, which the Mounties viewed as a possible breach of the criminal law. No charges were laid and, after a minor stir in the media, the episode was swept under the rug. Then, in April 1988, the *Toronto Star* published an exposé in which former RCMP officers claimed there was a strong case against Thornhill; the Mounties contended that the provincial government had used its political clout to block charges.

The revelations were a heavy blow to a government already tainted by scandal. Billy Joe MacLean, a cabinet minister and one of Premier John Buchanan's closest friends, had been forced out of office in 1986, after pleading guilty to cheating on his expense-account claims to the government. And a Tory backbencher had been sent to jail in 1988, after a sensational trial on similar charges of expense-account fraud. Thornhill, hoping to limit further damage to the government's tattered image, did the honourable thing. Days after the *Toronto Star* article hit the streets, he resigned his post as environment minister and minister of industry, trade, and technology. The RCMP later reopened its investigation; Thornhill was charged in 1991, but the case was eventually dropped.

Buchanan appointed a new industry minister on April 20. His choice was Donald Cameron, the man who had pledged two weeks earlier to make the Westray mine a reality. The choice was surprising; in political circles, word was that Buchanan disliked Cameron's squeaky-clean stance on patronage. But such an image was just what the battered government needed to project as Buchanan prepared to call a late-summer election. Another Pictou-area backbencher, Donald McInnes, replaced Thornhill as environment minister. With Pictou Centre MLA Jack MacIsaac already serving as minister of lands and forests, the county suddenly enjoyed a lot of political clout at the cabinet table. And one project all three ministers supported was the Westray mine.

At the provincial level, the pieces were beginning to fall into place. Curragh, through a subsidiary, Westray Coal, had applied to the Department of Mines and Energy for a mining lease in March; by midsummer, a draft had been hammered out. Then, Curragh vice president Marvin Pelley, who was handling the Westray negotiations, threw a wrench in the works, demanding several key clauses be "clarified or reworded." Among these were provisions dealing with surface coal rights, royalties, and damage to land above the mine workings. "It is unfortunate that many of these issues were not raised earlier," assistant deputy minister Richard Potter complained in a July memo to the new minister of mines and energy, Ken

Streatch. But Potter was optimistic the department's staff and the company could reach an agreement. There was another, potentially more serious, stumbling block. Curragh, reluctant to commit money to the project without an aid package from Ottawa, had yet to buy Suncor's Pictou coal rights. After several deadlines passed, the $7.5-million sale was expected to close in September 1988.

However, a federal funding package remained elusive. A series of meetings with Department of Regional Industrial Expansion officials in Ottawa and Halifax failed to break the logjam over production levels. Then, the Nova Scotia government stepped in to solve Clifford Frame's dilemma. Cameron's department—industry, trade, and technology—upped the ante with a $12-million loan. The provincial government also said it would buy an annual 275,000 tonnes of coal more than the Nova Scotia Power Corporation contract specified, for a total of 975,000 tonnes annually. That exceeded Ottawa's 900,000-tonne bottom line, and the provincial loan enabled Curragh to scale down its federal-funding request. On July 4, the company asked DRIE to guarantee 85 per cent of a $100-million bank loan and offset interest rates over seven years, rather than fifteen. The cost of providing the interest-rate subsidy was $26.7 million, less than half of Curragh's original funding request.

At the end of July, Buchanan called a provincial election. Opinion polls had the Liberals, under leader Vince MacLean, a fiercely partisan Cape Bretoner, running neck and neck with the Tories. A final deal to open the Westray mine was the kind of good-news announcement that could make people forget about RCMP investigations and phoney expense accounts when they voted on September 6. Buchanan wasted little time getting the word out. During an August 3 campaign swing through Pictou County, he assured an interviewer that Westray had a mining lease and a contract with the Nova Scotia Power Corporation. A mining lease was in fact ready for cabinet approval on August 8, even though surface-coal rights and other issues remained unresolved. The NSPC contract, however, was yet to be signed.

And to make matters worse, Nova Scotia Tories were getting little support from their federal colleagues. Cameron discussed the hold-up in a long conversation with the prime minister. Elmer MacKay did his part, asking DRIE to prepare a report on the Westray project for the federal cabinet. The document was ready by mid-August but was not put before cabinet; the sticking point, once again, was Devco. Cameron sent a letter to Robert de Cotret on August 24, assuring his federal counterpart that Devco sales to NSPC would not slip below current levels. De Cotret, however, replied that he needed "firm future commitments" that NSPC would buy Cape Breton coal.

That was not the answer Cameron and the provincial Tories were looking for, as the clock ticked towards election day. In his bid for a fifth term as MLA,

Cameron was portraying himself as a politician who could deliver. "Don Cameron has been fighting very hard for a coal mine in Plymouth," declared one of his campaign newspaper ads. "Give Don the job and he'll get it done." Cameron's Liberal opponent in Pictou East, Wayne Fraser, was calling on the government to end the speculation and announce whether or not there would be a mine. At that point, the provincial loan promised to Westray earlier in the summer had not been made public—nor had the government's commitment to exceed the NSPC contract with Westray, and buy an extra 275,000 tonnes of coal per year. By the final week of the campaign, Curragh and provincial officials were feverishly working out the details. The New Glasgow *Evening News* reported that Cameron had interrupted his re-election campaign to join the negotiations. Reporters were told to gather in Stellarton at half-past nine on the morning of September 1 for an announcement.

But problems surfaced during a last-minute meeting with Curragh officials. Cameron showed up as scheduled, but only to cancel the press conference. The industry minister refused to reveal what was wrong, other than a vague reference to "a misunderstanding in the language" that was holding up an announcement. NSPC had already awarded a $32-million contract to build Trenton 6's boiler, he pointed out, and it was designed to burn Pictou County coal. "That may be an added incentive to continue to press for an agreement," he conceded. A reporter asked who was to blame for the hold-up. "Well, maybe you can point a finger at me," Cameron said, "but I will not announce a project that is not fully completed."

Later in the day, Curragh forged ahead and announced the project anyway. Clifford Frame issued a press release from his head office in Toronto, announcing that construction of the Westray mine would begin in early 1989. The following morning, reporters tracked down Cameron and Elmer MacKay at MacKay's constituency office to find out what was going on. Cameron confirmed that the Nova Scotia government was lending Curragh $12 million, but stressed that the province had struck a hard bargain; the loan was at the full rate of interest and was repayable within five years. He made no mention of the extra coal that the Nova Scotia government was promising to purchase, over and above the NSPC contract. MacKay said that Ottawa would provide interest-rate subsidies and guarantee $85 million of the $100-million bank loan needed to finance the mine. "I guess it's a deal," said Cameron.

Today, Cameron says he is fed up with the opposition politicians and journalists who have repeatedly accused him of pushing through the Westray announcement on the eve of the 1988 election in a bid to win votes. "I didn't need that mine for my re-election," he insists. Cameron says he was reluctant to make an announcement about the mine so close to election day, but he was being pressured by the company. For proof, he says he cancelled the press conference as soon as he learned that Curragh and NSPC had yet to finalize a contract. And, he maintains, he only discussed Westray when reporters sought him out to comment on the Curragh press release. "The bottom line is, if you're trying to make the case that I needed this for my election campaign, it seems very strange to me that I would cancel an announcement."

There was another bottom line. Days before Nova Scotians went to the polls, the papers trumpeted the news that a multimillion-dollar project and hundreds of jobs were coming to Pictou County. And Pictou County proved to be one of the few bright spots for the Tories in the 1988 election. All three county ridings dutifully returned their Tory incumbents, albeit with sharply reduced majorities. In Pictou Centre, the riding that included Westville, the future Westray miner Eugene Johnson voted for the sitting MLA, Jack MacIsaac. Johnson's family had long voted Tory, but the mine announcement was an extra incentive to back the Conservatives. Next door, in Pictou East, Cameron's margin of victory over his Liberal opponent dropped from 2,500 votes in the 1984 election, to 753 votes only four years later.

Across the province, the Buchanan government took a drubbing; the Tories dropped from forty seats to twenty-eight. The Liberals soared from six seats to twenty-one, capturing all but one of the eleven ridings on Cape Breton Island. But it was not enough to end Buchanan's ten-year reign. Two New Democrats and an independent were also elected, giving the Tories a four-seat majority in the House of Assembly. If three more ridings had defected to the resurgent Liberals—say, for argument's sake, the three Pictou County seats—Buchanan would have been back on the opposition benches.

That scenario was fodder for the pundits. The reality was that the Tories had limped back into office, and it looked like smooth sailing for the Westray mine. There was just one glitch—MacKay's announcement of a federal funding deal was news to de Cotret and DRIE. Despite what MacKay was saying down in Nova Scotia, DRIE told Frame on September 9, there had been no decision on Curragh's application for assistance. And there would be no deal until Ottawa and the Nova Scotia government resolved the Devco issue. The fact remained that Clifford Frame was still looking for his hand-out from the federal government. And Cameron, MacKay, and Buchanan still had a lot of work to do.

LET'S MAKE A DEAL

ICK LOGAN'S VOICE PURRS over the telephone line, as smooth as silk, dripping with sincerity. It's a voice honed by years of wheeling and dealing in the back rooms of Ottawa, where Logan has put out his shingle as a lobbyist. Of course, Logan says, he would love to talk about how Curragh Resources, one of his clients, got federal funding for the Westray coal mine. There's just one problem, you see; he's the wrong guy to ask. "My role in this was very, very minimal," the genial president of R.B. Logan Consultants assures the reporter on the other end of the line. "You're asking the guy that maybe set up four or five meetings . . . I'd really like to help you, but I'm so far down there, you can't see me."

Logan is a legendary figure on Parliament Hill, a renowned raconteur and prankster. He is also the consummate insider, friend to just about everyone involved in helping Clifford Frame wring money for Westray out of Ottawa's reluctant bureaucrats. But he would rather see someone else take credit for cutting that particular deal. "It sort of took on a life of its own with the politicians, I suppose, that wanted it," Logan says, choosing his words with care. "It was in Elmer MacKay's riding and in a sense Don Cameron's riding. I suppose that they must have felt, I don't know, that it was a good project." The people to talk to, he tells the reporter more than once, are the politicians.

Logan could have mentioned one other politician who was solidly in the Curragh camp—Robert Coates. In the 1970s, when Coates and the federal Tories were languishing on the opposition benches, Logan was Coates's top aide. Among other things, they shared a conviction that Joe Clark was the wrong leader for the Tories, a view confirmed after Clark squandered the party's 1979 election victory. Coates was passed over for a cabinet post in Clark's short-lived government, and the snub made him a bitter enemy. He was one of the first MPs to plot against Clark, paving the way for the leadership review that ultimately brought Brian Mulroney to the Tory helm.

Logan did his part to undermine Clark, holding court in a parliamentary cafeteria each morning at a gathering of assorted Tory MPs, political aides, and journalists. It came to be called the Breakfast Club, and Logan served up a steady stream of malicious jokes and one-liners at Clark's expense. "In Logan's hands," journalist John Sawatsky wrote in his biography of Brian Mulroney, *The Politics of Ambition,* "Clark always emerged as a bumbling kid with his mittens on a string, or as a big turkey, or just as a plain jerk." A classic Logan line that made the rounds on Parliament Hill was prompted by a photograph in the *Toronto Sun* showing Clark standing with two Sunshine Girls, the women who posed for the pin-ups that have long been a regular feature of the paper. Logan used it to conduct something he called "the Clark loyalty test." The question was: "How many boobs do you count in this picture?" Anyone who counted more than four failed the test.

After Mulroney led the Tories back to power in 1984, Coates was rewarded with the defence portfolio. Logan, naturally, became the new minister's chief of staff. Nova Scotia was given a second voice in cabinet–Elmer MacKay, who became solicitor general. MacKay, another early backer of Mulroney's leadership bid, had resigned his Central Nova seat in 1983 so that Mulroney could enter the House of Commons; in 1984, the prime minister cleaned up in the Quebec riding of Manicouagan, while MacKay reclaimed Central Nova. It was a heady time to be a Nova Scotia Tory; the party controlled the levers of power on the federal and provincial levels, and anything seemed possible. Within a month of Mulroney's September 1984 victory, Coates and MacKay flew to Halifax for a meeting with Premier John Buchanan. As they smiled and joked for the cameras, Coates boasted that Mulroney was "sort of an ally of mine and Elmer's." The province's two cabinet posts, he said, showed the new prime minister was giving them "lots of clout to get things done."

One of Coates's first moves was to order new uniforms for the military, giving the army, air force, and navy the distinctive garb they wore before the forces were unified under the Liberals. He also promised a bigger role for Canada in the North Atlantic Treaty Organization. Coates, right-wing even by Conservative standards, got himself in hot water when he ridiculed peace protesters who were jailed after a demonstration in Halifax. The biggest complaint about Coates, however, was his free-spending ways. In five months as defence minister, Coates cost the public $70,000 for airfare and expenses, while his chief of staff, Logan, submitted another $10,000 in claims.

But Coates's high-flying career as a cabinet minister came to an abrupt halt on February 12, 1985. *The Ottawa Citizen* reported that Coates had paid a late-night visit to a German strip club, raising concerns about a possible breach of national security. The incident occurred in November 1984, while Coates and Logan were on a twelve-day tour of NATO countries. Their first stop was the Canadian Forces base at Lahr, Germany. After dining with the military brass, Coates, Logan, and another aide, press secretary Jeff Matthews, headed to a club, Tiffany's, for a nightcap. Coates had a drink at the bar with one of the strippers, a German divorcée who called herself Micki O'Neal. Innocent stuff—unless the patron happened to be a minister of the Crown, presumably carrying military secrets around in his head. Coates resigned the day the story hit the press, taking Logan with him and paving the way for Yukon MP Erik Nielsen, who took over as defence minister. It was the first of a string of resignations and slip-ups that rocked the fledgling Mulroney government.

Mulroney remained supportive, dismissing the ill-conceived nightclub visit as an "error in judgment." The prime minister declared publicly that "all doors" remained open to Coates in the future. Coates stayed on as an MP, but rarely showed his face in the House of Commons. To get even with the *Citizen*, he launched an expensive libel suit that dragged on for three years, until they settled out of court. He remained active in the Canada-Korea Parliamentary Friendship Society, a group he founded to foster closer ties with the Southeast Asian country. Behind the scenes, he made sure the folks back home were getting their fair share. In 1986, when a newspaper reporter tracked down the MP for a puff piece marking his twenty-ninth year in Parliament, Coates said he was too busy to give an interview. He was in the midst of a meeting at his Parliament Hill office with promoters of what he called "a potential new industry" for Nova Scotia. But he did not specify.

Coates's departure from cabinet thrust MacKay to the forefront. MacKay, a lawyer first elected to the House of Commons in 1971, was a no-nonsense politician whose idea of relaxation was wielding a chainsaw on one of his many Pictou County woodlots. MacKay caused his own share of headaches for the Mulroney government. The week Coates resigned, the opposition was out for MacKay's head over something more serious than a chat over drinks. MacKay, the minister responsible for the RCMP, had met privately with Richard Hatfield shortly before the Mounties charged the New Brunswick premier with possession of marijuana. MacKay held onto his post as solicitor general, despite that revelation, but then earned a public rebuke from Mulroney in May 1985 for downplaying a sexual harassment complaint lodged against their Tory colleague—Coates. Within a year, both MacKay and Coates were in hot water; they admitted to lobbying the federal fisheries department to grant a licence to a clam-harvesting firm part-owned by Coates's former aide—Richard Logan. It was indeed a small world.

As Nova Scotia's only remaining cabinet minister, MacKay made it his business to oversee the flow of government money to favoured areas and projects, just as Liberal power broker Allan J. MacEachen had done in the Trudeau years. MacEachen's riding, which straddled the Strait of Canso, separating Cape Breton Island from the Nova Scotia mainland, was located next door to MacKay's riding. For years, Central Nova had been passed over, as MacEachen heaped government largesse on his constituents; the blatant favouritism provoked much of the Cape Breton-bashing that surfaced in the debate over the Pictou County mine. With the Tories in power, MacKay did his part to redress the imbalance, pushing for government money to prop up failing industries and encourage new investment in his area. He helped engineer a rescue package—including a $100-million investment from the Department of Regional Industrial Expansion—to enable Montreal-based Lavalin Industries to buy the Trenton car works, salvaging hundreds of jobs. He even managed to wrangle $1 million in federal money for a new court building in Pictou, although courthouse construction was a provincial responsibility.

Logan went into business for himself after the Tiffany's debacle, opening a lobbying firm in Ottawa. When he was not promoting his clam-fishing scheme, Logan was trying to snag federal government contracts for a handful of steady clients. Among them was the upstart mining firm Curragh Resources. Like Coates, Logan had met Clifford Frame years before, through the Stephen Roman–Denison Mines connection. "All the corporate world is out there rubbing elbows with politicians," as Logan later put it. "You get to know a lot of people that way." Logan became a fan of Curragh's brash chairman; long after the Westray explosion, he was still promoting Frame as "one of the greatest mining people that Canada has ever produced." Logan also lobbied on behalf of Kilborn, one of the two engineering firms that produced the final plan for the Westray mine.

When Curragh came to Ottawa in 1988 to make its play for public money, it fielded a close-knit team of veteran Tory power brokers. In the background was the legendary Logan, greasing the wheels and setting up meetings with bureaucrats and politicians. In the back rooms were Coates and MacKay, who had been on hand for Frame's first face-to-face meeting with Industry Minister Robert de Cotret in May 1988. Besides opening doors in Ottawa, Coates had given his old friend Frame the inside track with the provincial government through another good friend, Buchanan. MacKay, who made it his business to see that Pictou County and Nova Scotia enjoyed the spoils of power, was an influential minister who could shepherd a funding deal through the federal cabinet. Despite some past "indiscretions," Coates and MacKay still had the confidence of Mulroney, the man they had helped elect as prime minister. Cliff Frame had some heavy hitters in his line-up; it was time to play ball with the federal government.

* * *

IN THE DAYS after the Buchanan Tories squeaked back into power on September 6, 1988, the Westray project won swift approval from the Nova Scotia government. On September 9, as scheduled, Westray Coal purchased Suncor's coal rights, including all feasibility studies and information compiled by that firm and Placer

Development. On the same day, the Department of Mines and Energy gave Westray first right of refusal to mine surface coal from the Wimpey pit in Stellarton. With those loose ends cleared up, cabinet passed an order September 13 granting Westray a mining lease covering a 1,150-hectare block of land fanning out in a northeasterly direction from the village of Plymouth.

The Nova Scotia Power Corporation, meanwhile, signed a contract on September 9 to buy 700,000 tonnes of Westray coal annually over fifteen years. Shipments were to begin in July 1991, in time to build a stockpile for the planned opening of the Trenton 6 plant four months later. The contract offered a two-tiered price based on the grade of coal delivered. Coal processed in Westray's wash plant to reduce ash content would fetch $74 per tonne, while higher-ash coal—the bulk of the mine's production—would sell for $60 per tonne. At those prices, Westray would gross close to $50 million annually. The contract provided for price increases tied to inflation and allowed the utility to impose penalties if the coal was not of the agreed quality. The overall price paid to Westray was comparable to, if not lower than, the approximately $65 per tonne Devco was receiving under its NSPC contract. On the same day, September 9, the Nova Scotia government put into writing its commitment to purchase an additional 275,000 tonnes of coal. With the mining lease and NSPC contract in place, cabinet approved Nova Scotia's $12-million loan to Westray on September 20.

It was a different story in Ottawa. Despite Elmer MacKay's wishful thinking on the eve of the Nova Scotia election, there was no agreement to provide federal assistance to Westray. The Department of Regional Industrial Expansion notified Clifford Frame in a September 9 letter that federal aid still hinged on protection for Devco. By the time Frame got the letter, Curragh was stuck with the Pictou coal rights and had a binding contract to supply NSPC. Those steps had been taken in the belief that federal money was in the bag, and Frame was not a happy camper. He fired off a pointed letter to federal Industry Minister Robert de Cotret on September 12, complaining that he had committed himself to the Westray project on the basis of the pre-election announcements in Nova Scotia. "It is quite clear . . . that there was consultation and agreement on financial assistance to Westray Coal between senior representatives of the federal and Nova Scotia governments prior to these announcements appearing in the press," Frame grumbled. He mentioned two of the "senior representatives" by name—Elmer MacKay and Donald Cameron. DRIE's response was to remind Frame not to believe everything he read in the papers. The level of "possible" federal support, de Cotret insisted, could not be discussed until Devco's coal sales to NSPC were set in stone.

Ottawa's hard line did nothing to dampen the enthusiasm of Curragh and its political backers. On September 22, a reception was held at the Heather Hotel in

Stellarton to once again announce that the Westray mine was going ahead. Frame blew into town for the occasion, joining MacKay, Premier John Buchanan, Robert Coates, and provincial Industry Minister Donald Cameron. Everyone, even Frame, was decked out in suits of Tory blue. Frame announced that construction would begin within a month, and promised that Curragh and Westray would be "good corporate citizens." Then the politicians took over, congratulating each other for their fine work. Buchanan called Coates "a champion of this from the beginning," revealing how the MP had set him up with Frame. Coates, who had just announced his retirement from politics, jokingly referred to himself as "the marriage broker in the deal." MacKay was still talking as if a federal deal were already in place. "We're bringing back coal [mining] to the mainland in a particularly elegant way," he told the gathering.

The reception also marked the return to Pictou County of a familiar face—Gerald Phillips. Curragh had sprung him from his temporary exile in the tar sands of northern Alberta to become Westray's on-site manager, with salary and bonuses in the $110,000 range. Phillips appeared a logical choice for the job; he had spent two years developing Suncor's plans for the mine and arguably knew as much as anyone about the project and its risks. Phillips wasted no time making his presence felt, promising reporters that Westray would hire locally and "make sure all employees are trained to work safely." Safety, after all, was the responsibility of the mine manager.

* * *

THE BACK-SLAPPING in Stellarton was in marked contrast to the arm-twisting going on behind the scenes in Ottawa. Funding negotiations hit a new snag in early October, just after Brian Mulroney called a federal election for November 21. Federal officials began making noises about aspects of the project other than its direct impact on Devco. Crown-owned Canadian National Railways was worried about its branch line to Sydney, which relied on Devco's coal shipments to Trenton. Direct aid to Westray would also set a precedent. Previous federal assistance to privately owned mining ventures had been limited to providing port facilities or rail lines—the prime example being Clifford Frame's last foray into the coal industry, the ill-fated Quintette project.

But the most serious objections arose after a review of Westray's business plan; officials of the Department of Regional Industrial Expansion concluded that the mine stood to reap handsome profits over the life of the NSPC contract. This raised questions about the limited amount of money Curragh was willing to invest—the company was proposing to sink $15 million of its own money into the project, including the $7.5-million cost of obtaining the coal rights—let alone the need for federal assistance. Curragh representatives and DRIE officials locked horns for three days in Ottawa in mid-October, but the marathon meetings ended without an agreement. "To be quite blunt," a senior civil servant later told *Maclean's*, "there was not a bureaucrat in the federal government who wanted the project to go ahead."

Curragh responded by going over the heads of the bureaucrats with a direct appeal to their political masters. The company sent a letter to Elmer MacKay on October 16, pleading for the minister's help in overcoming what it considered the "unfavourable biases" of DRIE officials. The same day, an accountant working for Curragh went straight to the top. William Redrupp of Price Waterhouse in Toronto, who was advising Curragh on the financial aspects of the Westray project, called Peter White at the Prime Minister's Office. White, the principal secretary at the PMO since the summer of 1988, was one of the prime minister's closest advisers, a charter member of Mulroney's inner circle. The two men had been classmates at Laval law school, and White had helped organize Mulroney's leadership campaigns in 1976 and 1983. And he was no stranger to the intricacies of corporate finance, having moved to the PMO from a top executive post in Conrad Black's business empire. To bring White up to speed on Curragh's application for financial assistance, Redrupp sent along a synopsis of what had been discussed at the meetings with DRIE.

The high-level contact brought results. White immediately waded into the complex financing issues surrounding Curragh's application for assistance. The day after Redrupp's call, White met with DRIE deputy minister Harry Rogers and two other senior department officials to discuss the disputed issues. As Mulroney crisscrossed the country, campaigning for re-election in the month that followed, his principal secretary kept close tabs on the Westray mine application. Curragh made sure both White and MacKay received copies of letters and other documents related to the funding request. And the PMO heavyweight played a direct role in the negotiations, attending key meetings between Curragh and the bureaucrats at DRIE.

White also acted as a liaison between the federal government and the Nova Scotia Power Corporation. The NSPC president, Louis Comeau, faxed three letters to White in October, each insisting that the Westray mine would not reduce the utility's purchases from Devco. As well, Pictou County's low-sulphur coal was a cornerstone of NSPC's effort to limit sulphur dioxide emissions, in keeping with a recently signed federal-provincial agreement designed to reduce acid rain. "Nova Scotia Power needs the Westray coal. We need it for environmental reasons," Comeau wrote. "If not, Nova Scotians will have to pay more for their power." White, who was also briefed by Comeau over the phone, forwarded the letters to

DRIE. The fate of a mining project in an out-of-the-way corner of Nova Scotia was commanding an inordinate amount of attention in the nation's capital.

* * *

WHILE COMEAU WAS trying to ease Ottawa's long-standing concerns about Devco, DRIE was taking a closer look at other aspects of the Westray project. The department's Halifax office retained Peter Hacquebard, a respected Nova Scotia geologist, to critique the geological work on the Foord seam. Since Curragh was relying on Placer Development's 1987 feasibility study, Hacquebard concentrated on those findings. A second report was commissioned from the Canada Centre for Mineral and Energy Technology, or CANMET, a division of the federal Department of Energy, Mines, and Resources. CANMET was asked to examine the entire Westray project to determine if it was technically sound.

In his October 14 report, Hacquebard told the federal government what most experts had been saying for years—numerous faults and the steep angle of the seam would make mining difficult and expensive. One known fault plunged the seam downward some sixty metres, he said. When miners reached that fault, they would have to drive a long tunnel through rock to once again reach coal. "These faults will pose major mining problems, and will be costly to cope with," he predicted. This was in marked contrast to Kilborn's opinion that only "minor" and "normal" faulting would be encountered.

Hacquebard also warned that the ash content of Foord coal fluctuated widely within the seam, casting doubt on Placer's estimates of the amount of good-quality coal available to be mined. Some forty holes had been drilled to obtain core samples during the exploration phase of the Pictou County development, he noted, but only nine were within the area staked out for mining. To confirm coal quality, Hacquebard told DRIE, Curragh should invest more money in exploration by drilling four new holes. Geologists on the staff of Nova Scotia's Department of Mines and Energy, working independently of Hacquebard, had come to a similar conclusion six months earlier.

CANMET, meanwhile, dispatched one of its engineers to Halifax on October 13 on a fact-finding mission. It was a rushed effort, based on a mere eighteen hours of meetings and research. The engineer discussed the project with Nova Scotia government officials and Westray's manager, Gerald Phillips, for about eight hours, then spent ten hours ploughing through a stack of feasibility studies and consultant's reports dating back to 1986. When the unidentified engineer handed in his nine-page report, even he admitted the effort was "limited and somewhat superficial."

The report touched on the salient points of the Kilborn study, offering a running commentary on various aspects of the project. The CANMET engineer, also working independently of Hacquebard, concluded that "a sound and logical argument" could be made for further drilling to confirm coal quality. The planned roof support methods met only "minimum" requirements, he said, "and will likely need upgrading in places." CANMET said its superficial study could not determine whether Kilborn's cost estimates took such problems into account.

CANMET's engineer took a dim view of Suncor's laboratory tests for methane, which were still being used to downplay the gassiness of the Foord seam. The finding of low methane content "does not give one confidence," he wrote with a hint of sarcasm. The risk of spontaneous combustion, dismissed in the Kilborn study and its predecessors, was taken seriously by CANMET. "Based upon experience elsewhere in the world, zones around faults may have elevated spontaneous combustion risk," the report noted; Westray was certain to encounter faulting. The report added that the thickness of the seam "also constitutes an increased spontaneous combustion risk." The planned monitoring system to detect carbon monoxide—which would indicate the presence of burning coal—was described as "prudent, sensible and necessary." But the diesel-powered vehicles to be used underground would also produce carbon monoxide, CANMET warned, making "early detection more complicated."

The mining method selected for Westray made sense, CANMET said, but time would tell if the room-and-pillar approach used in Western Canadian mines could be transferred to Pictou County. Experienced miners would have to be hired to make sure the technique was applied properly to conditions in the Foord seam. "There is always some risk associated with thick seam room and pillar mining," the report warned. "The real question is whether this property can wear the cost of the learning curve to get to a routine development/extraction practice." In other words, Westray's miners had to master the art of digging Pictou County coal as quickly and cheaply as possible—if the project was to succeed. The CANMET report was preoccupied with Westray's economic outlook, but that warning applied with equal force to the health and safety of workers. And it was the miners who could pay the price if Westray did not "wear the cost" of the learning curve.

The report ended with a lukewarm endorsement. "Technically, the proposal has no major failings. . . . On the surface [it] is a project with some expectation of financial success in spite of numerous technical uncertainties." CANMET recommended three steps be taken before Ottawa assisted the project. DRIE was advised to conduct an in-depth review of the project's geology, mining methods, and economics in order to "accurately assess" Westray's prospects. In addition, Ottawa needed to see the NSPC coal contract; it sounded "extraordinarily attractive" to CANMET's engineer, but he was not privy to the details. Finally, the industry department was advised to ask tough questions about Curragh's failure to secure financing from the banks without government help. "What reasons did the lenders give for not accepting Westray's business?" CANMET asked. "Is public sector funding in fact really necessary?"

The Hacquebard and CANMET reports gave the Westray project a passing grade at best. Both reports identified geological and technical problems that threatened the safety and viability of the mine. But Ottawa neglected to share all its findings with the Nova Scotia government. Only the Hacquebard report, the least-critical of the two, was forwarded to the Department of Mines and Energy in Halifax. The CANMET report, with its ominous warnings about dubious methane tests, steep learning curves, and inadequate roof support, remained buried deep within the federal government's burgeoning files on the Westray mine. Both reports recommended further exploration and study before mining began, but that advice went largely unheeded. Westray drilled eight additional holes between 1990 and 1992, but it remains unclear whether that exploration program was carried out in response to Hacquebard's report, or addressed his concerns. The in-depth federal study envisioned by CANMET never materialized.

* * *

WITH THE PRIME MINISTER'S OFFICE in the picture, there were two more meetings to try to hammer out a deal with Curragh. The first convened in Ottawa on October 26, 1988. Deputy minister Harry Rogers represented DRIE, while Clifford Frame headed the Curragh delegation. There was a follow-up meeting on November 7 at Curragh's offices in Toronto, with Nova Scotia Industry Minister Donald Cameron added to the guest list. Peter White of the PMO attended both meetings. DRIE set the agenda, demanding a number of conditions be placed on any assistance package. The bureaucrats wanted assurances that profits would be used to pay off bank loans, not just to line the pockets of Westray's shareholders. And if it cost more to develop the mine than expected, Westray's owners would have to pick up the tab, not Ottawa. The thrust of DRIE's position was simple—Curragh stood to make a bundle off Westray and hoped to pad its bottom line with public money.

The talks failed to break the impasse, and Frame was losing his patience. On November 9, two days after the Toronto meeting, he wrote a lengthy letter to Rogers at DRIE under the heading: "Pictou County Coal Project—Risk Assessment Analysis." Frame's frustration was palatable as he lectured Rogers about the perils of taking on the Westray project. "A lot can go wrong in the development of this mine. In particular the hazards of developing a deep coal mine of the nature of Pictou are many and can come without warning." Frame offered a partial list, in no particular order: unforeseen geological faults, flooding, poor roof and floor conditions, excess methane, unskilled and inexperienced workers, breaking in new equipment, and underground fires. The list read like a postmortem of the problems that would eventually haunt Westray, with one notable exception–Frame made no mention of the explosive potential of coal dust.

Although no mention of such hazards was made in Westray's business plan, Frame said, they had been factored into calculations of the money needed to bring the mine into production. It was Frame's "hope" that Curragh's mining expertise and his commitment to hiring the best mining experts and geologists available would reduce "to an absolute minimum the chance of these problems developing. . . . However, should the problems exceed those projected, the impact would be serious to the development of this project." Curragh and the Nova Scotia government, through its $12-million loan, would have a total of $52 million tied up in the project, he said; all Ottawa had to do was share the risk through a loan guarantee.

And Frame was tired of hearing Rogers and DRIE gripe about the huge profits the mine would generate. "You and your associates have repeatedly told us of the high level of profit illustrated in our business plan. Westray Coal sincerely hopes and expects that it will receive these levels of profits because without them we would not contemplate this project." Even if everything went smoothly, it would be 1995–eight years after the Pictou County coal rights were acquired–before the project began to make money. That was a long time to wait for a payback. "Surely," Frame protested, "these profits should be rewards to the developer which undertakes a project of this degree of risk."

The message to DRIE was blunt: Clifford Frame, twice Canada's mining man of the year, was not about to be pushed around by a bunch of number-crunchers in Ottawa. Westray was, to use Frame's word, a "marginal" project that had been passed up by two other companies. Frame could find better uses for his time and money, but he was willing to assume the risks, build the mine, and put people to work. In Frame's view, the federal government should be giving him some financial encouragement, instead of begrudging him a few million bucks of profit for all his trouble.

The November 9, 1988, letter stands out for reasons other than its preachy tone. It showed that Frame was walking into the project with his eyes open to the dangers. The letter outlined the risks of mining the Foord seam with more clarity than all the feasibility studies and reports that had been generated by the Westray project. Years before Westray went into production, Frame recognized that excess methane, faults, cave-ins, and the use of untrained workers could spell disaster. While the letter was preoccupied with the prospect of financial disaster, it was clear that explosive gases and underground fires would also put lives at risk. Frame made sure his warning received wide circulation in the corridors of power in Ottawa and Halifax; copies of the letter to DRIE were forwarded to Peter White at the PMO, and to Elmer MacKay and Donald Cameron. Thanks to Frame, the politicians were also walking into the Westray project with their eyes open.

The difference of opinion over Westray's profitability spilled over to the DRIE submission to the federal cabinet. A draft copy was faxed to Curragh's head office on November 10, but the company strongly objected to the department's portrayal of the project. The company's vice president, Marvin Pelley, took the liberty of redrafting the cabinet report and sending it back in a form "acceptable" to Curragh. That drew the PMO back into the negotiations. On November 14 and 15, in telephone conversations, White discussed the deadlock with DRIE officials and Curragh's accountants. A new report was prepared to incorporate some of Curragh's concerns, but Pelley complained that the document was still "very negative." On November 15, he demanded more changes, sending a letter to Rogers at DRIE, with copies to MacKay and White. "This could be a success story that may serve as a model of government-industry cooperation," he claimed.

As Curragh quibbled with bureaucrats and the PMO over the wording of the report to cabinet, the federal election was just around the corner. After a rocky campaign dominated by the controversial free-trade deal with the United States, the Mulroney Conservatives were returned to power on November 21, 1988. It was the second consecutive defeat for John Turner, the Liberal leader and future Curragh director. And in Central Nova, Elmer MacKay was re-elected with a healthy 3,999-vote margin over his Liberal rival. Despite the intervention by the PMO and all the frenzied negotiations behind the scenes, there was still no federal deal to assist Westray. But once again, the promise of a new coal mine had helped keep Pictou County within the Tory fold.

* * *

AS THE SMOKE CLEARED after the federal election, the Westray project finally entered the construction phase—even without the funding from Ottawa. In December 1988, earth-moving equipment cleared and levelled the mine site in Plymouth, the first step before tunnelling began. But there was no movement on Curragh's quest for federal help until the new year, when a new face came on the scene in Ottawa. Peter White began referring Curragh's calls to Stanley Hartt, a Montreal labour lawyer, who had just become Brian Mulroney's chief of staff after four years as deputy minister of finance. By January 1989, the Westray funding application was "in the PMO," according to a Curragh document setting out the sequence of events. The stubborn officials of the Department of Regional Industrial Expansion were relegated to the sidelines, as Hartt dealt directly with William Redrupp, the Price Waterhouse accountant working for Curragh. During January and February, Hartt and Redrupp discussed Westray at least twice, but they were unable to resolve a nagging question—why did such a profitable project need Ottawa's support? As Hartt explained in a February 8 fax to Redrupp, "the federal government is not satisfied as to the sharing of risk and rewards associated with the project."

There were still nagging questions about Devco. The new federal industry minister, Harvie Andre, advised Premier John Buchanan in early February that Ottawa needed an ironclad guarantee from the province that Westray would not reduce Devco's sales to the Nova Scotia Power Corporation. This had to include an assurance that the additional 275,000 tonnes of Westray coal to which the province had committed itself, over and above the NSPC contract, would not be diverted to replace Devco coal. At Buchanan's request, NSPC President Louis Comeau wrote back on March 1, promising that purchases of Devco coal would increase, despite competition from Westray.

In a letter to Andre, with a copy to Peter White at the PMO, Comeau said that Devco alone could not keep up with the utility's growing appetite for coal. The new power plant planned for Point Aconi on Cape Breton Island, using new technology to reduce acid-rain emissions from Devco's high-sulphur coal, would compensate for sales lost at Trenton. Then there was the so-called Bluenose Project, the Buchanan government's dubious scheme to export coal-generated electricity to the United States via underwater cable. If feasible, that project would require more power plants. And it was already too late to question the wisdom of using Westray coal. The utility "has ordered its equipment for the new Trenton power plant based on using low-sulphur coal from the Pictou County mine," Comeau grumbled. "The matter is becoming quite serious and, in our opinion, must be resolved quickly."

It was time for another appeal to the top. At Donald Cameron's request, Buchanan brought the Westray logjam to Brian Mulroney's attention when the prime minister visited Nova Scotia in late February 1989. As Buchanan recalled the meeting, Westray was one of several issues he raised with Mulroney, and they discussed it for no more than ten minutes. Mulroney was "not completely up to date on what it was all about," so Buchanan explained that the delay in federal funding was causing concern within the provincial government and Pictou County. "The answer he gave me at the time, I can recall well, was that he'd have his people look at it," Buchanan said, "which is the standard answer you'd get, but I know he made a note of it."

A few days later, on March 2, the vice president of Curragh, Marvin Pelley, arrived in Halifax to meet with Buchanan, Cameron, and Jack MacIsaac, another Pictou County MLA and by then the minister of mines and energy. Pelley was in town to put the finishing touches on an $8-million loan from the provincial industry department—money Curragh said it needed to begin tunnelling. It was essentially an advance on the $12-million loan the province had pledged back in 1988. Upon his return to Toronto, Pelley provided a blow-by-blow description of his Halifax meetings in a confidential memo to Clifford Frame and other Curragh executives. That memo, dated March 3, 1989, was later made public, and it provides a rare glimpse of the interplay between politicians and their corporate suitors.

At the initial meeting, according to Pelley's memo, Buchanan suggested the $8-million interim loan would be counterproductive for the province and for Curragh. It would "relieve some of the pressure" for a federal deal, the premier felt, further delaying the project. Cameron, however, took Pelley aside and assured him the opposite was true. Once construction began, and the Nova Scotia government had money tied up in the project, Pelley quoted him as saying, "the province cannot politically allow the project to shut down." Later, in the course of a candid three-hour discussion over dinner, Cameron made Pelley a promise: "Marvin, we are going to get this mine going with or without the feds."

Still it seemed the feds could be brought onside. During the meeting in the premier's office, Buchanan had briefed Pelley on his recent chat with Mulroney. In a later interview, Buchanan claimed he simply told Pelley that the discussion with the prime minister had been "positive," and that "hopefully things would be proceeding" towards a federal deal. But that was mild compared to the way Pelley recorded the conversation in his memo. In Pelley's version, Buchanan described Mulroney as "disappointed and surprised" to discover that no funding agreement had been worked out. The Westray deal, Mulroney was quoted as saying, was "one of the reasons" Elmer MacKay had been made minister in charge of the Atlantic Canada Opportunities Agency in January 1989. "The premier," Pelley noted in the memo, "stated that the prime minister will not likely give a direct order to get this thing done but rather will continue to play the role of the facilitator." Pelley came away from the meetings confident Curragh had a lot of friends in high places. "Everyone including the premier, Peter White, Elmer MacKay, and Don Cameron seems to be on top of this deal and are working towards a solution," he assured Frame.

Pelley may have been embellishing slightly to impress the boss. "Facilitator," Buchanan insisted later, was Pelley's word, not his. "There was no such comment along that line," he contended when asked about the memo. But "facilitator" was as good a word as any to describe Mulroney's contribution to getting funding negotiations back on track. Two of his top aides had spent hours in meetings and on the phone, fine-tuning the funding proposal with the bureaucrats and Curragh officials. And there was no denying Pelley's assertion that Tory politicians in Nova Scotia and Ottawa were "working towards a solution" that would get federal money into Curragh's hands. "Shortly thereafter, Peter White did make some calls to see what was going on," Buchanan conceded, "which is normal, of course; there's nothing unnatural about that."

Unnatural or not, the prime minister's staff was investing a lot of time and effort to make sure Ottawa did right by Clifford Frame and the Westray mine. In the wake of the Westray explosion, Mulroney assured Parliament that his interest in the mine was "no more nor less than in other cabinet decisions regarding financial assistance for regional development projects." Buchanan suspects Mulroney felt "a bit of an obligation" to keep tabs on the project, because he had once been the area's MP. In 1989, Mulroney came under renewed pressure from Pictou County's municipal leaders, who orchestrated a letter-writing campaign complaining that little had been accomplished since their trek to Ottawa a year earlier. They invited the prime minister to intervene personally to cut through the bureaucratic red tape delaying funding. "We urge you to expedite the mechanics of government," as Stellarton Mayor Clarence Porter put it.

After a round of meetings in Halifax in March 1989, Curragh vice president Marvin Pelley assured Clifford Frame that Nova Scotia's top Tories were "on top of this deal and are working towards a solution." [Courtesy of the *Chronicle-Herald* and the *Mail-Star*]

The prodding from Buchanan and former constituents had the desired effect. Mulroney instructed his staff to make inquiries, and that got the wheels of government moving again. On March 13, little more than a week after Pelley wrote his upbeat memo, Harvie Andre assured Curragh its position on Westray's profitability "will be properly presented in any document that is put before cabinet." At the same time, officials of his department—recently renamed Industry, Science, and Technology Canada—were asked to explain the delay on Curragh's funding application. Word came back that the profit issue remained the major stumbling block. But the bureaucrats promised to reopen talks with the Nova Scotia government to try to resolve the question of what the Westray project implied for Devco.

* * *

BACK IN NOVA SCOTIA, Donald Cameron was losing his patience. The industry minister went public in mid-March with a scathing attack on the bureaucrats standing in the way of federal support for Westray. Some officials in the Department of Industry, Science, and Technology were biased in favour of Devco, he claimed. He accused them of feeding "falsehoods" to the federal cabinet; he accused them of privately assuring Devco that they would scuttle the Westray project. If he were federal industry minister, Cameron fumed, heads would roll. Years later, Cameron would make no apologies for dressing down the bureaucrats in public. One moment they were questioning the project's viability, he says; the next they were arguing that it was too profitable. "These folks kept changing the goal posts," he says. "They were clearly biased."

Thanks to the Nova Scotia government's interim loan, work was moving ahead at the mine site in Pictou County. On April 10, 1989, Cameron and Louis Comeau, president of the Nova Scotia Power Corporation, set off a ceremonial blast to begin tunnel construction, five months behind schedule. "I'd like to put a few sticks of dynamite under some of my friends in Ottawa, to try to get this job done," Cameron quipped after hitting the plunger. The Devco chairman, John Terry, was in attendance, and Cameron tried to sound conciliatory. "There is no Devco-Westray competition. There is no Cape Breton–Pictou County competition. It is sad that some are trying to make it a competition," he told reporters, as if he had played no role in fuelling the regional rivalry. The no-shows among the politicos invited to the ceremony included Premier John Buchanan, Elmer MacKay, Pictou County's two other MLAs, and the man who had got the ball rolling in the first place, Robert Coates. It was starting to look as if the Westray project was Donald Cameron's project.

Cameron's unwavering support for Westray put him on the hot seat when the Nova Scotia House of Assembly opened its spring session. Since nearly half the Liberal caucus was from Cape Breton, the opposition was openly hostile to the Pictou County mine. Bernie Boudreau, an intense lawyer from Sydney who was among the new crop of Liberals elected in 1988, led the attack. Day in and day out, Boudreau peppered Cameron with questions about the province's loan to Westray, the Nova Scotia Power contract, and the lack of federal funding. Each time, Cameron insisted that burning Westray's low-sulphur coal made economic and environmental sense. The only alternative, he maintained, was spending some $200 million on scrubbers—on-site plants that reduced sulphur dioxide emissions—so the Trenton stations could burn Cape Breton coal.

Boudreau hit the bull's-eye with his criticism of the $14-million tunnelling contract, awarded without tender to Canadian Mine Development, although a provincial government loan was financing the work. Boudreau contended that Walker Mining, a Cape Breton company that had done work for Devco, was better qualified to build coal mines. But it was a case of Clifford Frame looking after his allies in the mining industry, as he had when he selected Kilborn to produce the latest Westray feasibility study. Canadian Mine Development, it turned out, was a division of Toronto-based Hillsborough Resources, which enjoyed a cosy relationship with Curragh Resources. The firm owned a chunk of one of Curragh's lead-zinc properties in the Yukon, and Frame sat on Hillsborough's board of directors.

The Liberals kept up the heat, inviting David Rose and Allister MacKenzie of Coalcor Resources to appear before a legislature committee. Rose made the most of the opportunity, complaining of political interference in the sale of the Pictou County coal rights to Curragh. Cameron, who could not contain his laughter during portions of Rose's testimony, dismissed the performance as "a farce." But Cameron, touchy at the best of times, became increasingly testy under the constant goading from the opposition. His frustration boiled over on May 18, during an exchange with Boudreau in the legislature. "Unless something unforeseen happens—an act of God—and the whole thing blows up," he declared, "I believe in my heart that this will be one of the better deals for the people of the province of Nova Scotia in the years to come." No one, least of all Donald Cameron, could have seen the irony in that glimpse of the future.

* * *

IN THE DYING DAYS of July 1989, after three months of incessant blasting, the Westray site fell silent. Canadian Mine Development laid off its thirty employees, leaving the main tunnels far from the Foord seam. Westray officials pleaded poverty—the money had run out, and without federal assistance, there would be no mine. By then, the company had drawn about $7 million of its $8-million interim loan from the Nova Scotia government. Westray had another $10 million of its own money tied up in the project, but $7.5 million of that had been the price of the coal rights. Suspending work was a dramatic way of pressuring Ottawa to bail out the project, but the company piously denied any ulterior motive. "It's not a bargaining ploy," claimed the Westray president, Kurt Forgaard. "We don't have money to continue." Only a skeleton staff was kept on.

The shutdown set off a political storm in the media. As Donald Cameron had predicted, the Nova Scotia government had too much at stake to stand back and watch the project die. Cameron trotted out his allegations of pro-Devco bias, and called on the federal government to bargain with Westray in good faith. Elmer MacKay joined Cameron in blaming the setback on a cabal of stubborn federal civil servants. "You have a very good mining operation led by Cliff Frame, trading opinions with a couple of bureaucrats who won't know a coal mine if they fell over one or fell into one," MacKay complained. "They are using a bit of bureaucratic licence and taking very hard-line positions on this."

Next to Cameron, MacKay was the politician with the most to lose if the project fizzled. Cameron had done his part at the provincial level, advancing the money needed to get Westray rolling. So it was up to MacKay, who now held the public works portfolio as well as being minister responsible for the Atlantic Canada Opportunities Agency, to make sure Ottawa came through. "I've dealt with Elmer

MacKay a long time, and I've known him all my life," Cameron told reporters. "He's not a wimp." MacKay was under the gun to make good on his repeated assurances of federal money for Westray, and it was not just Cameron who was putting on the pressure. Even the normally docile New Glasgow *Evening News* said MacKay deserved "a blast of reproach" from his constituents. MacKay promised to intervene personally with federal Industry Minister Harvie Andre and his other cabinet colleagues to get Westray going again.

Westray's critics in the Liberal party were quick to jump into the fray. Cape Breton MP David Dingwall predicted Westray would be a "devastating economic blow" to Cape Breton's mining industry. At the provincial level, MLA Bernie Boudreau accused Cameron of taking an "$8-million gamble" on his "pet project." If Cameron lost the bet and there was no federal deal, he said, a couple of holes in the ground would be all Nova Scotians had to show for their money. Cameron dismissed this as "silly politics," pointing out that the Westray loan was a drop in the bucket compared with the $1.3 billion Ottawa had lavished on Devco over the years. And there was no risk of the province losing its money, because Westray would go ahead. "As sure as the sun came up this morning," Cameron said, "we'll have a coal mine in Pictou County."

Then, one of Andre's aides acknowledged that Devco was not the only reason for the impasse. His press secretary, Jodi Redmond, was quoted in the *Chronicle-Herald* on August 10 as saying that "the rate of return to the promoter" was too high to justify the $60-million assistance package being sought. Cameron went ballistic. "This is typical of the kind of nonsense coming out of that department," he ranted. "Why do they give General Motors millions and millions in Ontario? That's far more than this mine would ever get." Besides a loan guarantee, he assured reporters, Westray was seeking a maximum of $26.7 million from Ottawa to reduce interest payments. Andre himself refused to be drawn into the debate, saying he did not want to negotiate "through the media."

Meanwhile, there were complaints within the coal industry about Curragh's effort to wring money out of the public purse. Dick Marshall, head of the Coal Association of Canada, said there was no precedent in Canada for direct government aid to a coal-mining venture. "A normal private-sector company would arrange its own financing and live or sink with it." That view was shared by Gary Livingstone, the president of Edmonton-based Luscar Coal. "We're very free entrepreneurs. We believe in viable free-standing operations that make economic sense." The last coal mine to attract government backing—other than Crown-owned Devco—was Quintette, Frame's previous foray into the coal business. Quintette, Faro, and now Westray—in less than a decade, Frame had rewritten the rules of government assistance to the mining industry. In the words of Nick Tintor, assistant editor of the

Northern Miner newspaper, the Curragh chairman was "a master of finding politically driven situations." And there was little doubt that Westray was a project driven by politics.

The public free-for-all over the work stoppage at Westray paid dividends, at least for Frame and Curragh. Almost overnight, the phantom bureaucrats whom Cameron and MacKay had dragged through the mud were warming up to the idea of a Pictou County mine. On August 8, Harry Rogers, the deputy minister of industry, science, and technology, was on the phone with Frame, suggesting revisions to the Westray funding application. Frame followed up a week later with a letter saying he was encouraged by the "cooperative attitude" that now prevailed within the department. According to Frame, the federal government had agreed to guarantee 85 per cent of the $100-million, fifteen-year bank loan needed to finance construction.

The only loose thread was the amount of money Ottawa would supply to reduce interest rates. Under the guidelines of the Atlantic Enterprise Program, the federal government would pay half the interest on the bank loan for seven years, up to a maximum of 6 per cent. That meant the level of federal assistance would be determined by the interest rate on the loan. At a low rate of 7 per cent, the interest-rate buydown for Westray could cost the government as little as $4 million. But if rates shot up, say, to 14 per cent, Ottawa could end up paying nearly seven times more–$26.7 million. Since interest rates were on the way down, Frame predicted the government subsidy was likely to be at the lower end of the scale. Whatever the interest rate, he was willing to "cap" Ottawa's contribution at $26.7 million, "in the interest of seeing this project go forward."

The year-long cold war between the bureaucrats and Curragh appeared to be entering a new era of detente. On August 15, the day Frame was making his latest pitch for funding, MacKay was confidently telling the media that Ottawa and the company were "fine-tuning" a final agreement. "It's important the project not get too much, but on the other hand it is equally important that the risk-taking element of this be borne in mind, too," MacKay explained. "Mining is not necessarily an enterprise you can always predict."

* * *

AS THE WAR OF WORDS raged in the media, Halifax *Chronicle-Herald* reporter Stewart Lewis was taking a closer look at how Curragh dealt with those risks. Back on May 9, 1989–three years to the day before the Westray mine exploded–the *Herald* published a story about Curragh's "atrocious" safety record at its Faro open-pit mine. Trevor Harding, a union official and a native of Shelburne, Nova Scotia, told Lewis there had been thirty-nine accidents at Faro in the previous month alone.

He also complained about the company's refusal to deal with a series of dust fires. "I hope the Nova Scotia government doesn't allow the lax health and safety we've encountered here," Harding, grievance chairman for the United Steelworkers of America, said in a telephone interview from the Yukon. "I hope the company puts health and safety as a priority right off the bat."

That story was buried on page 35. But Lewis followed up in mid-August with a story splashed on the front page of the *Herald*, under the headline: "Caution urged in working coal mine." The article delved into the Pictou coalfield's history of explosions and fires, and the Foord seam's reputation for giving off excessive amounts of methane. Former Devco president Derek Rance was quoted as saying that gas would be Westray's biggest challenge. "In order for that mine to be safe, everything is going to have to work right all of the time," Rance predicted. "The workforce is going to have to be really well trained and really observant. They're going to have to make sure their safety procedures are up to snuff and the ventilation is right on." Westray manager Gerald Phillips brushed aside the criticism as more Devco propaganda. The explosions and fires of the past had "more to do with the old mining methods," he told the *Herald*, parroting what mining consultants had been saying for years. With the $400,000 electronic monitoring system planned for Westray, Phillips said, "you can detect a problem before it becomes a real problem."

The safety warnings raised in Lewis's articles were drowned out by the political wrangling over whether Ottawa should assist the project. The sniping between MacKay, Cameron, and the Ottawa mandarins during the summer of 1989 upstaged concerns about the safety of a mine that might never be brought into production. "The safety of the proposed development has been questioned," a *Herald* editorial noted in mid-August, but its author said that issue was best left for "another day." It would be years before that day arrived, and safety at Westray once again captured the media's attention.

Out of the public eye, one woman who understood the dangers of mining Pictou County coal tried to make her voice heard over the din. On September 29, 1989, Jane MacKay of Oakville, Ontario, typed a two-page letter to Brian Mulroney, asking why his government was considering aid to Westray. She was worried not only about the project's environmental impact, but also about the safety of the miners. "The coal seams have extremely high methane gas content, which causes explosions," MacKay warned the prime minister. There were no longer trained miners in Pictou County, she noted, and Westray's imported workers would not know the area's sorry history of mining disasters and death. MacKay did—her grandfather was among those killed in Pictou County's mines.

The letter was passed on to Harvie Andre's office for reply. The industry minister wrote back two months later, thanking MacKay for her interest. Andre agreed that the Foord seam "does have a high methane content"—an odd concession, given that every consultant's report prepared since the early 1980s had been based on the premise that the gas content of the seam was low. But, Andre assured MacKay in the same breath, "modern mining and methane detection methods have significantly reduced mine hazards." Like Gerald Phillips, the federal government had complete faith in technology; the mistakes of the past would not be repeated at Westray's state-of-the-art colliery. But in 1989, the debate over safety was still academic. There would be no mine until Ottawa and Clifford Frame settled their protracted squabble over money. And there was still a lot of dickering ahead.

DOWNTIME

F CLIFFORD FRAME AND WESTRAY'S POLITICAL BACKERS managed to wrangle funding from Ottawa, there would be a major underground coal mine on the Nova Scotia mainland. And this posed a challenge for the government body in charge of enforcing mine safety laws—the Nova Scotia Department of Labour. Until 1986, mine inspection had been the realm of the Mines and Energy Department. But after the passage of a new provincial labour-safety law, the Occupational Health and Safety Act, the government's mine inspectors were transferred to the Labour Department. With a staff of eight, the inspection section was responsible for monitoring the operations of all underground and open-pit mines, as well as stone and gravel quarries. In a typical year, about four hundred inspections were carried out at sites scattered across the province.

The inspectors also enforced the Coal Mines Regulation Act, a thick statute consisting of some 150 provisions. The act regulated all aspects of underground coal-mining, from the qualifications of employees, to safety measures such as methane detection and stone dusting. Many of the sections originated in the late 1800s, when union lobbying and a spate of disasters forced the government to draw up safety regulations. The act in force as Westray came on stream still contained outdated references to hoisting ropes and steam engines, and using horses underground. The last major revision had been made in the late 1950s, based on the findings of an inquiry into the 1956 Springhill explosion. During the 1970s and 1980s, Labour Department officials took four or five runs at revising the act to reflect modern coal-mining methods, but none of the changes was ever translated into law. A new set of mining regulations was still in the draft stage when the Westray mine exploded.

Until Westray came along, there was no need for drastic changes to the Coal Mines Regulation Act. Most of Nova Scotia's coal was produced in Cape Breton by Devco, whose federally controlled collieries were inspected by officials of Labour Canada. Devco not only had its own inspectors, but also operated under a separate set of safety regulations, put in place by the federal government. After Westville's

Drummond mine closed in 1984, provincial mine inspectors were responsible for only one underground coal operation—Cape Breton's St. Rose colliery, operated by Evans Coal Mines. Evans was a marginal outfit, employing about forty miners to raise 30,000 tonnes of coal a year. The workforce at Westray would be six times larger, and produce that much coal every ten days. And Westray would be highly mechanized, using mobile coal-cutting machines that were new to Nova Scotia.

The Coal Mines Regulation Act was fine for an outdated operation like Evans but insufficient for a modern mine like Westray. The coal-dust provisions were a case in point. Requirements for watering or stone dusting to control dust in mines had been updated after the 1956 Springhill explosion, in which both methane and coal dust ignited. The new regulations were modelled after those being enforced by the United States Bureau of Mines—state-of-the-art statutes in the 1950s, but ill-suited to deal with use of the underground machinery and long lines of conveyors that characterized Westray's plans to mine the Foord seam. Continuous miners could dig coal faster than ever before; the by-product, of course, was more coal dust. And it was a different kind of dust. When the Nova Scotia regulations were enacted, the dust produced was the consistency of sand. The dust given off by modern machinery was more like talcum powder, so fine as to be almost invisible in the mine air.

By the time Westray came on stream in the late 1980s, Nova Scotia's Labour Department was still armed with antiquated mining laws. To make matters worse, the mine inspection unit was understaffed. There was talk within the department about hiring another inspector to deal exclusively with Westray, but no-one had been hired by the time Curragh Resources broke ground at Plymouth in the late fall of 1988. In October of that year, officials of the mine safety division met with Westray manager Gerald Phillips to get an overview of the project. Westray seemed to think the project would not be classified as a coal mine until it actually struck coal, but the province's director of mine safety, Claude White, disagreed. Even though the main tunnels would be driven hundreds of metres through rock, White told Phillips in a November 25 letter, the project would be subject to the Coal Mines Regulation Act. White's main concern was the potential for encountering methane as the tunnels passed through shallower seams. The department would consider waiving some regulations during the tunnelling phase, he said, "provided that certain precautions are taken and that methane gas does not become an undue problem." Privately, White felt Westray was not taking the gas problem seriously. His notebook for the period was filled with cryptic references to methane emissions and Pictou County's reputation for having gassy seams. "Lots of gas," White jotted down on November 29, 1988. "Why is G. Phillips . . . so complacent?"

A week later, Phillips assured White that the tunnelling work would comply with the coal mines act. "Safety precautions will be taken to ensure that the conditions underground are safe and monitored continuously," he wrote on December 5. And Phillips wanted to make sure the department knew who was boss. "I would like to stress that with my experience in coal mining, I fully recognize the potential hazards in all aspects of coal mining," he noted, barely hiding his annoyance. Phillips also stressed his extensive experience with the Pictou mine project; while with Suncor, Phillips had worked out a set of exemptions in 1986 with Walter Fell, White's predecessor as chief inspector. Westray, Phillips told White, would be seeking the same privileges.

Westray sought three written exemptions from the Coal Mines Regulation Act: the company needed to use dynamite to blast out the tunnels, planned to use electrical equipment not normally intended for coal mines, and wanted to substitute hard-rock miners for the trained coal-miners required under the act. White granted the request on February 17, 1989, but outlined strict conditions for monitoring methane and carbon monoxide levels. Labour officials also took a hard line on the key positions of mine examiners—employees with at least three years of experience in underground coal mining, who are responsible for gas testing and ensuring that safety rules were followed on their shifts. Phillips wanted to appoint "competent hard rock miners" to the posts, but the department refused to budge. At the end of the letter, White made a point of reminding Phillips that once Westray's tunnels reached coal, the exemptions would expire.

Phillips, whose mine manager's certification qualified him to be a mine examiner, ended up taking on gas-testing duties himself. White and John V. Smith, the department's Stellarton-based inspector of underground equipment, underscored the methane risk during a site visit at the end of March. "It was stressed that the importance of not operating any of the mobile equipment in the presence of methane was paramount," Smith noted later. The methane hazard was to be explained to all underground workers, Phillips was told, and gas-testing procedures set out in the department's February 17 letter were to be followed "religiously." In Smith's opinion, the mine inspectors had their work cut out for them. "It will be necessary to keep a close watch on this project," he confided in a memo to White, "to ensure that any safety related problems are dealt with promptly."

* * *

WESTRAY HAD ONE MORE favour to ask of the Nova Scotia government before tunnelling began, this one directed to the Department of Mines and Energy. Although the Labour Department was now in charge of on-the-job safety, Mines and Energy still approved mine plans and underground layouts before issuing a permit to a mine. In February 1989, Westray applied to Mines and Energy for permission to redirect the twin access tunnels about to be blasted towards the Foord seam. Back when Suncor was in charge of the project, the tunnels had been envisioned as sloping downward to the coal deposit in a northeasterly direction, then turning due north to enter the seam. As a precaution, Suncor had drilled nine exploratory holes to obtain core samples and make sure the route was free of major faults. But Phillips wanted to realign the tunnels about twenty degrees, so they would run in more of an easterly direction. That would increase the angle of the left turn into the Foord seam, leaving a configuration that resembled a sharp dogleg on a golf course. The new route shortened the distance to the seam, allowing Westray faster access to the area of coal known as the southwest block.

The changes did not sit well with Donald S. Jones, manager of mining engineering at Mines and Energy. "Our review has shown that the new tunnel alignment will generate conditions that are substantially different than were anticipated in the original tunnel alignment," he informed Phillips by letter on March 17. Jones foresaw two major problems. First, rock formations along the original route had been thoroughly examined, while only two test holes had been drilled in the area where Westray proposed to sink its tunnels. In "a complex geological environment such as Pictou," he noted, information gleaned from existing core samples could not be used to predict rock conditions along the new route.

There was enough data to predict the second problem—Westray's new tunnels were headed straight for an area crisscrossed by faults. Using geological findings incorporated into Westray's own feasibility studies, Jones produced a map showing where each fault would be intersected. Phillips's revised route would pass through four known faults, while the original heading would have encountered only one. The additional faults could create "extensive zones of . . . bad ground" along the main tunnels, Jones warned, and this would make roof support difficult. The areas of fractured rock and the lack of exploratory drilling, he told Phillips, "should be investigated and discussed with our department prior to approval of the revised plan."

But there was little time for Phillips to go back and rethink his plans. Westray was on a tight construction schedule to meet the 1991 deadline for supplying the new Trenton power plant. According to the timetable established by Westray's consultants, tunnel development should have begun in November 1988. By March 1989, thanks to the $8-million loan from Donald Cameron's industry department, the company was ready to proceed. To ease the last-minute concerns of officials at Mines and Energy, Westray called in Associated Mining Consultants Ltd. to check the new route; the consultants predicted underground conditions would be "substantially the same" as along the old alignment.

That was good enough for the department. Director of Mining Engineering Patrick Phalen, Jones's boss, was satisfied the new route would permit "easier and safer tunnel development" and provide "an improved point of access to the coal resource." Westray was given verbal approval to begin construction, and not a moment too soon. On April 10, 1989, little more than three weeks after Jones put his concerns in writing, Cameron arrived in Plymouth to set off the ceremonial blast marking the start of tunnelling. By the end of May, when Mines and Energy Minister Jack MacIsaac formally approved the realignment, seventy-five metres of tunnel had already been blasted along the revised route. But once Westray's tunnelling crews began burrowing through the faults in their path, Phalen and other officials at Mines and Energy would come to regret their hasty decision.

* * *

BY SEPTEMBER 1989, just over a month after work ceased at the Westray site, Ottawa was in a mood to bargain. In response to Clifford Frame's overture in mid-August, Industry, Science and Technology Canada put an offer on the table. Industry Minister Harvie Andre refused to divulge the details, but sources told the *Chronicle-Herald* that the package consisted of a loan guarantee and an interest-rate subsidy worth $8.75 million. That was a far cry from the more than $60 million Curragh had sought two years earlier. Curragh declined to discuss the details in public, but in Elmer MacKay's opinion, the offer was "appropriate." The news did not appease Donald Cameron, who vowed to have a few words with Andre about Ottawa's handling of the whole affair. In the meantime, the provincial government gave the company some breathing space by approving a two-month extension of the September 30 deadline for repaying the $8-million loan. Several more extensions would be granted while federal funding remained up in the air.

As usual, the public statements of Westray's Tory backers proved optimistic. On November 2, an infuriated Cameron revealed that Ottawa was still demanding guaranteed sales of Cape Breton coal to the Nova Scotia Power Corporation. The federal government, he charged, was using Westray's funding as a bargaining chip to boost Devco's sales. "As polite as I can be, that's blackmail," he grumbled to a scrum of reporters, as he left a cabinet meeting in Halifax. "And we're not going to give in to that." Then Cameron tried his own form of blackmail, vowing that Devco would not profit from Westray's misfortunes. If the Pictou mine did not go ahead, NSPC would import low-sulphur coal from outside the province for its Trenton plants.

In Ottawa, Andre denied he was blackmailing the Nova Scotia government. But he made it clear that money for Westray hinged on protection for Devco. "One of the things we want to ensure is that assistance we provide to Westray doesn't indirectly damage Devco by cutting into Devco's potential markets," he said, when confronted with Cameron's comment. "Once that is done, we will finish the deal." Small

Business Minister Tom Hockin, the federal minister responsible for Devco, backed Andre's stance. It was an ugly spectacle–two Tory governments publicly feuding over a coal mine that Nova Scotia desperately wanted, but Ottawa seemed content to live without. John Buchanan stepped in to play peacemaker. After meeting with Andre in mid-November at a first ministers conference in Ottawa, the premier was confident the Devco issue had finally been laid to rest. Within a month, Andre announced that a final deal was "in the typewriter."

Bernie Boudreau, a rookie member of Nova Scotia's Liberal opposition, emerged as Westray's most vocal critic. Financial and safety concerns, he argued, warranted "a professional, independent assessment" of the project. [Courtesy of the *Chronicle-Herald* and the *Mail-Star*]

Amidst the Tories' political infighting, an opinion piece by Liberal MLA Bernie Boudreau appeared in the *Chronicle-Herald* on November 16. Boudreau, who had become Westray's harshest critic, argued that politics and regional rivalry were obscuring far more serious questions. "My information leads me to believe that the mine is a questionable proposition technically and may even be unsafe because of gas and geological configuration." he wrote. If Curragh Resources could not bring Westray into production "safely or in a timely fashion [or] if there are cost overruns, or incompetent project administration, or technical difficulties, we will be unable to remove ourselves from the project in mid-stream."

It was time, Boudreau said, for "a professional, independent assessment," free of political interference, "so we may get a thorough and unbiased review of the technical and safety aspect[s] of this project." A handful of government geologists and engineers had been saying the same thing for more than a year–the project required more study and exploration. And Boudreau's plea, like their advice, was cast aside in the headlong rush to get Westray into production. In early December, Cameron complained that Ottawa's foot-dragging could prevent Westray from meeting the fall 1991 target for supplying Trenton 6. "They say that if they don't have a lot of problems, they might be able to meet that yet," he said. "But it's going to be difficult."

* * *

Labour Department inspectors kept tabs on Westray during the shutdown. The dozen or so employees still on the payroll were busy shotcreting completed sections of the tunnels and erecting a steel roof over the entrance at the surface. John V. Smith and another mine inspector, Albert McLean, toured the mine at least ten times between April 1989 and the summer of 1990. Smith and McLean knew their business–they had spent their entire careers in and around coal mines. Smith, a product of Britain's National Coal Board training system, had immigrated to Nova Scotia in the early 1960s to take a job as a mine inspector. McLean had been hired in 1978, after working for almost twenty years in Cape Breton collieries. There was a minor roof collapse in one of Westray's main tunnels in July 1989, around the time of the shutdown, but otherwise the inspectors found everything in order. Despite the lull in construction, little exploration work was carried out. Westray spent four thousand dollars to drill a single, shallow hole in July 1989 to test rock conditions along its new tunnel alignment.

The downtime gave Westray a chance to formalize its training program. The company was under pressure from the Nova Scotia government to hire locally, but faced a shortage of experienced coal miners; of approximately seven hundred people in the Pictou area who had submitted job applications by the summer of 1990, only twenty-four were certified to work as coal miners. Westray planned to import as many as fifty experienced miners from Alberta and overseas, and use that core group to train local people to work underground. In July 1990, Phillips submitted a proposal to the Labour Department that envisioned a combination of classroom instruction and on-the-job training. The coursework included one day of instruction on the hazards of mine gases such as methane. A single day was also devoted to fires and explosions, including coal-dust explosions and how to spread stone dust.

The Coal Mines Regulation Act sets out in detail the duties and training requirements for various categories of employees–from the miner at the face to the manager on the surface. Under the act, so-called "certificates of competency" for various posts are issued by a provincially appointed body, the board of examiners. Chief mine inspector Claude White and Patrick Phalen, the province's director of mining engineering, were among the members. Board members reviewed Westray's training plan in August, and were unimpressed. Westray proposed to streamline the certification process by establishing its own three-member board of examiners, consisting of a provincial mine inspector, Phillips, and a certified miner, presumably one employed at Westray. The board's response–thanks, but no thanks. Responsibility for certification testing of coal miners and rescue workers was delegated to Labour

Department inspectors, but the department's board of examiners would continue to examine those seeking certification for other underground posts. And, Phillips was told, the board reserved the right to monitor how Westray trained its workers.

* * *

THE WHOLE TIME Clifford Frame was begging the federal government for cash to get Westray going, Curragh Resources was raking in huge profits from its Faro mine. The company made $61.6 million in 1988, and almost an identical profit—$60.7 million—in 1989. "I could have made a piss-pot and walked away," as Frame later put it, in his colourful way. But his goal was not simply to be rich—he already was, anyway. No, Clifford Frame wanted to build Curragh Resources into a mining empire, just as his mentor Stephen Roman had done with Denison Mines. So profits were ploughed back into the company. The only exception was a $10-million dividend paid in 1989, split almost fifty-fifty with Giant Resources, the Australian company that owned almost half of Curragh. Roughly $2.5 million went to Curragh's largest shareholder, Clifford Frame.

The Australian connection, forged in 1987, had been part of Frame's initial expansion drive. But after a couple of failed bids to buy other mining firms, and after a fair amount of internal bickering, Giant and Curragh began to go their separate ways. One venture that enticed Frame, but not the Aussies, was the Pictou County coal project. But that acquisition was peanuts compared to the $140 million Curragh plunked down in May 1989 to buy one-fifth interest in Asturiana de Zinc, a Spanish firm that operated one of the largest zinc smelters in the world. Under the deal, ADZ, as the company was known, poured $61 million back into Curragh and became part owner of one of the company's zinc deposits. Curragh gained a foothold in the European Community and guaranteed sales to ADZ, providing some relief from the whims of the open market. Every major zinc producer operated its own smelters or had similar arrangements, Frame noted at the time, and now he was "part of the club."

Frame was already gearing up for the next step towards joining the big boys—taking Curragh public. Five million shares were put on the market in Canada and the United States in May 1990, at a price of $11.87 per share. Curragh aimed to raise about $60 million through the offering, enough to retire leftover debt from the ADZ deal and untangle itself from Giant, which had slipped into receivership. The share structure was heavily weighted in Frame's favour. He claimed title to a block of shares worth close to $100 million at the issue price; in return, he kept in the company $6 million he was entitled to take. Frame's shares also carried higher than usual voting power, allowing him to retain full control over appointments to the board of directors and all other aspects of Curragh's operations. The new shareholders would have to be content with collecting dividends while Frame called the shots.

The switch from private to public status required a reorganization of Curragh's stable of subsidiaries and affiliated firms. Among them was a Frame-owned outfit called C.H. Frame Consulting Services, which had been the nominal employer of Curragh's executive team. Under the restructuring, Frame Consulting was taken over and the executives officially became Curragh employees. For his trouble, Frame received $1.5 million from Curragh's coffers. Frame Consulting had also done work on the Westray project, and stood to collect $6 million in management fees when the mine was completed. Curragh absorbed those fees when it bought the company, and that arrangement raised a few eyebrows in Ottawa, because Curragh counted the deferred fees as part of its equity in the project; even with that $6-million boost, federal bureaucrats viewed the company's $15-million investment as too modest.

Westray itself was caught up in the reorganization. In December 1989, just before going public, Curragh assumed full ownership of the project. Boliden International Mining of Sweden, which marketed Curragh's lead and zinc overseas, converted a minority interest in the project into a 20-per-cent royalty, payable once the mine was in production. Until then, Curragh's strategy had been to keep Westray at arm's length, and let the project sink or swim depending on the level of assistance anticipated from Ottawa. With the federal government offering a stingy $8.75 million to ease interest rates, the takeover gave Westray the financial clout it would need to swing a bank loan. It also gave Westray deeper pockets to deal with any cost overruns that cropped up during the construction phase.

By 1990, Curragh was ready to take on two new mines, at opposite ends of the country. One was Westray; the other was Sä Dena Hes, an open-pit lead and zinc operation in the Yukon, near the British Columbia border. Another lead and zinc property, Stronsay, in northern B.C., was nearing the development stage. And Curragh was also looking at tapping deposits beside its flagship Faro mine, which was running out of ore. All the while, Frame's dreams grew bigger and bigger. He boasted publicly of having the resources and the Wall Street connections to finance a half-billion-dollar spending spree to bring more properties into the Curragh fold. But Frame got a chilly reception on the stock market. In a little over six months, the share value dropped more than one-third, to the $7 range. By the end of the year, the market had taken a $40-million bite out of Frame's $100-million windfall.

Investors had reason to be wary. For all his bravado and ingratiating charm, Frame could not shake his chequered past or his reputation as a promoter who thrived on government handouts. Frame and his team of executives were outsiders in Toronto's tight-knit mining community, noticeably absent from the St. Patrick's Day party at the Engineer's Club of Toronto and the annual convention for prospectors and developers, held at the Royal York Hotel. In the press, brokers and mining analysts invariably used the same words to describe Frame—"ambitious," "aggressive," "a bit of a maverick." Not to mention "flamboyant." One Bay Street money man described a memorable encounter with Frame for an article in *Report on Business Magazine*. In the middle of a meeting to discuss the 1990 share offering, a woman came in to present the chairman with a cigar and a generous portion of single-malt Scotch. It was barely ten o'clock in the morning.

Frame's refined tastes and big talk aside, Curragh had all the trappings of a big-time mining company. The company's ornately furnished offices, complete with full kitchen and private dining room, occupied the entire nineteenth floor of a downtown Toronto skyscraper, and rented for almost one million dollars a year. When not at the wheel of one of his Jags, the chairman was chauffeured around town in the company's sleek limousine, a battleship-grey stretch Lincoln. But looks were deceiving. Even Frame's desk was not what it seemed; the leather-topped monster that spanned one end of his spacious office may have looked every inch an antique, but it was a replica. Image and style alone could not change the reality that Curragh was a modest operation, rich in promise but reliant on the ups and downs of the base-metal market. To break into mining's major leagues, Curragh needed to diversify. Coal was a step in the right direction. And Westray, with its lucrative long-term contract and minimal drain on Curragh's finances, looked like Frame's next cash cow.

* * *

THE WESTRAY PROJECT cleared a major hurdle during the first month of 1990. An independent environmental assessment ordered by the federal government concluded that side effects of the project—noise, water pollution, surface damage above the workings—would be manageable. The downside, according to the consulting firm Acres International, was the potential impact on employment at Cape Breton's mines. The report also cast doubt on the environmental rationale for Westray—that its low-sulphur coal was vital to curbing acid rain. Devco, noted Acres, claimed it could meet all the coal needs of the Nova Scotia Power Corporation and adhere to sulphur-dioxide emission guidelines. Gerald Phillips was quick to condemn the report for giving short shrift to the benefits of low-sulphur coal. "If people sit down to look at the real facts, it's crazy not to develop the mine as quickly as possible," he told the New Glasgow *Evening News*. "No-one with any common sense would say no to it."

But to Bernie Boudreau, common sense dictated that the project be scrapped. When the Nova Scotia House of Assembly reconvened in the spring of 1990, the Liberal MLA renewed his attack on Cameron and Westray. Waving around documents filed in the United States as part of Curragh's bid to raise money on the stock market, Boudreau asked why a company so flush with cash could not pay back the $8-million advance on the provincial loan. He seized on the $6-million management fee tied to Westray, expressing outrage that Clifford Frame stood to pocket a "bonus" once the mine was completed.

And Boudreau once again played the safety card. "Serious technical and safety questions have never been adequately canvassed on this project," he told the legislature on March 22, repeating his warning of the previous November. "These questions include the gaseous nature of the coal formations, the accommodation of the fault structure, the method and the depth of the mining process, to name just a few." Sidestepping the safety issue, Cameron accused Boudreau of spouting "political nonsense" about the project's finances, and took delight in the fact that some Liberals, including provincial party leader Vince MacLean, were speaking in favour of the mine.

The federal funding offer made to Curragh in the fall of 1989 was still awaiting approval from the Treasury Board in Ottawa. Westray and Curragh had the inside track on the deal's progress. Almost every day, Cameron pestered Harry Rogers at the federal industry department for updates, then passed on his findings to Phillips. Phillips, in turn, fed a steady supply of memos to head office, outlining Cameron's assessment of the situation. The memos, some of which fell into the hands of the CBC-TV news program *the fifth estate*, suggested the provincial industry minister was growing increasingly frustrated with Ottawa's stalling. On March 29, 1990, for instance, Phillips described "Don" as "absolutely disgusted" that funding would not be approved that week.

So disgusted, Phillips said, that he was talking about cutting a new provincial deal to give Curragh its long-sought loan guarantee. "[Cameron] says he has enough support among his cabinet colleagues to finalize a provincial agreement," Phillips reported on April 2. "He says the project has been shut down long enough and he will do whatever it takes to have a quick start-up of the project." Phillips played hardball, insisting on the same 85-per-cent guarantee as Ottawa was considering. "Don wants to know if we can go with a smaller amount of loan guarantee," he told his superiors. "I told him that I am not sure, but if we have to give something up in one area we will expect to gain something in another area."

Regardless of what Phillips was telling his bosses, Cameron now says there was no move afoot to give more provincial aid to Westray. "We weren't going to get in any deeper. We didn't think we had to," he maintains. "They asked us to do more on many occasions; they asked us to do more on interest rates; they asked us to do more on the length of [re]payment. We said no. We told them we wanted the mine for sure, but we weren't going to sell the taxpayers out." John

Buchanan was no tightwad when it came to public money, but even the premier felt his government had done its share to help Westray. "There was talk about the province going further, providing much more," Buchanan recalls. In his view, the $12-million loan and guaranteed coal sales already in place were "a little more than maybe we should have [done] at the time. We certainly weren't prepared to go any further than that."

Westray's need for more provincial aid vanished within a month. Ottawa finally made a formal offer of assistance in early May 1990. The package was along the lines worked out the previous fall. Ottawa would back 85 per cent of the $100-million loan needed to complete construction. If the project shut down for any reason, taxpayers would have to pay Curragh's bankers up to $85 million to honour the partial guarantee. Curragh would be responsible for repaying the remaining $15 million. As security for the guarantee, the federal government had first call on the mine's assets, insurance policies, and mineral rights. Nova Scotia was relegated to the back seat; Ottawa had the right to recover all its money before the province received one cent of its $12-million loan. The offer also called for Ottawa to subsidize the interest rate on the bank loan with an outright grant, capped at $8.75 million over four years.

The deal came with several conditions. Curragh would have to comply with the recommendations of the environmental assessment. The loan would have to be repaid at an annual rate of at least $6.7 million, starting the last day of 1992. The first $20 million in surplus cash from Westray would have to go towards paying down the loan. And Curragh would have to maintain equity of $27 million in the project. But the term "equity" was defined loosely; Ottawa allowed Curragh to count the $12-million provincial loan and the $6 million in management fees from Frame Consulting as part of the $27-million requirement. In other words, Curragh would only have to shell out $9 million of its own money for the Westray project. With cost estimates hovering near the $140-million mark, that worked out to about one-fifteenth of the money needed to bring the mine into production.

It was a good deal, at least for Curragh. Clifford Frame had convinced the federal government to assume most of the risk; if the project went under, his own financial exposure on the loan was limited to $15 million. He had also achieved his goal of developing the mine with little drain on the Curragh treasury. The only disappointment was the interest-rate subsidy, which was far below the initial $63.9-million request. But the company was willing to settle for $8.75 million. Curragh's board of directors approved the deal in mid-July and promised to resume construction as soon as the bank loan was negotiated. "We're very satisfied with the federal package," Curragh's vice-chairman, George Whyte, told one newspaper. "Sometimes you ask for more than what is actually needed."

Curragh may have felt that Ottawa was driving a hard bargain, but that was not the opinion of the federal government's financial watchdog. The auditor general of Canada, Denis Desautels, reviewed all federal loan-guarantee programs in 1992 and found that the Mulroney government was ignoring its own guidelines for such assistance. Ottawa was shouldering too much of the risk, while bankers and project developers reaped most the gains. The Westray project was the worst offender. After reviewing the federal government's books, Desautels concluded that public assistance exceeded the projected cost of developing the mine. Based on a cost estimate of $130 million, the auditor general determined that the federal and provincial governments, through subsidies, loans, guarantees, and tax breaks, covered 103 per cent of start-up costs. Even if Curragh's own cost figure of $140 million were used, government money still accounted for virtually all Westray's financing. Little wonder Curragh was satisfied with the final federal deal.

But for the efforts of the bureaucrats at the Department of Industry, Science, and Technology, Curragh would have dipped even deeper into the public purse. In that sense, the two-and-a-half-year funding battle was a partial victory for the public interest. And it was won even though the political odds were stacked in the company's favour. Westray's detractors within the federal civil service had to endure public censure at the hands of Elmer MacKay and Donald Cameron. Behind the scenes, their dealings with Curragh were monitored—and at times usurped—by Prime Minister Brian Mulroney's closest advisers. While there was no evidence that Peter White and Stanley Hartt sided with Curragh in the negotiations, they did help clear many of the roadblocks holding up a funding deal. And their intervention sent a clear message to the bureaucrats: whatever its shortcomings, Westray was a project near and dear to the heart of the prime minister. And that may well have been the deciding factor in Ottawa's decision to come on board.

* * *

THE WEB OF FINANCIAL and legal agreements required to finalize the Westray deal produced a stack of documents several feet thick. Among these was a formal contract committing the Nova Scotia government to buy 275,000 tonnes of coal annually, on top of Westray's contract with the provincial utility. That was the amount of additional coal the province had agreed to buy in 1988, to appease the bureaucrats in Ottawa. At the end of August 1990, cabinet authorized the industry department to enter into a so-called take-or-pay contract for the additional coal. The provincial Crown corporation Novaco, which was now out of the coal business, was designated as purchaser.

On paper, it sounded like the "sweetheart deal" the opposition critics later made it out to be. Stripped to its essentials, the contract committed the government to pay $74 per tonne—the price NSPC was paying for Westray's premium coal—for up to 275,000 tonnes of coal each year, if the company was unable to find a buyer. That worked out to some $20 million per year over fifteen years. Westray, however, was required to beat the bushes to sell coal—to New Brunswick Power, utilities in New England, wherever—before coming cap in hand to the government. If the company failed to find any customers, the government had the option of buying any unsold coal, up to a maximum of 275,000 tonnes per year—that is, "take" it—or "pay" for it anyway, letting Westray leave it in the ground for future use. At the end of the contract's term, about the year 2008, Westray would have to pay back, interest-free, all money advanced for coal the government did not want.

The Liberals wasted no time attacking the deal once it became public knowledge in late 1990. Boudreau claimed it was tantamount to giving Clifford Frame "a blank cheque" and "a direct pipeline to our treasury." Westray, he charged, would enjoy an interest-free loan in the tens of millions of dollars for not digging coal that belonged to the people of Nova Scotia in the first place. Cameron said the government would never be called on to honour the contract, because Westray would have no problem selling all the coal it produced. And Nova Scotia Power would need an additional 100,000 tonnes per year at its Trenton power plants by the mid-1990s, when a contract with Pioneer's strip mine in Westville expired. Cameron remains adamant on that point. "Do you really believe there was some risk that they couldn't sell low-sulphur coal in this environmental climate?" he asked in a 1994 interview. "Curragh was told on many occasions that there was no way that the government would ever give them one red cent" under the take-or-pay deal. "It was to help put their financing in place. Period."

With the federal guarantee and the take-or-pay contract in its pocket, Curragh had no trouble securing a $100-million loan from the Bank of Nova Scotia. As part of the financial arrangements, the bank outlined the risks in a September 14, 1990, letter to the federal government. According to Robert Boomhour, the bank's assistant general manager, they were "typical of mining projects in general." Westray workers might unionize, raising the possibility that strikes could disrupt production; the fact that unions played a crucial role in making coal mines safe was ignored. Geological conditions—the depth and steep angle of the seam, "complex faulting," and varying seam thickness—"may adversely affect production costs." The letter also listed the project's strengths, including the coal's low sulphur content and the "significant coal mining experience" of Curragh's executives. The bank was apparently unaware that, with the exception of Westray manager Gerald Phillips, their experience was limited to mining surface coal.

Boomhour's letter dealt exclusively with financial risks. There was no mention of methane, or whether the geological conditions he described would put lives—not simply dollars—in danger. It was typical of the approach to the Westray project. Clifford Frame confronted the politicians with ominous talk of fires and explosions and excess gas, but that was merely a tactic to justify his inflated request for government backing. Federal officials were also preoccupied with Westray's finances and bottom line; Ottawa commissioned two reports on the project, yet ignored recommendations for an in-depth technical study. The only politician raising the red flag about safety was Bernie Boudreau, but even he tended to focus most of his attention on the financial implications.

The protracted battle between Curragh and Ottawa left the Westray project many months behind schedule. Tunnelling resumed in October 1990, after a fifteen-month shutdown. The access tunnels, which should have been ready by the summer of 1990, were less than one-fifth complete. Coal storage silos, the wash plant, a rail bridge over the East River, and other surface facilities remained to be built. A few kilometres away in Trenton, meanwhile, the Nova Scotia Power Corporation's new power plant was on schedule to open in the fall of 1991, and it would need Westray coal.

That gave Gerald Phillips a year to complete a mine that was supposed to take about three years to develop, according to the timetable set out by Curragh's consultants. It was a daunting task, but the Westray manager had an edge; his decision in February 1989 to realign the main slopes meant tunnelling crews would break into the Foord seam by the spring of 1991. A reporter for the *Northern Miner Magazine*, who toured the Westray site in mid-1990, predicted the last-minute design change would "save Curragh's bacon." Phillips confidently told the magazine he could meet the terms of the NSPC contract, delivering 300,000 tonnes of coal to Trenton by the end of 1991—no sweat. "You only have one chance to make a good impression," he said. "And we plan to do that by bringing the mine in on schedule and on budget." A lot of people were crossing their fingers and hoping Phillips could pull it off.

Westray manager Gerald Phillips in front of the wash plant building under construction in early 1991. "You only have one chance to make a good impression. And we plan to do that by bringing the mine in on schedule and on budget." [Photo by Steve Harder, courtesy of the *Chronicle-Herald* and the *Mail-Star*]

RECIPE FOR DISASTER

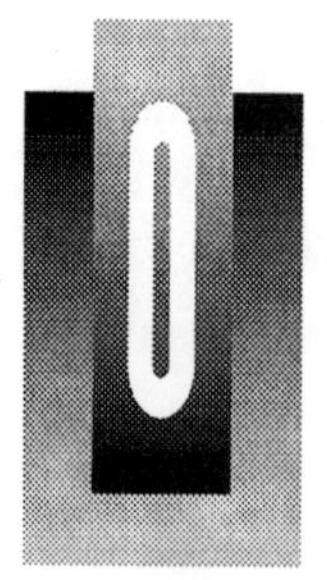

N A JUNE DAY IN 1990, a former high-level Nova Scotia bureaucrat named Michael Zareski calmly walked into Province House in Halifax and took his seat before an all-party committee of the Nova Scotia legislature. His testimony over the next hour or so set off a chain reaction that toppled a premier and changed the course of Nova Scotia's political history. And the unlikely beneficiary of Zareski's decision to blow the whistle on the Buchanan government was the province's industry minister, Donald Cameron.

Zareski had been the wunderkind of the provincial civil service, rising through the ranks to the powerful post of deputy minister in the Department of Government Services while still in his forties. Government Services was the government's purchasing arm, responsible for everything from buying toilet paper to signing multimillion-dollar leases on office buildings. That gave Zareski a front-row seat as the Conservative government's cronies lined up for contracts, jobs, and other favours. Zareski ended up being fired, partly because some of his colleagues in government seemed to feel that his increasing fascination with mysticism somehow raised questions about his mental stability. But the biggest problem for the government had been Zareski's refusal to play the political game. The invitation to testify before the public committee gave Zareski a chance to get even. As opposition politicians peppered him with questions and reporters furiously scribbled notes, Zareski rattled off a series of allegations of government patronage and corruption, some leading straight to the premier's office.

The Buchanan government tried to discredit Zareski, going so far as to question his sanity, but it could not control the political damage. The media pounced on Zareski's allegations and dug out other instances of patronage and shady deals. The RCMP launched an investigation. John Buchanan, whose ability for steering clear of previous scandals had earned him the nickname "Teflon John," was forced to run for cover. In September 1990, as the final paperwork for the Westray deal was being drawn up, Buchanan jumped at Brian Mulroney's offer of a Senate seat. After twelve years under Buchanan's rule, Nova Scotia was in the market for a new premier.

Cameron was a leading contender for the job. At the Tory leadership convention set for early February, Cameron faced two serious challengers. One was Roland Thornhill, a major force within Tory ranks, despite the RCMP investigation into his finances that had forced him out of cabinet. The other was Tom McInnis, the attorney general, who had been making headlines by implementing reforms to the justice system. In the wake of the Zareski scandal, reform was the buzz word in the leadership race. Cameron's campaign was based on a pledge to eradicate patronage, but no-one could accuse him of jumping on any bandwagons; he had been an outspoken critic of the patronage system since the 1970s. A Cameron government, he promised, would clean up Nova Scotia politics.

But the candidate from Pictou East had an Achilles heel. For all his talk about a new approach to politics, Cameron's coveted Westray mine looked to his critics like the product of old-style politics. The federal and Nova Scotia governments were putting millions of dollars of taxpayers' money on the line to bring jobs to Cameron's constituency—and Cameron had helped bring about the deals that were putting Pictou County back in the coal business. The would-be premier took a different view, maintaining Westray would be good for Nova Scotia and good for the environment, at minimal cost to the provincial treasury. The mine was "an important element in our overall energy strategy," he declared during the campaign, creating "a win-win situation for all Nova Scotians." Cameron's opponents, however, questioned how much time and effort he would have invested in the project had it not been located in his own back yard.

On the night of December 11, 1990, the Westray deal came under scrutiny on national television. CBC's *the fifth estate* aired a documentary by reporter Linden MacIntyre that delved into the political manoeuvring behind the mine's development. There was a passing reference to the history of fire and explosion in Pictou's mines, but the report focused on the public assistance lavished on the mine's promoter, Clifford Frame. The centrepiece was the take-or-pay contract, described as an "extraordinary deal" that promised to put tens of millions of dollars into Frame's pocket.

But when the newspapers hit the street the next morning, the headlines focused on another aspect of the CBC program. Curragh had given a Pictou-area company, Satellite Construction, an untendered $5-million contract for construction of a rail bridge and spur line to the Westray site. Satellite was owned by Eric Barker, one of Cameron's close friends and supporters, who had donated $3,500 to Cameron's leadership campaign. The documentary left the impression that Cameron had somehow steered the business to Barker. Frame saw nothing wrong with the appearance of favouring Cameron's friends. "I'd give a contract to the devil if he could do the job cheaper than somebody else," he said in an interview with MacIntyre.

Cameron, who was running third in the polls with two months to go in the leadership race, met the allegation head-on. He convened a press conference in Halifax to attack the CBC's "hatchet job" and defend the Westray deal. Cameron made no attempt to hide his anger as he lashed out at the "sleazebags" he claimed were using the media to drive him from the leadership race. Cameron denied doing anything to land the contract for Barker. The bridge contract had to be in place before the Bank of Nova Scotia granted its loan, he explained, and there had been no time for a tender call. Satellite had won three other Westray contracts put out to tender, he noted, in each case by submitting the lowest bid. These contracts included about $190,000 to excavate the site and supply cement.

For all his denunciation and denials, Cameron appeared ready to throw in the towel. "I'd be lying to you if I said it hasn't crossed my mind whether it's really worth trying to change the system, whether or not people really care," he admitted at the press conference. In the end, Cameron chose to hang tough. On February 9, he squeaked out a victory over Thornhill to win the Tory leadership. In less than two years, he had catapulted from the obscurity of the backbenches to the pinnacle of power. Although Cameron was the choice of the Conservative party, he still had to win the endorsement of the Nova Scotia electorate. He had plenty of time to launch his promised reforms and put his stamp on the government; an election did not have to be called until the fall of 1993. By then, people would have forgotten all about the bad old days of the Buchanan government. And the Westray mine would be in full production, silencing Cameron's critics once and for all.

* * *

AS TUNNELLING RESUMED at Westray in the fall of 1990, Department of Labour inspectors stepped up their visits to the site. On December 12, mine inspector Albert McLean gave the mine a clean bill of health, but issued an order under the Coal Mines Regulation Act requiring the use of a certified mine examiner to test methane levels. McLean's own tests showed no methane, but the department remained concerned about the potential of blasting in the presence of gas. The main tunnels, McLean noted, were passing through thin bands of coal as they neared the Foord seam.

At the same time, the department had set the ground rules for the use of diesel-powered vehicles in the mine. Legislation dealt exclusivley with rail-mounted diesel equipment, the only kind in vogue when the act was last revised, in the 1950s. Coal mining had changed dramatically in thirty years, and tractors were in common use in American collieries. Westray wanted to use a variety of rubber-tired machines to ferry men and materials in and out of the mine—front-end loaders, utility trucks, converted farm tractors. Not all the machines were equipped with the flame-proof engines usually required for underground use. Besides the fire hazard they would create, the machines would pollute the mine air with carbon monoxide. The Department of Labour had to start from scratch to devise rules for use of the vehicles; a series of strict conditions was imposed. Tractors could be used on a trial basis, for one year, but they were to remain at least one hundred metres from the coalface. Ventilation was to be well maintained in all areas where diesel equipment was operated, and stone dusting was to be thorough.

In early 1991, McLean sounded the alarm about a serious safety hazard. On February 15, he sent a memo to Colin MacDonald, the department's supervisor of mine inspection, saying he was "very concerned" about the condition of the mine's roof. McLean wanted MacDonald and the chief inspector of mines, Claude White, to see for themselves, and a tour was arranged for February 20. McLean's brief report on the follow-up inspection with his superiors suggested roof conditions had improved. Crews were erecting steel roof supports, he noted, and the tunnels were entering an area of "hard stone" that would make the roof "somewhat better."

Westray was also concerned about the roof, and had retained Golder Associates, a Vancouver-based engineering consulting firm, to monitor conditions. Golder had a long association with the project, having devised a roof support plan for Suncor in the mid-1980s. After a tour in mid-April, Golder engineers said the roof was in "good" condition, "in spite of the relatively economical support currently installed." Westray was relying on wire mesh held in place with roof bolts, augmented with steel beams and arches where needed. Golder was pleased to see that Westray was cutting smaller intersections at each crosscut. The roof above the first intersection had collapsed in July 1989, requiring the installation of an "unconventional" concrete pillar to hold up the roof.

On April 1, with the tunnels at the thousand-metre mark and inside the Foord seam, contractor Canadian Mine Development was let go. Blasting ceased, and Westray began using its mining machines to dig out the main slopes. "Roof condition seems to be improving," McLean noted after an inspection in early April. A month later, he reported that "stone dust has been ordered and should arrive very soon." Westray, it seemed, might even come close to meeting the deadline to supply Trenton 6. By May, Associated Mining Consultants Ltd. estimated that the project was 90-per-cent complete. AMCL, which had already done work for Westray, was now monitoring the project for the Bank of Nova Scotia. "It's extremely fast to achieve an underground mine in ten months. Normally you're talking years," AMCL engineer Alan Craven told the *Chronicle-Herald*. Maybe too fast, as it would turn out. But that was not Craven's view, as Westray dashed towards the finish line. "It's looking more and more like a very good project."

Then, on the night of May 23, there was a close call. A thirty-metre section of bolted roof in the No. 2 main tunnel collapsed moments after a work crew sensed the danger and cleared out. That was the night Eugene Johnson and the other shaken members of B-crew had their showdown with Gerald Phillips. Phillips called McLean to report the incident, stressing that there had been no injuries and that no equipment had been lost. McLean passed the information on to his superiors, but it was a week before he went to Westray to take a look. The rock was piled close to seven metres high, and McLean estimated that some ten thousand tonnes of stone had come down.

It took two weeks to put up heavy steel arches, rebolt the roof, and clean up the mess. McLean and John V. Smith, the electrical/mechanical inspector, met with Phillips on June 5; they were told that Golder's engineers would be investigating the collapse. The two-metre-long bolts used to secure the area had obviously been insufficient, and Westray had already switched to longer bolts. Despite the precautions, Smith remained uneasy about the situation. "The security of the roof in these two slopes appears to be the subject of some speculation," he reported in a June 17 memo to White. "It is hoped that any potential problems which are found will be corrected as soon as possible."

* * *

WHEN THE HOUSE OF ASSEMBLY opened its spring session in 1991, Bernie Boudreau still had his sights set on Westray. It was Donald Cameron's first session as premier, and the Liberals were determined to undermine the Tory government's newfound commitment to reform and fiscal restraint. Their chief weapon was the government assistance showered on Clifford Frame and Curragh Resources; namely, the take-or-pay agreement guaranteeing the sale of 275,000 tonnes of coal annually. Under questioning from Boudreau in May, Cameron's replacement as industry minister, Tom McInnis, said the government could end up paying Curragh $14 million per year to leave coal in the ground. It was a conservative estimate—the contracts called for payments as high as $20 million—but since the government refused to make the contract public, no-one could challenge McInnis's figure.

That still worked out to a $210-million liability over the contract's fifteen-year term. Boudreau made the most of the revelation, conjuring up the image of Frame rubbing his hands as he pocketed millions of tax dollars for doing nothing. McInnis countered that the province would never be called on to honour the contract; buyers were lining up for Westray's low-sulphur coal, he claimed, and the mine would be "a tremendous success." The premier waded into the debate, demanding that Boudreau "put up or shut up" about Westray. "I'm telling the public and the province that as long as I'm premier, this contract won't cost them anything," a defiant Cameron declared on May 28. If the contract cost taxpayers $14 million a year, he promised to resign; if not, then Boudreau should put his seat on the line. Boudreau begged off, saying he wanted to see the fine print of the contract before taking the offer. But the dramatic episode raised the stakes for the new premier. In one burst of rhetoric, Cameron had pinned his political future to the success of the Westray mine.

Boudreau's reputation as Westray's harshest critic made him a lightning rod for complaints about the mine. So it was only a matter of time before word of the massive May 23 cave-in reached his ears. Boudreau asked about the incident in the House of Assembly on July 2—but neither Natural Resources Minister Chuck MacNeil nor Labour Minister Leroy Legere, knew anything about the cave-in. MacNeil quickly queried his staff, and Patrick Phalen assured him the roof-fall was "an isolated incident" and under investigation.

The cave-in was confirmed privately to Boudreau, who was outraged that the ministers responsible for mine safety were so ill-informed. "Are we monitoring this mine or are we treating them with kid gloves?" he asked when the House resumed sitting on July 3. Legere, a former teacher who had been labour minister for just five months, insisted the matter was in hand. His department had investigated the cave-in, and the company had retained a consulting firm to study the roof problem. Boudreau pressed the point, calling on Legere to shut down Westray until the cause of the collapse was determined. The combination of "a dangerous mine and apparently a lax operator," he charged, created the "potential for a lethal incident" at Westray. Legere said Boudreau and the Liberals were only interested in scoring political points. "All they want to know is who is at fault," he shot back. "They are not concerned about the safety of the miners."

Surrounded by reporters outside the legislature, Legere did not inspire confidence as he discussed the cave-in. "I'm sure that the people in the department that looked into it felt matters were being taken care of," he sputtered, "and that there was no immediate, I guess, safety problem with the mine." Legere, who did double duty as minister of fisheries, was just as shaky when asked what caused the collapse. "I am, at this moment, not quite sure whether it's a fault in the rock formation, or whether it's a structural fault," he said.

It was painfully obvious that Legere did not have the answers. And he was in no hurry to find out what had happened at Westray. It was July 4, two days after Boudreau first asked about the cave-in, when he received a detailed memo from chief inspector Claude White. The department had investigated, White told his minister, and would be receiving a copy of Golder's report on the incident. Gerald Phillips went public on July 6, calling on politicians and the media to give

Westray "fair treatment." Boudreau had not demanded the closure of Devco mines after cave-ins, he noted in a newspaper interview. Phillips also contended there had been ample warning of the May 23 collapse, and brushed it off as a minor setback. "We let it come down, resupported it, and then went back to work." Phillips made it sound like just another day at the office. But Eugene Johnson and the other men on shift that night knew how close they had come to being under the roof when it came crashing down.

Boudreau kept up the heat. On July 10, he revealed during Question Period that Westray had been tunnelling through coal for four months. Boudreau demanded to know whether the company was still operating under the Department of Labour's temporary easing of blasting regulations and other safety rules, which had been granted during the rock-tunnelling phase of the mine's development. For the second time in a week, Legere was forced to admit he had no idea what was going on at Westray. "I do not on a daily basis get reports from the mines," he said. If he had, he would have known that blasting had ceased months earlier. "The people in the Department of Labour are not treating the Westray mine any differently than they are treating any other mine in the province," Legere insisted, "despite what the honourable member is trying to imply." Legere once again laid down a political smoke screen to avoid tough questions about safety at Westray.

* * *

IT WAS THE END of July before Golder Associates sent an engineer to Plymouth to review roof conditions. Westray had been changing its roof support methods on the fly, noted consultant Richard Brummer, and it was time for a more scientific approach. Brummer recommended a comprehensive program of research and monitoring, supervised by Golder, to determine the strength of the rock layers directly above the mine and to develop "a more rational method of support."

Westray had been relying on roof bolts, Brummer noted, although steel arches would be used for the entire length of the main slopes in "a more traditional coal mine." While he felt Westray's approach was sound, there was one proviso—the roof had to be closely watched to ensure it was holding up. Brummer had seen some evidence of deterioration since his last visit to the mine in April: shotcrete applied to the upper sections of the tunnels was cracked and broken in dozens of places. Among the explanations was a build-up of pressure on the roof. It was also possible that shale behind the cement coating was swelling from exposure to moisture. Golder's recommended study would pinpoint the cause and suggest solutions.

The Golder report did not reach Westray until early September. In the meantime, the Department of Labour continued to keep an eye on roof conditions. In mid-July, McLean and Smith sat down with Phillips to review safety conditions. Phillips said his staff was regularly checking the roof, and monitoring equipment had been ordered. When he griped about the poor press Westray had received over the May cave-in, he got little sympathy from the inspectors. According to Smith's notes, Phillips was told that a roof support plan was needed; if there was a "refusal" to comply, he was warned, the department "would have no alternative but to become involved." The inspectors, it appeared, were not prepared to write off the roof problem as anti-Westray propaganda.

The inspectors raised one other matter. The company was well into the Foord seam, and mining machinery was producing coal dust. In one of the tunnels, Smith reported, "dry conditions combined with fine coal dust from the continuous miner," plus a methane reading of just under 1 per cent, "was a matter of some concern." But a supply of stone dust had just arrived, and the company was dealing with the problem. Smith was pleased to note that a stone-dusting machine was being readied for underground use, and that two more were on order.

Within a month, there was another major cave-in. On August 13, about ten metres of roof collapsed in the No. 1 main tunnel, deep in the mine. McLean was on vacation at the time of the cave-in, and was not informed until he showed up for an inspection on August 29. McLean's concern about the roof was turning to alarm. "This is the second fall at this mine within a few months," he told his supervisor, Colin MacDonald, in a memo. "Roof condition at this mine is a soft stone (shaley), not ideal for this type of mining." During his inspection, McLean said, he urged the miners to ensure their workplace was "safe and secure."

The Department of Labour decided that the same message should be taken to Westray's senior management. Four officials—White, MacDonald, McLean, and Smith—toured the mine on September 3, then sat down with Phillips and underground manager Roger Parry. White led off the discussion, taking the company to task for not reporting the August cave-in. Phillips explained that his secretary had tried to get in touch with McLean and White but had not left a message. The oversight, he said, would not be repeated. Then Phillips distributed copies of the Golder report, and assured the inspectors that a roof-monitoring program would be implemented. On the subject of coal dust, Phillips promised to have a plan for stone dusting and dust sampling in the department's hands by the end of September. The Labour Department had reviewed the stone-dusting plan Phillips had prepared back in 1986, when he was still with Suncor. But that was only a hypothetical approach; the department wanted to see how Phillips intended to carry out dust sampling and stone dusting in a working mine.

The inspectors also raised some new concerns. Large drums of flammable hydraulic oil had been spotted underground, and garbage was strewn about the workings. Phillips agreed to remove the oil and tidy up the mine. A small diesel-powered bulldozer belonging to Satellite Construction had also been found underground. Westray was thinking of buying the 'dozer, and was using it on a trial basis. That was not good enough for the inspectors, who told Phillips that all diesel equipment had to be modified for underground use and approved by the department. Phillips was cool to the idea of a miners safety committee being allowed to tour the mine on a monthly basis, and had to be reminded that it was required by law.

That meeting came a week before the grand opening. By the time Clifford Frame and the politicians showed up on September 11 for the celebration, the mine was clean and stone dust had been spread everywhere. The roof, however, refused to cooperate. There was another serious cave-in on the night of September 20, as crews drove the No. 2 main tunnel through a fault. A pile of rock about ten metres high came tumbling down, but no-one was injured. When he surveyed the caved-in area a week later, some miners told McLean that the unstable roof was getting on their nerves. "I feel we were fortunate that no one was under the falls," he told his superiors on September 26. McLean was apparently losing faith in Golder–he recommended an "independent consultant" be brought in to study roof conditions.

Until then, cave-ins had been confined to the main slopes. But the roof problem also affected new tunnels being driven into the first production area, the southwest section. Between September 29 and October 12, there were three rock-falls in the southwest. The largest covered an area about fifteen metres by six metres. On each occasion, the roof had been reinforced with bolts or arches, but came down anyway.

The spate of cave-ins added weight to McLean's call for an independent review. The department, however, took a different approach. White, the chief inspector, arranged a meeting for mid-October to clear the air and find out what was wrong at Westray. Phillips, Parry, and senior engineer Kevin Atherton attended from the company; Brummer was on hand from Golder; White, Smith, and McLean represented the department. The discussion was frank. Labour officials asked why the number of cave-ins was increasing despite the efforts of Golder and Westray. Golder's engineer said roof support in the mine was already "robust." The company insisted there had been ample warning of each fall, and that workers had not been in danger.

A new strategy was hammered out. Westray agreed to hire a full-time engineer to check the roof on a daily basis, and to draw up plans for roof control and monitoring by the end of the month. Cave-ins of supported roof more than ten metres from the working face were to be reported to the department. All roof-falls, reportable or not, were to be investigated by the company, with the results available to provincial inspectors upon request. Finally, Westray was to prepare written rules to ensure that miners did not work under areas of unsupported roof as mining advanced. Almost as an afterthought, Phillips was asked near the close of the meeting about the overdue stone-dusting plan. That plan, he promised, would be forwarded to the department by the end of October.

November rolled around, and the Department of Labour had yet to see Westray's roof control plans. On November 4, McLean issued an order under the Occupational Health and Safety Act, giving the company until the middle of the month to file the required documents. That deadline also passed. Phillips finally forwarded the roof-support and monitoring plans to Claude White at the end of November. There were two more roof-falls in early December, but White and his colleagues were satisfied with the company's proposals. The stone-dusting plan, however, remained elusive. When he met with Labour officials just before Christmas to review the roof situation, Phillips assured White he would file a stone-dusting scheme by the end of January 1992. Since the roof was the department's primary concern, the new deadline was accepted without protest.

* * *

AS WESTRAY STRUGGLED to shore up the roof during the fall of 1991, some miners decided they had seen enough. A few voted with their feet and quit; others asked

Mine inspectors Albert McLean, left, and John V. Smith were the Department of Labour's eyes and ears at the Westray mine. During one visit, mine manager Gerald Phillips was warned to get a roof support plan in place. [Photo by Andrew Vaughan, courtesy of the Canadian Press]

Bernie Boudreau to sound the alarm. Boudreau went public with their concerns, but the only reporter to dig into the safety issue was Betsy Chambers, who covered Nova Scotia politics for the Thomson newspaper chain. It was her business to keep on top of the Westray story—Chambers filed reports for the New Glasgow *Evening News* and the Sydney-based *Cape Breton Post,* Thomson-owned dailies in the thick of the debate over the province's coal strategy. On October 23, both papers published long accounts of her interview with a Sydney man who had quit after four days at Westray. "Some things are more important than money, I guess," explained the miner, who spoke out on the condition that his name not be used.

The man, an experienced hard-rock miner, painted a grim picture of conditions underground. Training was almost non-existent, the roof was unstable, and dust was a foot thick on the floor. Chambers seized on the dust problem and put in a call to Claude White at the Department of Labour. "Well, if it's the right kind of dust, that's good," he said—the right kind being stone dust. "If it's the wrong kind of dust, we would be concerned. . . . Certainly if it was a foot [deep], it would exceed the standards that would be expected in that mine." The department was working with the company "to get them up to acceptable standards" for monitoring dust conditions, he said. White had good reason to doubt the miner's story—McLean's latest inspection reports showed stone dust was being spread throughout the mine.

Boudreau, however, viewed the allegations as serious enough to warrant shutting down the mine. "With conditions unchecked," he claimed, "it is only a matter of time before a major catastrophe happens." Phillips pointed out that hard-rock miners found it tough to adapt to digging coal. "Any of the people who have quit who have been here a week—they have no idea what they are saying, and they surely aren't experts in this type of mining," he told Chambers. Phillips accused the United Mine Workers of America of stirring up trouble over safety as part of its union drive. As for Boudreau, Phillips again wondered aloud why the Liberal MLA had said nothing about recent cave-ins at Devco mines. Pictou County's largest paper, the New Glasgow *Evening News,* backed the company against what it saw as yet another malicious attack from Cape Bretoners. "We have no reason to suspect at this time they are not paying close attention to every possible safety precaution at the new mine," the paper noted in an editorial that seemed contradictory to the evidence uncovered in its own pages. "We find it hard to believe Mr. Phillips would allow his miners to work in any area that threatened their life!"

Chambers followed up with an October 24 story, carried by the *Evening News,* revealing that Albert McLean's son, Wayne, was a welder doing surface work at Westray. The inspector was in a conflict of interest, Boudreau charged; McLean had the power to shut down the mine as unsafe, but that would mean putting his son out of work. When the article appeared, the labour minister, Leroy Legere, agreed that Wayne McLean's hiring could raise eyebrows. But, he added, there was no need for Albert McLean or any other inspector to close down the mine. "I'm satisfied that the safety precautions are being followed to the limit," he said. Phillips publicly denied that McLean had asked him to give his son a job, and he insisted the company avoided "political requests" to hire people.

Westray's public image continued to take a battering during the fall of 1991. As the bad press over safety died away, the Nova Scotia Power Corporation revealed that it was receiving very little coal from the mine. Trenton 6 had started up on October 7, but Westray was supplying only one-quarter of the plant's fuel requirement. To compensate, the utility was buying additional coal from Devco and from the Pioneer strip mine in Westville. There was no coal to fire the older Trenton 5 station, which had been shut down for repairs. Westray promised to have enough coal by the new year to meet its contract to supply both plants.

In the meantime, Westray would be bailed out by its old nemesis, Devco. It was an embarrassing situation for the Nova Scotia government, which had been thumbing its nose at Cape Breton's coal industry for years. For Curragh, Westray was becoming a financial liability at a time when the parent company needed all the money it could get. After turning a $32-million profit in 1990, Curragh began a long run of bad luck. Metal prices entered a downward slide. Then the marketing arrangement with Asturiana de Zinc, the Spanish smelting firm, turned sour; it cost Curragh millions of dollars to disentangle itself from the deal. At the end of 1991, Curragh shares were trading on the stock market for a little over $3, a quarter of their 1990 issue price, and losses for 1991 would total $28 million.

If Westray could not produce coal, Curragh hoped it could produce cash. In September 1991, around the time of the gala opening, Curragh floated a $55-million bond issue, using the mine as collateral. Part of the proceeds were to be used to pay off the $12-million Nova Scotia loan, so the company could get out from under the hefty 11.75-per-cent interest rate. But there were few takers, and the bond scheme was abandoned. On November 28, the company announced the sale of "non-essential assets" in an renewed effort to pay down debts and expand the Faro mine. The asset Curragh now considered most expendable was Westray, and all or part of the mine was up for grabs. Would-be buyers were assured that production problems were temporary; the mine would deliver 130,000 tonnes of coal to NSPC by the end of March 1992 and would be at full production by the summer. "This is an extremely good property, one of the best that Curragh has," Curragh's Marvin Pelley, who now held the title of president of the company's coal division, declared in a television interview. "We're just suffering normal start-up problems." A major American coal-mining firm took a look, but did not put in an offer. No-one, it seemed, was eager to take Westray off Frame's hands.

Boudreau used Westray's financial woes as a platform to once more press the safety issue. At the end of a long interview about the mine's production shortfall on the CBC-TV news program *First Edition,* he noted that there had been six reported cave-ins at the mine in six months. "My information, from people working in the mine, is that it is not safe," he said bluntly. "I've already asked that until the Department of Labour can assure the people in that mine that these roof-falls will not cause injury or death, then they should close it temporarily."

Despite the seriousness of Boudreau's charges, journalists let the cave-ins and other safety issues fade into the background. When Westray made the headlines or the newscasts, the story angle was invariably about politics or economics or environmental concerns. Boudreau's persistent cries of wolf were ignored; most reporters, like the Tory government, were quick to write off his dire warnings as political mischief. "We didn't think Boudreau had any credibility," explained Mike Hornbrook, a CBC radio reporter. "People would roll their eyes when Boudreau got up on his favourite sawhorse again." The November 1 interview was the last public reference to unsafe conditions at Westray until disaster struck six months later.

* * *

OUT OF THE PUBLIC EYE, Westray was headed for a showdown with provincial officials responsible for approving mine plans. Engineer Donald Jones of the Department of Natural Resources–the new name for Mines and Energy–visited the mine in October 1991 and filed a progress report. The department had earlier approved the realignment of the main slopes to speed up production, but the company's decision to mine the southwest section first was proving to be a serious mistake. Coal quality was poor, more than 40-per-cent ash in some places; after the wash plant removed the impurities and waste rock, Westray was producing between ten and twenty tonnes of sellable coal for every hundred tonnes of raw material brought to the surface. That explained the meagre coal deliveries to Trenton 6. And despite all the crowing about the Foord seam's low-sulphur content, the southwest was yielding coal containing up to 1.3-per-cent sulphur, compared to the projected 0.8 per cent.

Roof conditions, Jones noted, were just as poor as the coal. There had been three cave-ins in the southwest section during October, and some areas of bolted roof were taking a heavy load. The company, along with Golder Associates, was working with the department to deal with the problem, he said. Jones was more concerned about Westray's intention to tunnel near the flooded Allan mine. Every consulting firm had recommended leaving a block of coal at least one hundred metres thick, since the workings were difficult to pinpoint from old maps. The barrier offered a margin of safety; if Westray inadvertently broke into the abandoned workings, a flash flood could inundate crews working the new mine. Jones told

Gerald Phillips the department would need "detailed engineering studies" before the change could be approved.

A month later, the department asked Jones to review a revised underground layout submitted by Westray. To his surprise, the company wanted to dig coal within 40 metres of the Allan workings. "I would not, under any circumstances, approve mining within the 100-metre barrier pillar without extensive work to prove it can be done safely," Jones told his superior, Patrick Phalen, in a November 20 memo. That was just one of several major changes to the original mining plan drawn up for Westray by Associated Mining Consultants Ltd. The government, which had issued Westray's mining permit based on the AMCL approach, had the right to veto the changes.

Jones did not like what he saw. Not only was Westray coming too close to the Allan mine, but the company was also throwing out the original approach to mining the southwest section in a bid to get the coal out faster, he concluded. Even as he was writing his memo, mining crews were slicing through the coal at a different angle than originally contemplated under the AMCL plan. That procedure, Jones felt, could be responsible for the pressure building up on the roof and, in turn, for the cave-ins that were disrupting production. "It is our concern that the realignment of the southwest block sections was done to expedite coal production without due consideration of the factors that may have been incorporated into the mine plans by AMCL," he warned. And Westray was about to do the same thing in other sections of the mine, even though the AMCL approach to extracting coal appeared "more logical." There was also a greater risk of subsidence–surface damage caused when rock collapsed into abandoned workings–to a section of the Trans-Canada Highway that ran above the mine.

Other woes could be traced to the last-minute decision to realign the main tunnels, Jones said. Westray had encountered poor roof conditions and faults along the new route, just as he had predicted in 1989. And the company had little to show for its troubles. The seam was reached earlier, but the poor quality of the coal wiped out that advantage. By the fall of 1991, Westray was in the same area of the seam as it would have been under the original plan, Jones noted. But the access slopes were longer and more twisted, which could interrupt the flow of coal to the surface along the conveyor system. "If the original tunnel alignment had been maintained, the major fault zone recently intersected (which necessitated the turning of the main access slopes) would not have caused the major ground control problems that were encountered," Jones explained in his memo. "Westray did not fully realize the implications of making these realignment changes."

Neither had the province's director of mining engineering. Patrick Phalen was kicking himself for approving the realignment of the main tunnels in the first place. "I approved this plan but, in hindsight, realignment of the tunnels has not been in the best interest of the company," Phalen admitted in a November 22 memo to the deputy minister of natural resources, John Mullally. Based on Jones's misgivings, however, Phalen was prepared to lower the boom and reject the new plans. "We believe Westray is making important decisions regarding the mine plan without sufficient input from professional mining engineers who have sufficient experience in this type of operation," Phalen advised the deputy minister. The company was "clearly in violation of the Mineral Resources Act" by not using qualified engineers, "which could lead to the forfeiture of their mining permit."

Phalen informed Westray of his decision the same day, November 22. Approval was withheld until the company submitted a full report to justify the changes sought. "In the absence of information explaining the changes you are proposing, it is difficult to provide specific comments," he told Phillips, "but our concerns relate to the safety of mine workers . . . and the efficient operation of the mine." Phalen did highlight a couple of reservations–subsidence under the highway and the plan to encroach on the barrier pillar surrounding the Allan workings. "Time is of the essence," Phalen noted, pledging that his department would make "every effort" to work out an acceptable plan. The government had allowed Westray to get into this mess; now it would have to help the company to find a way out.

Phillips did not share Phalen's sense of urgency. It was December 10 before he discussed the matter with Phalen, and another week before he responded in writing. In polite terms, he told Phalen to back off. "In our opinion, there are no major significant changes to the original mine plan," Phillips contended in a December 17 letter. It was the same mine, using the same room-and-pillar methods, with the same equipment as approved in 1988. Westray, he said, required some latitude to adjust its plans, as underground conditions changed and the company gained experience in mining the Foord seam. Phillips denied any intention of mining close to the Allan mine, and devoted two pages to a detailed defence of the changes sought. "The safety of our workers, subsidence, long-term viability, efficient operations of the mine, and the maximum practical extraction of the coal resource are amongst our primary concerns," he assured the department.

Natural Resources made an abrupt about-face. Within three days, Phillips was told the revised plan was acceptable. At the end of December, Phalen followed up with a letter laying new ground rules for the department's monitoring of future changes to the mining plan. The complex geology of the Foord seam made it "impractical" for Westray to follow a set game plan, Phalen conceded. "We believe that it is the operator's responsibility to plan and operate the mine," he added, "so we will not refuse a request to approve a change in the mine plan unless we believe the change will cause unsafe conditions, unreasonably affect other persons or unnecessarily reduce the life of the mine."

Part of the problem was the vague definition of what constituted a change to a mine plan under the Mineral Resources Act, which had been revamped earlier in 1991. In his December 31 letter, Phalen outlined a procedure for Westray to update its mine plan every three months, and promised that approval would not be withheld without first consulting the company. The tough talk about rescinding Westray's mining permit was long gone, replaced by a conciliatory approach. "You can be assured that we will discuss any concerns we have with you before making a decision not to approve a mine plan," Phalen told Phillips.

Critics would later point to the dispute over the mine plan as evidence that Westray received special treatment from government regulators in the months leading up to the explosion. But Phalen bristles at the suggestion. "The correspondence speaks for itself," he maintains. "As far as I'm concerned, they were treated the same as anybody else." The role of Natural Resources is to "audit" changes to mine plans, he explains, and ensure they are approved by a qualified engineer. If the plans appear to be in order, department staff can comment on changes, but are not in a position to "check or argue," Phalen says. The decision to make changes ultimately rests with the mine operator.

One thing was certain as 1992 dawned—officials at Natural Resources would think twice before they questioned the way Gerald Phillips ran his coal mine.

* * *

WESTRAY WAS NOT ALONE as it tackled the double challenge of increasing production while keeping the roof from falling in. The federal and provincial governments came to the rescue on two fronts. To help sort out the roof problem, Ottawa agreed in December 1991 to cost-share a $364,000 monitoring program, providing $214,000 under the Canada–Nova Scotia Mineral Development Agreement. Westray put up the remaining $150,000. The federal government and the company chipped in identical amounts to fund a separate study of subsidence at the mine, for a total federal input of $428,000. According to minutes of a December 5 meeting in Halifax, some government officials questioned how the public would react to news of these grants, coming as they did on top of the millions of dollars in loans and loan guarantees lavished on Westray. But the proposals were approved in the interests of improving productivity and "health and safety" at the mine. Despite Albert McLean's call for an "independent consultant" to study roof conditions, both contracts were farmed out to Golder Associates, which had been involved with the Pictou County project even before Frame came on the scene.

Meanwhile, the provincial government had paved the way for Westray to get its hands on more coal. On the morning of January 6, 1992, residents on the outskirts of Stellarton were jarred awake by the rumble of earth-moving machinery almost at their doorsteps. Workers were clearing brush and earth at the Wimpey pit, within a few hundred metres of some homes, as a prelude to strip mining. Three days earlier, the Nova Scotia Department of the Environment had given Westray the green light to dig 100,000 tonnes of coal at the site. The coal was deemed a "bulk sample," a loophole that saved the company from undergoing a time-consuming environmental assessment of the strip mine.

The Wimpey coal, easy to reach and cheap to mine—compared to the cost of getting at underground coal—had caught the eye of every mining company that had considered setting up shop in Pictou County since the early 1980s. Suncor had lobbied for rights to the surface coal, while Placer Development had demanded them outright. When Curragh came aboard in 1988, it secured the right of first refusal to the site. In January 1989, shortly before tunnelling began at Plymouth, Curragh sought a lease on the surface coal, but was turned down. As Mines and Energy Minister Jack MacIsaac noted at the time in a letter to Curragh, "it was our intention to ensure that this coal would be available in the event of a substantial disruption in the production from your underground mine." But by the fall of 1991, with underground production in disarray, the provincial government came through as promised. On November 28, barely a week after officials at Natural Resources were talking about pulling Westray's underground mining permit, the same department granted the company a lease to dig Wimpey coal. That meant the only barrier to moving into full-scale production at the Wimpey strip mine was environmental approval.

In January 1992, when the heavy equipment moved in, tempers flared in Stellarton, and a storm of political controversy erupted in Halifax. Strip mining had long been a sore point in the Pictou area, and landowners near the site complained they had no notice of the company's plans. "I call it 'the blitzkrieg,'" said Stellarton town councillor Stephen Kirincich, who lived nearby. The Liberals and New Democrats decried what they saw as a blatant attempt to circumvent environmental laws to cater to Westray and Clifford Frame. The "sample," they pointed out, represented one-tenth of Westray's projected annual production. Premier Donald Cameron refused to get "down in the gutter" with Westray's critics. He blamed the whole mess on the Liberals, who, he said, had held up federal aid and put the underground mine behind schedule.

There was a side benefit for the province—the Nova Scotia Power Corporation saved an estimated $1 million. Since an open pit was cheaper to operate than an underground mine, the utility was paying Westray about $10 per tonne less for

Wimpey coal than called for in its contract. But the Town of Stellarton and residents near the site balked at having to endure the dust and noise of a strip mine just so that Westray would be able to solve its production problems. Opposition increased in mid-April, when Westray applied to the government for permission to operate a full-scale operation at Wimpey, again without an environmental review. The company wanted to take about 200,000 tonnes of coal annually over ten years, creating about eighty jobs. "Because this area's been significantly mined in the past," Gerald Phillips argued in the press, "we feel from an environmental standpoint it's relatively simple to approve the project." The government had until May 13, 1992, to make a decision.

Coal from the strip mine, trucked to the Plymouth site for processing at the wash plant, boosted production by as much as 1,000 tonnes per day. The additional sales were vital for a company beset with mounting operating costs. An additional $7 million had been set aside for more steel roof supports and other unbudgeted expenses during the last four months of 1991. The April 1, 1992, target date for reaching commercial production levels at Plymouth had been pushed ahead to the end of July. Westray's bankers, who were receiving monthly updates on the roof conditions and production troubles, were becoming anxious. As early as December 1991, Associated Mining Consultants had warned the Bank of Nova Scotia that Westray's financial situation "is extremely sensitive and must be viewed with extreme caution."

Curragh's executives were painfully aware of the production shortfall and roof control problems. Gerald Phillips was in charge down in Pictou County, but he answered to head office. Internal memos flew back and forth between Toronto and Plymouth, as Westray struggled to fulfil its contract and increase production. The cost of additional roof support and engineering consultants at Westray was draining cash from a company that was already losing money at its other mines. But no-one in Curragh's upper echelons could claim much experience with an underground coal mine, let alone one as troublesome as Westray. Kurt Forgaard, Curragh's former president, had been the lone exception; his résumé included senior posts at three coal operations, including an underground mine in Alberta. But Forgaard left the company in 1990, after a run-in with Frame.

Westray, Curragh's only coal mine, was the responsibility of the president of the coal division, Marvin Pelley. Pelley had been adept at mining the federal and Nova Scotia governments for money, but his coal experience was limited to Quintette, a far bigger undertaking than Westray, and an open pit as opposed to an underground operation. About all Quintette had in common with Curragh's Plymouth mine was technical glitches and production problems. Pelley was chief engineer at Quintette for almost five years, until Frame saved him in the mid-

1980s with a call to join the Curragh team. Named head of the coal division at the time of Westray's grand opening in September 1991, Pelley oversaw the project as it moved from the development to the production stage.

For Westray, it was a time of growing pains and growing problems. All through the fall of 1991, and into 1992, the company was struggling to keep the roof from caving in, all the while trying to play catch-up to production targets that were no longer realistic. At times, relations between head office and Westray management became testy; in January 1992, Pelley fired off a memo to Phillips, threatening unspecified disciplinary action unless a layer of coal at least four feet thick was left at the top of the seam, to make the roof more stable. But by the spring of 1992, Pelley was assuring the *Northern Miner Magazine* that the bad roof and areas of faulted ground were "normal start-up problems," and they had been overcome. Westray was fine-tuning its roof support methods on the fly, but that was not a matter for concern. "As the mine matures," he noted, "changes to the support system are likely to occur."

But Pelley would not be around to see Westray mature. At the beginning of April 1992, Frame transferred him to Whitehorse to head Curragh's Yukon division, a job which put him in charge of the Faro and Sä Dena Hes mines. He replaced Colin Benner, who became Curragh's new president of operations. Benner was given direct responsibility for Westray, and he soon flew to Plymouth to take a look at his new baby. It was not a pretty sight. He relayed his findings back to head office in an April 16 memo that focused on the poor production figures and the grim outlook for repayment of the Westray bank loan. But Benner's challenge was to get Westray back on track in all areas, not just production. Among the goals of the internal task force he set up in mid-April, three weeks before the explosion, was the creation of "a safe mine plan." Unless drastic measures were taken, Westray seemed destined to become a financial disaster, a mini-version of Clifford Frame's last coal project, Quintette.

Westray's woes were not news to Frame. The Curragh chairman ran a tight ship, and he made it his business to find out what was going on throughout his far-flung mining empire. He routinely circumvented the chain of command to talk directly to the people running his mines. "Frame was always on the phone to his operating people; it was just his nature," says a Curragh official. When Westray's roof control problems were at their worst in the fall of 1991, the official noted, Frame was in daily telephone contact with the Plymouth office. But in interviews conducted after the explosion, Frame denied that he or anyone else at Curragh's head office had any inkling that safety was being compromised at Westray. "Never at any time were we in this office made aware that there was ever anything to worry about," he told *Maclean's* magazine in the spring of 1993. "Now, since the accident, we've found we're not perfect." Frame also maintained he never scrimped on safety equipment for his operations, and Westray was no exception. However, the responsibility for ensuring that mines are run safely, he insisted, is not assumed by company brass ensconced in an office tower in Toronto. "[It's] up to the people at the mine site to do the job."

* * *

AT THE WESTRAY MINE SITE, protests to management about working conditions or safety standards fell on deaf ears. Miners who came forward with complaints were told the company's files were bulging with applications from people who would gladly take the place of anyone who wanted out. That left two choices—quit, or stay on in the hope things would get better. A third option was to form a union, which would give the men the collective clout they needed to demand changes. During the fall of 1991, the United Mine Workers of America tried to organize the miners, but it was an uphill battle. The long-standing feud between Cape Breton and Pictou County left many wary of joining the UMW, the union that also represented Devco miners. After the UMW organizing drive fizzled, the United Steelworkers of America stepped in. The Steelworkers, which represented workers at Curragh's Faro mine, were still signing up members at the time of the explosion.

Without a union, the miners were on their own. At the Department of Labour's prompting, a safety committee was set up in the fall of 1991. During a tour of the mine with Roger Parry on October 7, the miners on the committee complained about the lack of stone dusting, and about the garbage that was piling up underground. The company took the view that the workers were responsible for removing garbage, and Parry said he would ask miners "to work between shifts to stone-dust." The committee held at least one meeting with Albert McLean, who urged its members to "become more involved with the workers" so complaints could be brought to "the proper authority." But the group's impact was minimal. Randy Facette, an experienced coal miner who served on the committee, dismissed it as "a joke." His recollections: "We never got any help at all from the company—having actual health-and-safety committee meetings, having a chairman; there was nothing like that."

While McLean was encouraging workers to come forward with complaints about safety, it was company policy to keep such grievances in-house. Westray had an eight-page health and safety policy designed not only to prevent accidents, but also to prevent workers from talking to outsiders about problems at the mine. The reporting of accidents to the Department of Labour and the provincial Workers' Compensation Board was the sole responsibility of management. The policy also imposed a gag order; it decreed that "under no circumstances may information [about problems] be released to any other person without the expressed authority of the vice president, general manager of Westray Coal," the title held by mine manager Gerald Phillips. The policy flew in the face of the Occupational Health and Safety Act, which granted employees the right to complain directly to mine inspectors and other government officials responsible for enforcing safety laws.

That policy could not prevent former employees from coming forward. Mike Wrice quit in October 1991, not long after he met Curragh director John Turner at the grand opening and made a crack about the mine's unstable roof. The cave-ins were only one of his reasons for leaving; Wrice was also worried about the coal dust, and he found it hard to believe the Department of Labour was letting Westray use farm tractors in the mine. Although the vehicles were modified for underground use, the starters sparked and the exhausts were hot, creating a fire hazard. Wrice, married with two sons, decided to take his chances and look for another job.

Wrice applied for unemployment insurance benefits, saying he had quit because the mine was unsafe. The claim was rejected. Wrice appealed the ruling but knew he needed evidence to show he had reason to fear for his safety. None of the men still working at the mine would back him up—they were too afraid of being fired for speaking out. Wrice turned to the Department of Labour. He telephoned the department's main office in Halifax in January 1992, and complained about the cave-ins, the coal dust, the tractors. Wrice does not remember the name of the official who ended up handling the call, but the response remains clear in his mind. "As far as we're aware," said the labour official, "this company is meeting all of the regulations that are required by the province."

The appeal came down to Wrice's word against the company's. Even though Wrice had no evidence to back his claim, one of the appeal panel's three members voted in his favour. The dissenting member, whose name was recorded only as B. Chisholm, knew of a number of experienced miners who had fled Westray for safety reasons. Wrice still lost the appeal, and that meant he was penalized for quitting without cause; he had to wait more than three months for his first unemployment cheque. Carl Guptill, who was injured after being forced to work without his helmet light, approached the Department of Labour at about the same time as Wrice did. Guptill had an even longer list of safety infractions, but the department took no action in response to his complaints. Many of the miners still toughing it out at Westray concluded that it was futile to complain to the Department of Labour, or anyone else. As the months wore on, Wrice checked in with his former colleagues from time to time to find out if conditions had improved. "No, Mike," said one. "Believe it or not, it's actually getting worse."

As experienced miners got fed up and bailed out, men with little or no mining background took their place. By late 1991, Westray had about seventy miners on staff, but only fifteen were certified to work in coal mines. The other fifty-five were trainees—everything from raw rookies to men whose experience was limited to hard-rock mining. A training officer was hired in the late summer of 1991—Bill MacCulloch, a former radio reporter and westling-ring announcer who had formerly administered Pictou County's industrial development agency. Since he had never worked in a coal mine, MacCulloch's role was limited to developing programs and organizing classroom sessions. Once newcomers set foot in the mine, many were quickly left to their own devices. By law, trainees were supposed to work under the supervision of a certified coal miner, but high attrition and the push for production often made such supervision impossible. Years earlier, a federal government engineer had warned about the "learning curve" Westray miners would have to climb to become adept at mining the tricky Foord seam; with each new arrival, the learning curve grew steeper and steeper.

Westray had one project under way that was supposed to bring miners up to speed on the hazards of coal mining and the proper use of equipment. In mid-1991, Wade Coates—the son of retired Tory MP Robert Coates, the self-proclaimed "marriage broker" in the Westray deal—was hired to compile a safety manual. Westray's existing manual was a three-inch binder containing a hodgepodge of pamphlets from manufacturers, handbooks borrowed from other mines, and information on the prevention of methane and dust explosions. Coates was handed a copy and asked to transfer the material onto computer disk. Coates, who had a master of education degree but knew nothing about mining, was allowed to make minor editing changes, such as rewriting technical passages to make them more understandable to the average miner. He spent the better part of a year on the project, working out of his Halifax home. No sense of urgency surrounded the work; Westray did not impose a deadline, and no-one from Plymouth called to monitor his progress. By May 1992, the manual was almost complete, and part of a rough draft was in the company's hands. Coates was still pecking away at his keyboard when the mine exploded.

Despite Gerald Phillips's proclamation that Westray avoided "political requests" to hire people, Coates had the inside track on the job. Rick Logan, Curragh's Ottawa lobbyist, was a crony of Robert Coates and a friend of the family; when he heard Westray needed someone in Nova Scotia to put together the manual, he immediately suggested Wade. The younger Coates had worked for Nova Scotia's Department of Advanced Education, helping to revamp the province's community college system. But when Donald Cameron came to power in early 1991, he cancelled the contracts of a number of employees hired during the Buchanan years, Coates included. He was unemployed until Logan steered the Westray job his way. Coates visited the mine site only once, to be officially hired by Phillips and to pick up the manual. Pay was set at $2,500 a month, but there was no formal contract.

Westray needed more than a revised policy manual to get on track. Conditions underground deteriorated as the months wore on. Safety took a back seat to production, as the company struggled to meet its contract to supply the Trenton power plants. Tractors and other diesel-powered vehicles edged their way closer and closer to the working face, contrary to the conditions imposed by the Department of Labour. Methane levels in some areas of the mine crept dangerously close to explosive levels, undetected by the remote sensors set up at various underground locations. Cave-ins went unreported. Oily rags and discarded fuel containers littered the tunnels. Mining machines operated with broken methane-detectors. A thick layer of coal dust accumulated underfoot.

The actions of some miners added to the danger. Matches and cigarette butts were found in the tunnels, despite the Coal Mines Regulation Act's ban on smoking underground. Mining crews often rejigged vent tubes to flush out methane in their working area, a practice that allowed the gas to collect in areas robbed of fresh air. In some instances, workers tampered with the methanometers on the continuous miners, either on their own initiative or because their superiors told them to, so that the machines would continue to operate when methane rose above the approved level of 1.25 per cent. "People got complacent there," one miner told the *Chronicle-Herald* not long after the explosion. "It started off where little things, safety-wise, weren't being followed. And it just escalated. They kept building and building and building." It was a recipe for disaster.

* * *

DURING THE EARLY MONTHS of 1992, mine inspector Albert McLean reported improvement in the condition of the roof. But he kept hounding the company to do something about the build-up of coal dust. "Stone dust needs to be spread on a more regular basis," he noted after a January 22 inspection. "Mr. [Roger] Parry agree[d] to see to this." Three weeks later, McLean told Gerald Phillips and Glyn Jones, the mine's assistant supervisor, that stone dust was needed "in different areas of the mine." The mine was in need of a "housecleaning," including the removal of coal spilled from the conveyor line. "Both agree to have these items corrected," he wrote in his February 13 report. At a meeting with Westray officials on February 26, McLean again raised the issue. Two new machines were being used to spread dust, he was told, and a system of taking dust samples would be in place by the middle of March.

Steps were taken to get more stone dust. The company's files contained a purchase order, dated March 3, for 2,400 twenty-five-kilogram bags of stone dust from a local supplier. But when McLean inspected the mine again on March 17, he found that stone dusting and "housecleaning" still needed to be done. Parry told the inspector he would "look after the items of concern. [He] also stated that a plan for stone dusting is being put in place." It had been six months since the Department of Labour first asked for the stone-dusting plan, and the company was still pleading for more time.

Then the roof began acting up again. As crews extracted coal from pillars in the Southwest One section, its main production area, the roof began to crack and the floor heaved. As pressure increased, pillars collapsed and sections of the roof caved in. Parry notified McLean, who went underground on March 31 for a look. Coal being crushed under the weight of the roof was releasing high volumes of gas. "The methane readings in this area have been recorded from 1-4%," McLean reported—perilously close to the explosive range. "The methane is coming in waves." The ventilation system seemed to be keeping the gas in check, however; about twenty metres away from the area of high gas concentrations, McLean noted, the methane readings were zero. "The manager has the situation under control and is monitoring it on a daily basis," he told Claude White. Westray later filed a report with White, stating that the afflicted area had been sealed off.

McLean filed scores of reports during three years of inspecting the Westray mine, but March 17 was the first time he complained of dangerous levels of methane. Nevertheless, the company was left to deal with the gas. Department of Labour records indicate there were no more inspections until April 29, when McLean recorded methane levels of 0.75 per cent "through the mine." Those readings were within legal limits—work was to cease at a level of 1.25 per cent, and areas with readings of 2.5 per cent were to be cleared—and suggested the gas had been confined to the sealed-off area. Alan Craven of Associated Mining Consultants, who inspected the mine on behalf of the Bank of Nova Scotia, also reported that the company had the situation in hand. The crushed pillars had given off gas "to levels that made ventilation to required standards difficult," he noted. But by the time he went underground in mid-April, the area had been "sealed off for safety reasons."

The methane may have been bottled up, but another hazard lurked right under the miners' feet. McLean had repeatedly warned the company to deal with the coal dust problem, and put a dusting and sampling plan in place. On April 29, he finally did something about the situation. After inspecting the mine with his superior, White, McLean ordered the company to clean up coal dust immediately and file a stone-dusting plan by mid-May. After months of prodding and reminders and polite requests, the Department of Labour was finally putting its foot down on Westray. Ten days after McLean issued his order, a deadly combination of methane and coal dust exploded, wrecking the mine and killing everyone working below.

* * *

THE DISASTER OF MAY 9, 1992, has thrust the Department of Labour's dealings with Westray under a microscope. Why was there a delay in issuing orders to deal with potentially explosive coal dust? Why did government inspectors give the company so much leeway? Over the previous three years, they had observed dozens of violations of provincial mining laws, yet no charges were laid. Failure to spread stone dust, failure to file a stone-dusting plan, storage of flammable materials underground, use of an unauthorized bulldozer, garbage strewn underground—all could have led to charges under the Coal Mines Regulation Act and the Occupational Health and Safety Act. Even the twelve-hour shifts worked at Westray were against the law; miners were supposed to work shifts no longer than eight hours. Formal orders, the first step in a prosecution, were filed on four occasions during Westray's short history, including the April 29 ultimatum on stone dusting. That was as far as the Labour Department was prepared to go to compel Westray to obey the letter of the law.

There was no agency to challenge that approach. Virtually everyone with a stake in Westray, from the miners working underground, to the bankers who had financed the project, counted on the Labour Department to ensure the mine was safe. The Bank of Nova Scotia retained Associated Mining Consultants Ltd. to keep an eye on Westray's expenditures and production levels, but that was all. Although AMCL engineers reported cave-ins and other safety-related problems to the bank, they relied on the provincial inspectors to enforce the law. As late as March 1992, according to one AMCL official, the Department of Labour assured the bank that Westray was complying with all safety laws. The federal government, in turn, relied on the bank. Officials of Industry, Science, and Technology Canada asked to see a few AMCL reports after the spate of cave-ins made headlines in late 1991. Although Ottawa had put a huge chunk of public money on the line through its loan guarantee, the federal government made no attempt to independently check safety conditions at Westray.

The Department of Labour was the first and last line of defence against safety violations at Westray. But the department's philosophy was to use the carrot of persuasion to enforce the law, rather than the stick of prosecution. From 1986, the year mine inspection became the responsibility of the Department of Labour, to the day Westray exploded, not a single Nova Scotia mining company was prosecuted for violating safety laws. During that period, hundreds of mining accidents were recorded, and four mine workers died. There was a simple explanation—inspectors had to jump through a series of bureaucratic hoops before laying charges. It was departmental policy to give three orders to most companies violating safety laws, just as police officers often issue warnings to speeders. As well, a series of conditions had to be considered before charges were laid, including the officials' time that would be tied up in a prosecution. "Even where there is evidence which could result in conviction," according to the department's internal policy guidelines, "not every violation or alleged offender will be prosecuted."

The lack of political will to prosecute was just one shortcoming. The inspection division, like most government services, was feeling the pinch of government cutbacks, and the inspectors were stretched thin. The mine safety division's seven inspectors, who had a budget of $650,000, were expected to monitor as many as five hundred job sites, inspect mine rescue equipment, conduct training sessions, and administer tests to those seeking certificates as coal miners. At the time Westray was being developed, the giant Rio Algom tin mine was coming on stream at the extreme western end of the province, further increasing the workload. The idea of hiring an additional inspector to handle Westray never materialized; Albert McLean and John V. Smith were left to cope as best they could, juggling their frequent visits to Westray with other inspections and work commitments. To compound matters, the ongoing effort to revise the mining regulations absorbed countless hours. And it all had to be accomplished within a department rife with internal problems. An independent review conducted after the Westray explosion revealed low morale among staff, sloppy record-keeping, and a dearth of leadership in the department's upper echelons.

In the fall of 1992, a disgruntled Labour official went public to complain about the no-win situation facing mine inspectors. "If you want to get a prosecution in this department, you've almost got to have a fatality on the site," the unnamed staffer told the Halifax *Daily News*. "There is a very narrow window to prosecute if nobody is hurt on the job. And there's very little support for you from people higher up. You had better have it nailed down, airtight, or they won't even look at it." And the lack of support from above put inspectors in a Catch-22 situation. "The man on the spot has to have the guiding word," the official contended. "If you're constantly having to second-guess yourself because you don't know if your decision in the field is going to be supported back at the department, it makes a difficult job that much more difficult."

The Department of Labour's handling of Westray should be seen in that context. McLean and Smith carried out more than fifty inspections at Westray between 1988 and 1992. When violations were noted—such as use of unauthorized equipment or storage of oil underground—the inspectors were satisfied with the company's promises to rectify the situation. That passive approach was consistent with the department's handling of most companies. But a handful of McLean's inspection reports noted serious safety hazards, such as coal dust and poor roof conditions, and one mentioned dangerously high methane levels. Those conditions, left unchecked, put lives at risk. And inspectors had the power, not to mention a duty, to close down the mine if they felt conditions were dangerous.

But McLean and Smith were not operating in a vacuum. Their reports and memos ended up on the desks of their superiors in Halifax. McLean took particular care to record in writing the times when he told his supervisors about cave-ins or other incidents. On a half-dozen occasions, the inspectors called in their superiors, Colin MacDonald and Claude White, to take a look for themselves. Former miners, such as Carl Guptill and Mike Wrice, had come forward with complaints about safety, but to no avail. Instead of cracking down on violations, the Department of Labour chose to give Westray more time to work out the roof control problems, more time to get a stone-dusting plan in place—more time to make the mine safe.

Smith spent May 8, 1992, at the Labour Department offices in Halifax, going over a diesel equipment code that was to be incorporated into the new mining regulations. Smith had been in hospital for tests on the day of the April 29 inspection, so he was not sure how bad the coal dust was at Westray. But he knew one thing—if Albert McLean had issued written orders, the situation must be serious. Over the lunch break, as he walked along a Halifax street with chief inspector Claude White, Smith raised the subject.

"By the way," he asked, "what are you going to do about these orders over at Gerald's place?"

"Albert will be over there on Tuesday or Wednesday of next week," White noted. That would be about the time Westray's fifteen-day grace period on McLean's demands was up.

"Well," Smith persisted, "what are you going to do?"

"I guess we're not going to have any other alternative. If [Phillips] has not brought it up to a reasonable standard, we're going to have to stop it."

Smith was left with the impression that there would be no more fooling around. He knew as well as anyone that the company had been given ample opportunity to clean up its act. If the coal dust problem had not been tackled by the second week of May, the Labour Department was going to shut down Westray. And the time for action was drawing near. But as they spoke, the men of A-crew were deep inside the mine, going about their work in the ever-present coal dust and methane. Within eight hours, the men of B-crew would take their places. And at 5:18 the following morning, for Glenn Martin, Eugene Johnson, Roy Feltmate, Myles Gillis, Mike MacKay, Robbie Fraser, and twenty other men nearing the end of their shift, time ran out.

AFTERMATH

FALLOUT

ROVINCE HOUSE GREETS THE EYE like an oasis of dusty-brown sandstone, its sleek Georgian elegance offering a welcome relief from the office towers of downtown Halifax. During the century and a half it has served as the seat of Nova Scotia's government, its walls have echoed with great speeches and rancorous debate. Province House was the birthplace of self-government in Canada, the setting for Joseph Howe's eloquent defence of freedom of the press. And it was here, in the dying days of May 1992, that collective shock over the Westray disaster gave way to tough questions about why twenty-six miners were dead.

Two separate investigations were already under way. The Department of Labour, which was responsible for securing the mine as an accident scene, was looking into possible violations of the Coal Mines Regulation Act and the Occupational Health and Safety Act. Given the allegations being bandied about in the media—high levels of methane, unchecked coal dust, and other safety problems—there was a lot to investigate. On May 15, the Cameron government had set up the Westray Mine Public Inquiry to examine the disaster and its causes. Headed by Mr. Justice Peter Richard of the Nova Scotia Supreme Court, the inquiry was given wide powers to call witnesses and gather documents to evaluate the way the mine had been planned, financed, and operated. The inquiry was certain to cover some of the same ground as the Labour Department probe; but in the rush to find answers, no-one questioned whether the parallel investigations could coexist.

Before these investigations got fully geared up, however, there was more-immediate political business to attend to. Opposition politicians broke their self-imposed silence on May 20, just one day after the memorial service at the New Glasgow arena. Their target was Minister of Labour Leroy Legere, the man responsible for mine safety in Nova Scotia. It had been eleven days since the explosion, but Legere showed up in the House of Assembly ill-prepared for the onslaught of questions from the Liberals and New Democrats. He was shaky when asked whether Westray inspection reports had crossed his desk. He was unable to confirm whether Westray workers would be part of his department's investigation. He said he "assumed" the company had complied with safety problems identified by government inspectors long before the explosion. Legere looked no more in tune with his department than he had the summer before, when he was surprised by opposition questions about a major cave-in at Westray. The performance left the opposition outraged and the government on the defensive. "Read the file, because there are more questions coming," the Liberal's Bernie Boudreau scolded Legere, waving a sheaf of papers. "Read the file."

Labour minister Leroy Legere was ill-prepared for the onslaught of questions thrown his way when the Westray disaster was debated on the floor of the Nova Scotia legislature. [Courtesy of the *Chronicle-Herald* and the *Mail-Star*]

Legere had read at least one inspection report. Within minutes of his mauling at the hands of the opposition, he emerged from the House and told reporters about the April 29 inspection, the last time the Labour Department had checked the mine. Coal-dust levels exceeded the legal limit, Legere admitted, and the company had been ordered to apply stone dust immediately to ward off an explosion. That was more fodder for the opposition. When the House resumed sitting on May 21, Boudreau accused the government of "pussyfooting" around with Westray, and demanded to know why the mine had not been shut down until the coal dust was neutralized. Legere's response may not have inspired confidence, but it underlined his department's passive approach to law enforcement. "The company is responsible for the safety, and we monitor it," he said. Legere was quick to deflect criticism away from himself as minister: "The mine inspectors have the authority and the responsibility to inspect, to issue orders, and to follow up," he said. "I was not apprised of the [coal-dust] order that was issued until after the explosion."

As if being left in the dark about potentially explosive conditions was not bad enough, Legere revealed that there had been no follow-up inspection. An underground fire had occurred at the Evans mine on Cape Breton Island, he explained; the inspectors there had been prevented from returning to Westray to ensure that stone dusting was being done. When Evans miners disputed that claim, Legere backtracked and said his staff had given him erroneous information. An inspector had in fact been at Westray on May 6, just three days before the explosion, to test employees seeking their coal miners' papers. Legere refused to explain why the mine was not checked again at that time, saying the inspector's actions would be dealt with at the public inquiry.

The miscue reinforced the image of a minister struggling to get a grip on his department. By the end of the week, Legere was still dodging questions in the legislature and refusing to speak to reporters. At one point, in his haste to get away from the barrage of criticism, he slammed the door of his car on a reporter's microphone. The Liberals and NDP called for his resignation. "The minister has been half asleep on the job," the *Chronicle-Herald* chided in an editorial, "and it's time he woke up." The Halifax *Daily News* described Legere's performance as "inept." Donald Cameron came to the aid of his beleaguered minister. Legere had struggled with questions in the legislature because key members of his department's staff, the people who had the answers, had been "working day and night" at the mine site since the explosion. "Leroy Legere is a very decent and honourable individual who just happens to be in the way of a political tornado," the premier told reporters.

Cameron was in the path of the same tornado. When not out of town attending constitutional talks, he endured his fair share of tongue-lashing in the legislature. Boudreau led the attack, rehashing Cameron's high-profile role in steering the Westray project through stormy political waters. "Will he admit that he invested so much political capital in this deal that he turned a deaf ear and a blind eye to any concerns raised about that mine?" demanded Boudreau. Cameron accused the opposition of creating "a new standard in politics," whereby politicians who supported a project were automatically responsible if death or injury occurred. Opposition to the project had focused almost exclusively on the financial impact on Cape Breton, he protested, and it was not for him to say whether Westray or any other mine was safe. "I am not a mining engineer or a safety expert."

Westray also topped the opposition agenda in Ottawa. The Mulroney government rejected calls for a federal inquiry into why Westray had been subsidized, despite the objections of senior bureaucrats. Federal ministers deflected questions about safety; mine inspection, they pointed out, was up to the Nova Scotia Department of Labour. Brian Mulroney rose in the House of Commons to defend the spending of public money on the project. "After careful consideration, the government agreed to provide assistance, as we do in many regional development programs across Canada," he said. Westray, the prime minister insisted, had been handled no differently from the Hibernia offshore oil project. It was also normal, he said, for provincial premiers to personally lobby for certain projects, as Cameron had done on Westray's behalf. There was no mention of the leading role

Mulroney's top aides had played in the drawn-out negotiations over a financial aid package. Other players in the Westray deal opted to take a low profile. Robert Coates, the retired MP who had snagged a high-paying international trade post from the federal government, was no longer bragging about bringing Clifford Frame to Nova Scotia. "What I had to do [with Westray] has nothing to do with an accident in a coal mine," he told a reporter who tracked him down a few weeks after the explosion. "There's no reason to talk to me."

In the midst of the partisan wrangling, Nova Scotia's three New Democratic Party MLAs travelled to Stellarton to meet privately with about sixty Westray employees and organizers for the United Steelworkers of America. During the three-hour meeting, miners told horror stories about working conditions in the mine before and after the explosion. Westray was still sending men underground to carry out routine maintenance, they said, although rising methane levels and coal dust could fuel another explosion. Miners who refused to re-enter the wrecked mine were told to "hit the highway." NDP Leader Alexa McDonough, upon her return to Halifax, warned Legere he would have "blood on his hands" unless his department closed down the mine.

The Westray employees also revealed that documents were being destroyed at the mine site. Garbage bags full of shredded paper were spotted at the mine's offices; one worker said he used the bucket of a front-end loader to haul away destroyed documents. When McDonough went public with these allegations on May 20, Legere responded that the company was under Department of Labour orders to secure its files, and said the documents were under "lock and key." Then McDonough produced an internal Westray memo that outlined security procedures in the wake of the explosion. According to the memo, prepared by Westray training officer Bill MacCulloch and dated May 17, "any shredding, if necessary" was to be done by two persons, using a shredding machine located in the mine's security office. Once again, Legere was upstaged by his critics. The government had "proceeded as best as we could" to secure the documents, was his lame reply.

Faced with evidence that shredding was being carried out, the government took action. On May 21, Solicitor General Joel Matheson called in the RCMP to take possession of the company's records. The Mounties, one step ahead of the government, had decided a day earlier to launch a criminal investigation. During the rescue effort, police had been content to play a support role, identifying bodies brought to the surface and controlling traffic to and from the mine site, to keep the media hordes at bay. But the accusations being discussed in the media—unsafe working conditions, destruction of documents—spurred the force into action. By May 23, a stack of Westray files and documentation was in the hands of the RCMP, locked in a cell in the Stellarton detachment for safekeeping until they could be forwarded to the Westray Mine Public Inquiry. Westray's lawyers handed over the records without asking for a search warrant; the Mounties made it clear that allegations of document shredding would form only part of a police investigation to determine whether crimes had been committed at Westray.

Westray responded to the allegations with a terse press release. An internal investigation, the company announced, had determined that "no papers relevant to the public inquiry, including documents concerning safety procedures at the mine, have been shredded at Westray." The carefully worded statement did not deny that shredding had taken place, only that the material shredded was not "relevant" to the upcoming inquiry. Westray officials, the people whose actions were the subject of a public inquiry and a criminal investigation, were in the enviable position of deciding what documents should be preserved. Shortly after the records were turned over, Westray's lawyers asked the RCMP to issue a public statement dismissing the shredding accusations as groundless; the Mounties refused, saying they needed more time to investigate.

Weeks later, Curragh executive Marvin Pelley claimed that the only documents shredded had been a batch of uncompleted press releases containing information that could have been upsetting to the families of the dead miners. It was a dubious explanation; Curragh and Westray had relied on Colin Benner's briefings, rather than written releases, to update the press and public in the days following the disaster. During the entire rescue effort, from May 9 to May 14, the company distributed only one written statement, two pages long. Besides, the destruction of a few press releases could not explain the bags full of shredded paper discarded before the police took possession of the company's files. MacCulloch, however, now contends that his memo was misinterpreted—that it was designed to make sure no documents relevant to the investigation of the explosion were destroyed. "It got all blown out of proportion," he says. The fact that Westray employees saw bags of shredded material, to his mind, proves nothing untoward was being done.

On May 25, five days after McDonough first sounded the alarm, the public inquiry moved in and took control of all records pertaining to the Westray mine. The inquiry commissioner, Justice Richard, had been in the midst of assembling a staff when the shredding allegation surfaced. John Merrick, a Halifax lawyer retained to act as the inquiry's chief counsel, contacted the company, seeking an explanation. "There were documents being shredded," Merrick later told a newspaper reporter, "but the company has assured me under oath that there were no relevant documents shredded." The inquiry, unlike the RCMP, was willing to take the company at its word. Justice Richard also ordered the provincial and federal governments and Curragh's head office in Toronto to hand over all documents relating to the mine. Those documents, and the records from the mine site in RCMP custody, were to be forwarded to the inquiry's Halifax offices.

Questions were also being raised about the security of evidence that lay underground. Dozens of Westray and Devco employees had gone down into the mine to monitor conditions after the rescue operation ceased. They were accompanied by mine inspectors, but Bernie Boudreau claimed that was akin to leaving "the fox . . . in charge of the chicken coop." Officials of the Department of Labour, who would also be under scrutiny at the inquiry, should not control access to the mine, he contended, and the RCMP or inquiry officials should step in. "In the case of any serious accident, whether it be a bus going off the road, or a mine explosion, there is always the potential for criminal charges to follow, so the prudent move by the RCMP is . . . to secure the site and preserve the evidence," said Boudreau. As soon as the Mounties launched their investigation on May 21, they posted a guard at the mine portal twenty-four hours a day, to keep tabs on everyone heading underground.

In the meantime, the Nova Scotia government scrambled to deal with the deepening political crisis. Legere's lacklustre performance in the House was bad enough, but it was nothing compared to the image of Westray officials busily destroying evidence right under the nose of the Labour Department. Over the weekend of May 23–24, a team of government officials, including some of the premier's top aides, gathered at the department's Halifax offices to sift through inspection reports and other Westray records. Legere was flown in on Sunday from his home in Yarmouth, at the western end of the province, for a badly needed briefing. Two days later, on May 26, Legere opened the first in a series of press briefings to outline his department's past dealings with Westray and to explain events unfolding at the mine. In the days ahead, the government—not the opposition—would control the Westray agenda.

The first task of the better-informed labour minister was to explain Westray's proposal to seal off the devastated working areas of the mine. The company wanted to erect thick barriers of reinforced concrete, one thousand metres down each of the main slopes. Once in place, Legere said, this bulwark would protect the upper reaches of the mine if rising methane levels set off a second explosion. The sealed-off area would still be accessible; heavy steel doors would be installed in the barrier, allowing rescue crews to re-enter to look for evidence or the bodies of the eleven men still in the mine. It was an ambitious undertaking, requiring that work crews be sent into the mine for up to three months. The proposal, which needed approval from the department, added weight to the assertions of Westray miners and the NDP that the pit was unsafe. So did a cave-in deep in the mine on May 27, which released explosive levels of methane; luckily, no-one was underground at the time.

As part of the damage-control effort, the government retained an American consultant, Don Mitchell, from the coal-mining state of Pennsylvania–at a fee of $150 U.S. per hour. Mitchell, who had fifty years of experience in the industry and a long list of degrees and awards, wasted little time making his mark, recommending approval for the explosion barrier. On May 27, Mitchell accompanied Legere to a media briefing and outlined his theory of the events at Westray on May 9. After reviewing autopsy results and reports from draegermen, he concluded that methane had ignited in the southwest section, killing the eleven men working in that area. He confirmed what some Westray miners and experts had been suggesting in the press for about a week–the gas explosion had stirred up coal dust, touching off a fireball that roared through the tunnels and killed the remaining fifteen miners. It was the kind of explosion that could happen "in any mine," he contended. "There is nothing that would show that this is anything more than a typical explosion, where a number of regrettable things all come together."

Mitchell had been on the job for only a few days, and had never set foot in the Westray mine, but he was already prepared to write off the disaster as a "regrettable" accident. Mitchell could not say where the methane came from, or why it had not been detected by the mine's monitoring system. He speculated that the gas pooled at the mine roof, as had happened in several recent colliery explosions in the United States. Despite the destructive force of the blast and the loss of life, Mitchell hesitated to describe it as a major dust explosion. Some stone dust appeared to have been spread in the mine before the explosion, he said, which at least limited the distance the coal-dust fire travelled through the mine. Whether there had been enough stone dust used to satisfy the requirements of the Coal Mines Regulation Act, he added, might never be determined. Under questioning from reporters, Mitchell conceded that some miners might have survived if the Westray disaster had been caused by the methane explosion alone–and not been exacerbated by the ignited coal.

Mitchell's hasty conclusions put the government on the defensive again. "If he were prudent, he would withhold his opinion until he gathered a bit of evidence," observed Bernie Boudreau. "Otherwise people will suspect that he's simply involved in a public relations exercise rather than a real attempt to determine the cause of the explosion." The editorial writers at the *Chronicle-Herald* denounced Mitchell's findings as a clumsy attempt at damage control and "an insult to the families of the 26 men who died in the mine." Mitchell, it turned out, had a reputation as a hired gun who defended mine operators in the wake of explosions and disasters. Joe Main, a safety official with the United Mine Workers of America in Washington, D.C., said his union had often locked horns with Mitchell. "For him to go up there raises serious questions about the integrity of the investigation."

The Labour Department stuck to its guns. To counter the appearance of conflict of interest, a retired RCMP superintendent, Harry Murphy, was hired to head the departmental investigation. The probe, separate from the public inquiry, was expected to involve scores of witnesses, and Legere predicted it could continue until fall. Time, however, was a critical factor. In order to prosecute Westray for violating a provincial statute like the Coal Mines Regulation Act, charges had to be filed within six months of the day the alleged offence occurred. If the government investigation was not finished on November 9, 1992, the whole process would be moot.

By the end of May, three weeks after the explosion, the search for answers was off to a rocky start. Allegations that evidence had been lost or destroyed hung in the air, casting doubt over whether the full story of what went wrong at Westray would ever be told. There was no shortage of attention being paid to the case—the disaster was the subject of three separate investigations. Labour officials had been the first off the mark, but the department's credibility suffered with each new revelation of unsafe conditions at the mine. A public inquiry was gearing up to look at all aspects of Westray's development and operations, but was not expected to open hearings until the fall. And then, the Mounties were also on the scene, gathering evidence that could lead to criminal charges against the company or its officials. It was inevitable that three investigations into the same incident would overlap—and, perhaps, clash. And that set the stage for a legal battle that threatened to keep Westray's secrets buried for a long time.

QUEST FOR ANSWERS

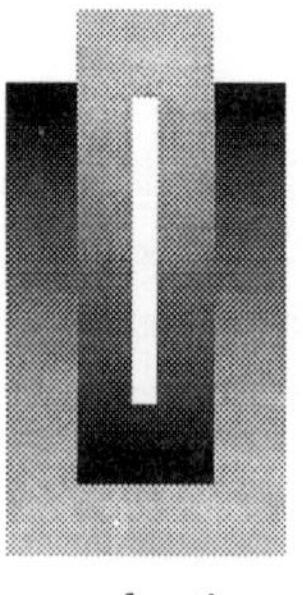

IN PLYMOUTH, NOVA SCOTIA, life slowly returned to normal after the explosion. The roadblocks had been lifted; the reporters and camera crews were gone; the funerals and memorial services were over. No surface sign of the tragedy remained at the mine site: welders had replaced the steel sheets blown off the portal roof in the explosion. Although the mine had been knocked out of production, three-quarters of the company's 160 employees were back at work by the end of May. A small amount of coal stockpiled on the surface was run through the wash plant, providing a few trainloads for shipment to Trenton. But prospects were bleak for the seventy-five surviving miners. Most of them resisted the idea of going back underground to do repairs and monitor conditions there, even though the Department of Labour was supervising the work. Those fortunate enough to have been off work on the morning of May 9 did not want to tempt fate. "I'm not sure if I could go back," explained Wyman Gosbee, who had worked the last shift before the blast and taken part in the rescue effort. "It's like being down in hell and waiting for Satan to jump out at you."

At the miners' request, the union drive under way at the time of the explosion was reactivated. The United Steelworkers of America had been on the verge of forming a union when the mine exploded; membership cards were being sent out at the time of the blast, and some widows received them in the mail within days of their husbands' deaths. On May 29, Westray employees voted 91–21 to form Steelworkers Local 9332. It was a hollow victory. Within a week, the idled mine announced the layoff of 104 employees paid by the hour. A skeleton staff of 18, and about 30 management-level employees, was kept on the payroll. Westray described the layoffs as temporary, holding out hope that some workers would be recalled if the Department of Labour approved the plan for erecting explosion-proof barriers in the crippled mine. That project would tide workers over until it became clear whether the mine could be repaired and put back into production.

But Westray was forced to rethink those plans after a major cave-in in late May, the second in a week. For the handful of Westray draegermen still willing to go underground, it was the last straw. "We don't want to be guinea pigs any more and lose any more of our men," one unidentified miner told the press. The Department of Labour, once it was notified of the collapse, barred anyone not equipped with draeger gear from entering the mine. The idea of installing concrete barriers was abandoned. In its place, Westray submitted a new plan to flood the North Mains, the deepest part of the mine. Pumping in millions of gallons of water would help keep the roof in place and prevent methane from building up to explosive levels.

If the Department of Labour approved the flooding plan, there would be new problems. The bodies of eleven miners would be entombed in a watery grave, possibly precluding a future effort to bring them to the surface. There was also the chance that evidence needed to establish the cause of the explosion would be washed away. While the area of the southwest section where the explosion began would not be submerged, access to it would be cut off. Westray, however, claimed a new tunnel could be dug to re-enter that area if investigators wanted to take a look. The company informed relatives of the dead miners, who had banded together in early June to protect their interests, that flooding was the first step towards repairing the mine and eventually recovering the bodies of Westray's remaining eleven victims. The families' spokesman, Kenton Teasdale, an Antigonish County teacher and the father-in-law of explosion victim Myles Gillis, served notice that the group would monitor the flooding operation to make sure the company kept its word.

The Department of Labour gave Westray the go-ahead in mid-June. But it would take time to lay a three-hundred-metre pipeline from the East River and extend it down the main tunnel, and several weeks to pump in enough water to close off the working areas. That meant there was still time to send investigators into the southwest section before it was cut off by flood waters, possibly for good. In late June, as water was being pumped in, Westray put forward a new proposal: flooding would stop once the North Mains area was submerged, allowing investigators to examine the southwest section. The company would send in its mining experts, along with representatives of the three public agencies investigating the explosion—the Department of Labour, the RCMP, and the Richard inquiry. The plan was risky; methane had been collecting in the sealed-up area for more than a month. Before the details were worked out, however, there was another serious cave-in in early August. Although the five men in the mine escaped injury, Westray refused to send more workers underground. The foray into the southwest section was scrubbed, and flooding was confined to the North Mains.

Even without a first-hand look at the southwest, the inquiry was to begin its public hearings in October. Staff members and a team of four mining experts faced the monumental task of sifting through about a quarter of a million pages of documents and interviewing hundreds of witnesses. An office was set up in a skyscraper in downtown Halifax, but speculation that the public hearings would be held in the provincial capital caused an uproar in Pictou County. The inquiry commissioner bowed to the pressure and agreed to hold most hearings at a museum in Stellarton, within sight of Westray's surface plant.

Justice Richard went public in mid-June to talk about how the inquiry would proceed. A former salesman and gas station owner, Peter Richard was appointed to the bench by the federal Liberals in 1978, after practising law for only eleven years. His résumé included a stint as lawyer to a royal commission set up in the early 1970s to investigate the sinking of the oil tanker *Arrow* off the Nova Scotia coast—valuable experience for the Westray post. At sixty, he was one of the Supreme Court's most senior trial judges, having presided over criminal and civil cases for fourteen years. At a press conference in Halifax on June 16, Richard described his mandate as "awesome" and promised to look into the politics behind the mine's development, the Foord seam's geology—"anything that bears on the explosion." But the judge said it was not for him to decide whether the mine should reopen. "That's a business decision," he insisted. Asked about the reports that documents had been shredded, Richard told reporters those allegations would be explored during the upcoming hearings. But he was confident that no papers relevant to his work had been destroyed. "As far as documents are concerned," he said, pointing to a room at the inquiry's offices filled with boxes of files, "our cup runneth over."

Peter Richard, a Nova Scotia Supreme Court judge, promised the public inquiry would delve into "anything that bears on the explosion," including the political decisions that led to Westray's development. [Courtesy of the Westray Mine Public Inquiry]

The first step was to sort out who would have the right to appear before the inquiry and question witnesses. Richard convened a two-day session in the gymnasium of a community college in Stellarton to hear submissions. There was no question that status would be granted to the people most directly affected by the explosion—the relatives of the dead men, and Westray's recently unionized miners and surface employees. Richard also agreed that the government should cover the legal fees for both groups, up to $1,600 a day for each lawyer. The miners and the victims' families used the hearing as a platform to take their first shots at Westray and the Nova Scotia government. "The big question that must be asked is whether or not the operators of the mine sacrificed safety for the sake of production," said Tony Ross, lawyer for the Westray Families Group. Ronald Pink, the lawyer representing the Steelworkers, blamed the government. "We believe that the Province of Nova Scotia, through its desire to have this mine in operation, has created a recipe which led to a disaster," he said. "It was production at any cost."

Mine owner Curragh and officials responsible for safety at Westray were also granted status, but would have to pay their own legal fees. Besides Curragh, mine manager Gerald Phillips, underground boss Roger Parry, and a group of about forty non-unionized middle managers were given the right to take part in public hearings. Albert McLean, the Labour Department inspector who had monitored Westray, caused a stir by asking that he and fellow inspector John V. Smith be given separate status and inquiry-funded lawyers. The provincial attorney general had already been given status to represent the provincial government, including McLean's employer, the Labour Department. But McLean told the judge he was not sure whether the government's lawyers were "going to protect the department or my interest" when he testified about his actions at Westray, and he wanted his own lawyer. McLean and Smith had good reason to fear that the government might hang them out to dry—for weeks, Leroy Legere had been dodging criticism about his own performance as labour minister by pointing fingers at the inspectors. After a quick huddle with McLean and Smith during the lunch break, a government lawyer announced that the inspectors had changed their minds, at least for the moment. The two men would stick with the government's lawyers and seek their own counsel only "if the need arises in the future."

Behind the scenes, inquiry officials and relatives of the men killed in the explosion were headed for a showdown. The Westray Families Group, which represented the families of all but one of the twenty-six miners—almost one hundred people—started as "a support thing," says Bert Martin, who served as secretary. As events unfolded, the organization became a means of ensuring that the families had a say in crucial decisions, such as flooding the mine. The group also pushed for recovery of the bodies of the eleven men still in the mine, including Martin's son, Glenn. But most of all, the group was a watchdog, making sure the inquiry, the Nova Scotia government, and the Mounties left no stone unturned in their investigations.

The issue that drove a wedge between the families and inquiry officials was money: the level of public funding would dictate the role the Families Group played in the hearings. Richard was determined to keep the inquiry's expenditures in check. No-one wanted a repeat of the royal commission into the Donald Marshall case, which spent more than $8 million in the late 1980s investigating the way the Nova Scotia justice system handled the case of a native teenager wrongly convicted and imprisoned for murder. When Tony Ross submitted a budget of about $300,000 to interview witnesses and review documents in preparation for the hearings, the inquiry countered with an offer of about $50,000, to cover legal fees in the pre-hearing stage. More disturbing, from the families' viewpoint, was Richard's contention that the Families Group need only take part during the early weeks of hearings, when the inquiry would deal with the cause of the explosion. After that, the judge felt, lawyers for the miners union could carry the ball for the families as well.

But the interests of the Families Group and Westray's miners were growing further apart by the day. The Westray local of the Steelworkers, headed by Randy Facette, supported the company's bid to operate a strip mine in Stellarton until the underground mine could be refurbished. The surface mine, if approved by the provincial government, promised to put scores of laid-off miners and wash plant employees back to work as early as the fall. But for the families of many of the miners killed in the explosion, distrust of Westray's actions and motives was turning to outright hostility. Many Westray widows were still bitter about having been informed of the explosion through anonymous early-morning phone calls. Media allegations of unsafe working conditions only strengthened their conviction that the company was to blame for their husbands' deaths.

Curragh took steps to soften the financial impact on the stricken families. The company offered counselling, sped up payment of life insurance benefits, and covered medical and dental bills for three months after the explosion. A support centre was set up in a vacant storefront on Stellarton's Foord Street. It was supposed to be a place where employees and bereaved relatives could drop in for information or referrals to counselling services, or just for conversation and a cup of coffee. The decor was not calculated to help people forget about the explosion; a wall of the one-room centre featured a huge aerial photograph of the mine site. Few family members used the facility, and it became a hang-out for unemployed workers who saw the company as their best bet to get back to work.

What was missing—and not just from the centre—was a human touch. Some widows thought Curragh and Westray should apologize for their husbands' deaths; at the very least, they expected some sort of official explanation of what caused the explosion. No apology or explanation was forthcoming. There were people in Pictou County who remembered how the operators of Stellarton's MacGregor mine reacted after a 1952 explosion. Company officials visited the family of each of the nineteen men who died, handed out envelopes containing $1,400, and expressed their condolences. Westray's approach, by comparison, seemed cold and distant. And the company's financial support was also found wanting. In the months after the explosion, donors ranging from schoolchildren to large corporations poured $1.8 million into the disaster fund set up to help the families. Neither Westray nor Curragh contributed a penny to the fund—despite Frame's assurances, in the middle of the rescue operation, that there would be compensation for the victims' families.

Months of frustration and bitterness finally boiled over in late August. The families broke their silence at a press conference in Stellarton, lashing out at the provincial government, Westray, Curragh, and the Richard inquiry. Kenton Teasdale, the group's chairman, read a statement demanding Westray's mining leases be revoked for "gross violations" of the province's safety laws. He attacked the Cameron government for failing to stop the "deadly combination of violations of law and safety practices" that led to the explosion. And he accused Richard and inquiry lawyer John Merrick of pinching pennies. The legal funding being offered was "so restrictive" that the families' participation in the inquiry would be "virtually eliminated," he contended.

Genesta Halloran, who had been left on her own at age twenty-seven to raise two young children, followed with an emotional appeal for the recovery of the miners' remains. One of the eleven men still in the mine was her husband, electrician John Halloran. "We need the bodies brought back to us," she said. "I personally need a place to go and shed my tears." Halloran charged that the company was still doing little to help the families, or keep them informed. "I want the average John Doe to understand that Westray has done nothing for us as far as moral support, or financial support," she said. Westray's drop-in centre and offers of counselling could not overcome the growing distrust—and in some cases, outright hatred—felt by those most deeply touched by the disaster.

Despite the united front presented to the press, some relatives took exception to the hard line of the Families Group. Bernadette Feltmate, for one, complained publicly about the statement's political overtones. But she agreed that the priority should be bringing the bodies of the remaining men—including that of her husband, Roy—to the surface. Another widow, Harriet Munroe, fired off letters to several newspapers, defending Westray and Curragh. "No amount of money or bickering over whether the mine reopens or not is going to bring back my husband," she wrote. "If any one or more persons are to blame for my husband's death, then they will be dealt with accordingly." Munroe, who lived a short drive from Stellarton and was the only widow to visit the support centre on a regular basis, praised the company's efforts to help people get on with their lives.

Richard made no apology for his effort to rein in costs. "I have a responsibility to the public to ensure inquiry costs are controlled, and one way to achieve that objective is to avoid duplication and redundancy as much as possible," he protested. But public sympathy was solidly behind the victims of the disaster; newspaper editorials and columns came down hard on Richard for putting the families in a financial straitjacket. Within days, the inquiry relented and offered more money. The move did not mollify the families. On August 27, Teasdale demanded that Richard and Merrick be replaced, and asked the RCMP to take over the mine and seize the inquiry's documents. "I really honestly believe there is no alternative but to ask the RCMP to come in," he told reporters. "We can't risk blowing this investigation."

* * *

AS THE DUST SETTLED from Westray, mining analysts were confident that any rumours of Curragh's death were premature. Although the company's already-battered stock had taken a tumble, the firm seemed solid enough to weather the storm. True, Curragh had just come off a disastrous year, racking up close to $100 million in losses in 1991. But lead and zinc prices were on the rebound, and as of the end of March 1992, the company had $70 million in cash on hand–a war chest left over from the unwinding of the Spanish smelter deal. Curragh's real challenge was to endure the political and media aftershock of the explosion. "It's a public relations disaster," explained one industry observer. "But in terms of their financial exposure, I believe it's limited." Another Toronto broker who dealt extensively in Curragh stock dismissed the allegations of unsafe working conditions, reminding one reporter that Westray had garnered the John T. Ryan award for its low accident rate. But miners were already telling the media that mine management had fudged accident figures to capture the award.

Image problems aside, the bills for Westray were piling up. The failed rescue effort, repairs, legal fees, and maintenance would total $15 million by the end of 1992. That did not take into account lost revenue, damage to tunnels, and the $14 million in mining equipment destroyed in the explosion. Westray was insured against property damage and interruption of business from an explosion, but the policy's $25-million limit was not enough to cover the massive loss. The financial mess deepened in early June, when the Bank of Nova Scotia called in the $95 million advanced to Curragh under its $100-million Westray loan. Insurance policies and the mine's assets, which had been used to secure the loan, were frozen. The bank's decision required the federal government, as guarantor, to fork over $80.75 million within six months. Curragh paid back the remaining 15 per cent of the loan, $14.25 million, in mid-June. Even after taking a multimillion-dollar hit in the pocketbook, the company reported earnings of $5.2 million for the quarter ending June 30.

Curragh continued to pour in money to keep Westray afloat, hoping to recover some of the payout from the insurance settlement. The company was better off than the Nova Scotia government. Triggering the federal guarantee took the bank off the hook and left Ottawa with the right to seize the mine's assets and insurance proceeds. The provincial government was entitled to nothing until the feds recovered their $80.75 million. "We're at the back of the line," admitted Tom McInnis, Nova Scotia's industry minister. Westray had been paying interest on its $12-million loan from the province, but had not been required to repay the principal until 1995. McInnis conceded there was no hope the province would ever see a penny of its stake in Westray. All told, the Westray disaster could cost Canadian taxpayers up to $96.6 million. That was the sum total of the federal government's $80.75-million loan payout, $3.9 million in federal interest-rate subsidies, and the province's $12-million loan. All taxpayers had to show for their money was a slice of the insurance settlement—and a wrecked mine.

Clifford Frame, the man whose political connections had brought Westray into being, took his share of knocks in the press. Government largesse and inside consulting deals were a lightning rod for criticism. Smarting from the attacks, Frame sought out Jennifer Wells, a writer with *Report on Business Magazine*, to plead his case. "We didn't take one nickel out of the Yukon for those mines," he protested. "I kept them going. Took my losses. . . . Only an asshole like me would do that. . . . I took on the shit projects in this country." Frame and his investors had little to show for their trouble. Low metal prices and the stigma of the explosion had driven Curragh shares down to $2.90, one-quarter of their face value. And Frame's dream of making Curragh a force on the international mining scene was fading fast.

The shock waves from Westray reached Curragh's operations in Western Canada. The company that once seemed able to get whatever it wanted from the government found its operations under intense scrutiny. In the summer of 1992, Yukon officials charged the company with four violations of safety laws at the Faro mine, where the Steelworkers had long complained of poor safety standards. The charges were relatively minor—a faulty brake on a front-end loader and lack of oxygen bottles at first-aid stations—but they made news across Canada. Curragh was also finding that the door to the public treasury had been slammed shut. In early 1992, the company had been negotiating loan guarantees worth $34 million with Ottawa and the Yukon government. The money was needed to develop a lead-zinc deposit near Faro, which was running out of ore. But after the Westray explosion, the two governments put negotiations on hold, creating another financial crisis for the company.

As Curragh licked its financial wounds, it also had to devote time and money to preparing for the public inquiry. In late May, the company announced it had retained a five-member team of respected mining experts from Canada and the United States to determine the cause of the explosion. One of the consultants was no stranger to the Westray project—the team's coordinator was Derek Steele of Dames & Moore in Ohio, which had identified the potential ingredients for a methane explosion as early as 1986. Gerald Phillips was reassigned to help put together the company's case for the inquiry. His replacement was Dan Currie, a former Devco official coaxed out of retirement to assume the post of mine manager. Roger Parry, the underground manager and Phillips's right-hand man, was not so lucky; he was laid off a few months after the explosion. Phillips, who, like Parry, was an obvious target of the Labour Department and police investigations, ducked all questions about his actions before the explosion. "We will make the true facts known during the inquiry," he promised in a July interview with the *Chronicle-Herald,* his only public statement on the subject. "And I think we'll surprise a lot of people."

Phillips was also point man on Westray's bid to rise from the ashes of the disaster by shifting its operations to a strip mine a few kilometres to the west, which had been tapped to supplement production from the underground as early as January 1992. As Frame had said within a few days of the explosion, the coal just waiting to be dug at the Wimpey pit on the outskirts of Stellarton was the best hope for getting Westray back into production. A strip mine would keep coal flowing to the Nova Scotia Power Corporation and provide the millions of dollars needed to repair the underground mine at Plymouth. If the federal government agreed to hold off on seizing Westray's assets, it would recoup some of the money Curragh was losing every day the mine remained idle. And it would put as many as one hundred Westray miners and surface employees back to work.

But before Westray could go back to the well at Wimpey, the proposal had to clear a major hurdle. Nova Scotia's environment minister, Terry Donahoe, had to agree to exempt the strip mine from a full environmental assessment. An assessment would require an in-depth study and public hearings, a process that could take up to a year to complete. But Westray could ill-afford to waste time on talk when it could be mining coal. Donahoe was asked to waive the full assessment, just as the provincial government had bailed out the company with the 100,000-tonne "bulk sample" earlier in the year. But this time, Westray wanted to scoop up ten times more coal—1 million tonnes of coal per year for two years. Enough, the company hoped, to tide it over until the underground mine could be repaired and reopened.

To bolster its case, Westray Coal released a consultant's report in mid-July that concluded the strip mine's environmental impact would be "minor." Dust and noise would be kept within permissible limits, and treatment ponds would be used to prevent run-off into adjacent streams. Once the coal was removed, the fifty-hectare pit would be refilled and landscaped for recreational use. Heavy machinery and dump trucks would be used to excavate the site, making blasting unnecessary. The scale of the operation, however, was anything but minor. The coal would be trucked to the Plymouth site, run through the wash plant, and shipped by rail to Trenton. And for laid-off Westray employees and other unemployed Pictou County residents, there was the promise of sixty jobs at the pit and another forty at the wash plant.

Westray also commissioned a poll, which showed that a slim majority of Pictou County residents supported the strip mine. Of the 430 people canvassed in late June, 53 per cent backed the proposal; 27 per cent were opposed, and the others offered no opinion. Virtually all those in favour of the mine said the area needed the jobs. Almost half of those opposed told pollsters their biggest concern was for the welfare of workers, reflecting the image of Westray Coal as unsafe. Few of those questioned would be directly affected by the dust and noise of the strip mine; less than one-fifth of the respondents lived in Stellarton.

The rosy picture painted by the report provoked a backlash from the project's opponents. Stellarton's municipal leaders branded Westray's tactics "sleazy" and rejected the report as "window dressing." Most of the criticism centred on the compensation being offered to people living near the Wimpey site. Westray promised cash, clothes dryers, vinyl siding, and air conditioners to the roughly thirty families in the immediate area; the company was also willing to pay their property taxes and water bills while the strip mine was in operation. The offer amounted to a tacit admission that the area would be choked with dust and dirt. Those were not the only inducements intended to win support for the open pit. Within days of the explosion, the Curragh chairman, Clifford Frame, had promised as compensation an unspecified payment to the families of the dead miners, based on the amount of coal to be dug at Wimpey. Westray extended a similar royalty offer to the Town of Stellarton.

As the September deadline approached for a decision, pressure mounted on Donahoe. Westray employees, working with local clergymen, circulated a petition calling on the environment minister to exempt the project from review; 5,500 people signed, including 500 Stellarton residents. Unemployed miners demonstrated outside Stellarton's town hall to show their displeasure with Mayor Clarence Porter, a vocal opponent of the strip mine. "How do you spell Porter?" they chanted. "U-I-C." Pictou County businessmen, who were out more than one million dollars a month—the amount Westray had been spending on everything from workboots to tractors—threw their support behind the strip mine. The Westray Families Group demanded a full environmental review. Genesta Halloran, one of those widowed by the explosion, denounced Frame's offer to the relatives. The families were being "used" by the company, she said, in an effort to line the company's pockets.

Westray stepped up its public relations campaign as the summer wore on. Phillips ventured into enemy territory to personally lobby Liberal and NDP MLAs. He failed to convert Westray's biggest critics, who complained that the company stood to reap enormous profits by mining Wimpey coal on the cheap—that is, in a less-costly surface operation—to meet its NSPC contract. Phillips did nothing to dispel that notion, and refused to disclose how much it would cost to strip-mine each tonne of coal. He even issued a mild threat, telling one newspaper that the company might abandon the project if the drawn-out environmental review was ordered. "We're not trying to put pressure on the government," he claimed. "What's important right now is jobs."

But leaving Phillips in charge of the controversial proposal was a tactical error by Curragh. As miners came forward with horror stories about working at Westray, the credibility of the man in charge at the time of the explosion sank lower and lower. As grief turned to anger, fingers were pointed at Phillips. Classmates taunted his younger children in the schoolyard. And Phillips could not escape the daily diet of pre-explosion revelations being served up in the media. In July, the *Chronicle-Herald* published a series of stories revealing that Phillips had altered the underground mining layout over the opposition of provincial bureaucrats; a few weeks later, two follow-up articles raised disturbing questions about Phillips's safety record at Alberta's Reiff Terrace colliery in the late 1970s. As Westray struggled to steer the strip mine through turbulent waters, Phillips, the project's chief proponent, was making his own waves.

Donahoe released his decision on September 3. Despite the petitions and the lobbying and the need for jobs, Westray would not be able to take more coal from the Wimpey pit without a full environmental assessment. Donahoe, who admitted he was in "a no-win situation," said his department's officials advised against pushing ahead with the project. Westray's plans to compensate neighbouring residents and manage noise and dust were not enough—other side effects of the strip mine, such as the potential for fire and explosions, required detailed scrutiny. "There are legitimate, significant, and vitally important environmental issues which have to be addressed here," Donahoe insisted.

With a strip mine tied up in environmental red tape until the summer of 1993, Westray was left with no short-term source of coal and cash. The company announced it would proceed with the environmental review rather than abandon the project. "We're sorry that the working people of Pictou County cannot get on with the jobs that would have been created by the mine," said Diane Webb, Clifford Frame's executive assistant, who had been seconded to Westray after the explosion. Within a week, the company laid off 17 management-level employees, bringing the number of people thrown out of work since May to about 150. Only a handful of employees remained on staff to act as security guards. The Nova Scotia Power Corporation again turned to Devco for coal to fuel its Trenton plants.

Westray miners lucky enough to find jobs at mines in other parts of Canada had been fleeing Pictou County for months; for those sticking around in hopes of snagging a job at the strip mine and eventually going back underground at Plymouth, the decision was a devastating blow. "We've got a lot of miners that just don't want to stay around and fight any more," said Teddy Deane, a laid-off miner in his early twenties from Westville. "They're mentally exhausted." In an area where jobs were few and far between, employment prospects for many of Westray's former workers were grim.

* * *

LONG BEFORE the Families Group called on the RCMP to take control of the Westray investigation, the Mounties had been quietly building a case for criminal charges. A team of a half-dozen investigators, headed by veteran Staff Sergeant Ches MacDonald, was working out of the RCMP's detachment in Truro, about seventy kilometres west of Stellarton. Between late May and the end of August, the police interviewed more than two hundred Westray employees, past and present, looking for evidence of unsafe practices to support charging the company with criminal negligence causing the miners' deaths. In June, detectives seized four plastic bags of coal dust that had been taken from the mine before the explosion. The samples were sent off for laboratory analysis to determine whether the amount of combustible material was within legal levels. Then, in early September, the RCMP investigation went public. A team of Mounties showed up at the Halifax office of the public inquiry with a search warrant demanding hundreds of Westray's internal documents and files—the same material the company had turned over shortly after the shredding allegations surfaced in late May.

The seizure netted the police a gold mine of material, everything from computer print-outs from underground monitoring equipment, to employees' notebooks and confidential memos exchanged between officials of Westray and its parent company, Curragh. It also provided the first public glimpse of the scope of the criminal investigation. In documents filed at the New Glasgow courthouse before the search warrant was issued, the RCMP accused Curragh and Westray of criminal negligence in the miners' deaths, "by operating a coal mine without due regard to [the] safety of its employees." The Mounties levelled a host of allegations. The company had made coal production "paramount" and neglected the safety of the workers. The level of coal dust was excessive, training was inadequate, and the ventilation system and methane detectors had been altered to keep mining machines operating. According to the police, illegal activity had continued after the explosion; maintenance crews sent into the wrecked mine after May 9 had modified mining equipment to make it conform with safety laws. The Mounties had not sealed off the mine until May 21, twelve days later.

The RCMP hoped to prove not only that the Westray mine was unsafe, but also that conditions were so bad, and condoned by management for so long, that the company's actions amounted to a breach of the criminal law. The Criminal Code of Canada defines criminal negligence as conduct that shows "wanton and reckless disregard for the lives or safety of others." The conduct can either be an accused's actions, or an accused's failure to do something that person or company was duty-bound to do. In other words, the company can be held criminally responsible if employees are forced to work in unsafe conditions, or if its officials fail to comply with safety laws.

By a macabre coincidence, Westray was not the only mine disaster the Mounties were treating as a criminal case in September 1992. A week after the police took possession of the inquiry documents, an underground explosion at the Giant Yellowknife gold mine killed nine workers. The blast had occurred during a bitter strike, and investigators quickly concluded that a bomb had been planted and the miners had been murdered. The act of deliberately planting a bomb, however, bore little resemblance to the long-standing safety lapses being used to build a criminal case at Westray.

Treating the Westray disaster as a possible criminal act increased the stakes for those who might be charged. If the Labour Department's probe led to charges under Nova Scotia's mine safety laws, the maximum penalty would be a year in jail and a $10,000 fine. But criminal negligence causing death is one of the most serious crimes on the books. A conviction can bring a life sentence in prison for an individual; a corporation faces a fine, with no limit on the amount a judge can impose. In search warrants, the RCMP named Westray and Curragh as the targets of its probe. Privately, however, Staff Sergeant MacDonald told lawyers for the mine managers that up to three former Westray officials could face charges.

Seizing documents was only a prelude to a bolder investigative move. On September 17, the Mounties showed up at the mine gate with a search warrant authorizing them to take over the site for two weeks. The police would use that time to send two specially trained officers underground to collect dust samples, take photographs, and scour the mine for other pieces of evidence. The officers would be escorted by teams of draegermen, many recruited from the ranks of Westray's unemployed miners. Using oxygen equipment, as they had during the rescue effort, the draeger teams would need days to re-establish ventilation and monitor gas conditions. Their ultimate goal was to enter the sealed-off southwest section, the area where the explosion was believed to have originated.

Returning to that area four months after the explosion was a risky business. The southwest section was blocked off with a plywood barrier, reinforced with hundreds of bags of stone dust, that had to be dismantled. Mining consultants warned that high concentrations of methane had built up behind the seal. When the section was reopened, fresh air from the mine would mix with the gas; the result would be areas where methane was diluted to its explosive concentrations, between 5 per cent and 15 per cent. And there was the ever-present danger of more cave-ins. Westray had backed out of the planned foray into the southwest section of the mine a month earlier, claiming the mission was too dangerous.

The company stuck to its guns as the RCMP took control of the mine site and began commandeering equipment. In a public statement, Westray attacked the RCMP's underground search as a "poorly conceived scheme" that risked lives and the mine's stability. The company washed its hands of the affair, serving notice that it would sue the Mounties if their actions caused further damage to the mine. The Department of Labour and inquiry officials also begged off, saying the risks outweighed the value of any evidence that might be recovered. As inquiry lawyer John Merrick put it, "the last thing the inquiry needed was a second Westray tragedy." The RCMP, however, was determined to go it alone. Sergeant Bill Price, a spokesman for the force, questioned the company's motives, speculating that "they don't want us to see what's down there."

Despite the dire warnings, the search went off without a hitch. After a week of preparation and repair work, the RCMP officers and their support teams finally entered the southwest section, on September 24. Gas levels topped 90 per cent in some areas, making the air unbreathable but posing no explosion risk. Working in four-hour shifts, the investigators spent days collecting dust samples and examining the continuous miners and other machines being used in the section at the time of the explosion. Corporal Harry Ullock of the RCMP's Pictou detachment, one of the Mounties who volunteered to go underground, examined and photographed ten pieces of machinery in the southwest section. A farm tractor of the type Westray used to transport men in and out of the mine, he reported, was the only piece of equipment that showed signs of damage from the explosion.

The RCMP relinquished control of the mine at the end of September, after collecting scorched mining helmets, damaged self-rescuers, a machine-mounted methane detector, and scores of other pieces of potential evidence. But one item removed from the mine during the two-week operation was not destined to end up as an exhibit at some future trial. Deep underground, the draegermen working with the RCMP located a metal lunchpail that had belonged to Eugene Johnson. After the search, they presented it to Johnson's eldest son, eleven-year-old Michael. It survived the explosion intact, a tangible reminder of a father who knew where to find all the good fishing holes, a father who would never again come home from work with that lunchpail in his hand.

* * *

THE RCMP's HEROICS upstaged the Richard inquiry, which was set to begin public hearings in mid-October. But the criminal probe threatened to do more than overshadow the inquiry's efforts. For months, legal experts had been warning that a Westray inquiry and a Westray criminal prosecution could not coexist. "It should be one or the other," Mr. Justice Samuel Grange, the Ontario judge who headed a 1983 inquiry into the deaths of twenty-eight babies at Toronto's Hospital for Sick Children, told the *Chronicle-Herald* in July. A nurse, Susan Nelles, faced criminal charges; to avoid prejudicing her case, the Ontario courts prohibited Grange from naming those responsible for the deaths in the inquiry's final report. Nelles was later exonerated. In 1990, the Supreme Court of Canada went a step further, shutting down an Ontario inquiry into the Patricia Starr fundraising scandal because it crossed the fine line between an inquiry and a criminal prosecution.

The Westray inquiry would not be able to walk that line without a fight. In mid-September, as the RCMP seized control of the mine, lawyers for the mine's top officials took legal action. The Nova Scotia Supreme Court was asked to delay the inquiry's public hearings, at least until any criminal charges cleared the courts. Former underground manager Roger Parry was first off the mark, but the challenge was quickly joined by Gerald Phillips and five others who had held high-level posts at Westray. Curragh shied away from the challenge, preferring to let its former employees take the heat for trying to prevent the long-awaited inquiry from beginning its work. The company, which was on record as welcoming the chance to clear the air at the inquiry, did its part by paying the managers' hefty legal bills. Inquiry staff vowed to put up a fight. "This has been a disaster that has affected deeply all Nova Scotians," inquiry lawyer John Merrick said. "It's time this thing was brought to the table so people can see once and for all what caused the Westray explosion."

The managers attacked the inquiry on two fronts. The first argument was based on the division of powers set out in the Constitution. The inquiry had been established by the Nova Scotia government, but its terms of reference threatened to spill over into the realm of criminal law, a federal jurisdiction. As a special examiner under the Coal Mines Regulation Act, Justice Richard was required to report any evidence or suspicion of "culpable neglect" in the mine's operation. The judge was also expected to determine whether neglect "caused or contributed to" the explosion. Culpable neglect, the lawyers maintained, was synonymous with criminal negligence. The inquiry was also mandated to decide whether Westray and its employees had complied with the laws and regulations governing mining, which could be interpreted as including the criminal law. To the managers, this meant the inquiry could point fingers and name names. If the courts accepted their argument, the inquiry was dead in the water. The provincial Supreme Court would have no choice but to strike down the probe as unconstitutional.

The other weapon in the managers' arsenal was the decade-old Charter of Rights and Freedoms. The charter gave teeth to an array of legal rights for those facing criminal or mine-safety charges; the inquiry's public scrutiny of events surrounding the disaster threatened to ride roughshod over those rights. With the RCMP investigation grabbing headlines, and the deadline looming for the Department of Labour to lay charges, the prospect of charges was real. If any of Westray's managers were named in the charges, they had the right to a fair trial, free of prejudicial publicity. The inquiry's high-profile hearings, on the other hand, promised to be front-page news in Nova Scotia for months. Those accused could be forced under a subpoena to testify at the inquiry, violating their right to remain silent. Incriminating evidence presented at the inquiry could also be used against them in the subsequent prosecutions. To preserve those rights, the court was asked to put the inquiry on hold until trials on all charges arising from the disaster, criminal or otherwise, were completed.

The legal challenge, if successful, would restrict the work of every public inquiry set up in Nova Scotia, and possibly across the country. If the courts accepted the jurisdictional argument based on the division of powers, future inquiries would be restricted in their ability to investigate events that could lead to a criminal prosecution. But the Charter rights being invoked by the managers could derail any inquiry into any matter that might lead to the laying of charges under the Criminal Code or other statute. Since inquiries were invariably set up in response to disasters or other controversial events, there was always the possibility that someone involved would face prosecution. In Britain, the practice was to let inquiries do their work and forgo prosecution of those to blame, no matter how strong the evidence against them. That was not the policy in Canada, but the Westray case would force the courts to decide whether a public inquiry and a prosecution could coexist, and on what terms.

The managers cleared the first hurdle with ease. On September 30, Chief Justice Constance Glube of the Nova Scotia Supreme Court's trial division ordered the inquiry to hold off on public hearings. It was a temporary injunction, designed to allow a full hearing of the legal challenge at the beginning of November. During its four-month existence, the inquiry had spent close to one million dollars in public money preparing for the start of hearings. It had not heard a single witness or presented a shred of evidence pointing to the cause of the explosion. And now the entire effort was sidetracked in a legal fight that could put the inquiry's work on hold—permanently.

* * *

WITHIN DAYS of the order halting the inquiry, the spectre of charges arising from the Westray disaster became real. On October 5, the Department of Labour wrapped up its investigation by laying more than fifty charges against Curragh and four former Westray officials. The charges, laid under the Occupational Health and Safety Act and the Coal Mines Regulation Act, covered a wide range of safety violations in the month preceding the explosion. The mine officials charged were four of the seven who had mounted the legal challenge to the inquiry, and all were British expatriates who had worked with Gerald Phillips at other mines. Phillips and Roger Parry each faced thirty-two of the counts, reflecting their key management posts at the mine. Parry's twin brother Robert, formerly in charge of maintenance of the mine's equipment, was named in eleven. Glyn Jones, who had held the post of assistant superintendent, one rung below the underground manager, faced twenty-two charges. Their former employer, Curragh, was charged eighteen times. Many of the accused were jointly charged with the same offence, bringing the total number of charges to fifty-two. Each offence was punishable upon conviction by a maximum $10,000 fine and, for the former managers, up to a year in jail.

The charges encompassed many of the allegations former Westray miners had been making for months—forcing men to work in unsafe conditions; allowing discarded fuel cans to litter the workings; using machinery not certified for underground use; failing to properly train workers; and improperly using acetylene torches in the mine. Coal dust and methane figured prominently in several charges. Curragh, Phillips, Jones, and Roger Parry were accused of failing to take steps to deal with the build-up of dust "to prevent an explosion." The same four were charged with failing to withdraw miners from the mine's southwest section, the ignition point of the explosion, when methane levels exceeded 2.5 per cent. Curragh and all four former managers were charged with allowing workers to operate a continuous miner while the methane detector was not functioning. The list went on and on.

The sheer volume of charges was a stinging indictment of the Labour Department's lax enforcement practices before the explosion. And there could have been more; the six-month time limit for filing charges meant infractions committed before April 1992 could not be prosecuted. As had been the case throughout Nova Scotia's long history of coal mining, it had taken a tragedy to prod the government into taking a tough stance on safety. "The charges that have been laid bring into question the department's own conduct almost as much as it brings into question the conduct of the company and its top officials," huffed the Liberals' Bernie Boudreau. Opposition critics and the Westray Families Group asked the obvious question: why was nothing done to stop unsafe practices before the explosion, when there was still time to avert disaster? Even Westray's lawyer, Bruce MacIntosh, criticized the department for failing to investigate the actions of its own staff, to determine whether charges should be laid "within the department itself."

Labour Minister Leroy Legere surfaced a day after the charges were laid, and defended his department's hands-off approach to Westray before May 9, 1992. "The department issued orders and expected compliance," he told a crowded press conference in Halifax. "The department does not, nor I believe is it expected to, or intended to, provide a constant daily surveillance of any operation." Appearing bored with the onslaught of questions, Legere claimed his staff had done their jobs "to the best of their abilities." No-one had been disciplined or fired for their handling of Westray, he confirmed, rolling his eyes as yet another reporter asked about the inspectors. And he denied suggestions that the Conservative government's support for Westray had made the inspectors reluctant to take a harder line on safety. Although the public inquiry could be on hold for months, possibly even years, Legere insisted that was the proper forum for determining whether labour officials had acted properly.

* * *

THE SECOND ROUND of the legal battle over the Westray inquiry opened on November 2 in a cavernous, oak-panelled courtroom on the Halifax waterfront. Constance Glube, an experienced judge and the first woman to hold the post of chief justice anywhere in Canada, was back on the bench. For the next three days, she was bombarded with legal arguments, charges, and countercharges. Bruce Outhouse, a Halifax lawyer acting for former mine manager Gerald Phillips, attacked the probe as "a criminal investigation masquerading as a public inquiry." The terms of reference, he said, required Justice Richard to single out those to blame for the disaster, crossing the jurisdictional line laid down by earlier court rulings. Robert Barnes, who represented five of the seven Westray officials, warned that the "media free-for-all" surrounding the inquiry's hearings would make it impossible for those facing Labour Department charges—and criminal charges, if laid—to get a fair trial. Testimony and documents presented at the inquiry would offer a windfall of evidence not normally available to police investigators. Prosecutors, Barnes predicted, would "reap the benefits of this inquiry in spades."

Lined up against the managers were lawyers for a half-dozen of the parties granted status to appear before the inquiry. John Merrick, the inquiry's lawyer, vowed that the probe, regardless of the wording of its terms of reference, would refrain from naming those responsible for the explosion. Pretrial publicity could be minimized through bans on publication of some evidence, he argued, and those facing charges would not be compelled to testify at the inquiry until their trials were over. The lawyer for the Steelworkers union, Ray Larkin, pleaded that the inquiry was needed to examine safety conditions and government regulation—issues that would be overlooked in prosecutions arising from the explosion. "We hope we'll find out [at the inquiry] what went wrong and why," he said, "and prevent its recurrence."

Caught in the middle of the legal debate were lawyers for the Nova Scotia government. The inquiry, whatever the shortcomings of its mandate, was the government's creation. Now, six months after the explosion, the Cameron government was doing its best to disown its troubled offspring. The Attorney General's Department, the government's legal arm, was trying to be all things to all people. As purveyor of legal advice to the various departments of government, it was supposed to defend mine inspectors and other government officials at the inquiry. As author of the inquiry's terms of reference in May 1992, it was called on to defend them in court. And as the department responsible for the administration of justice within in the province, it had a duty to ensure that the inquiry's work did not jeopardize the prosecution of charges arising from the disaster. The department resolved its dilemma by asking Glube to delay the inquiry. The "common-sense and appropriate" approach, argued government lawyer Reinhold Endres, was to put prosecutions ahead of the inquiry's hearings.

The concerns raised by the attorney general's lawyers were valid, but the timing was suspect. The Tory government was in the final year of its mandate, and rumours that an election was imminent had been circulating since the summer. Cameron, who commanded a slim majority in the House of Assembly, was running out of time to prove he could win the approval of the electorate—and not just the support of delegates to a party convention. The Westray explosion, and the constant opposition attacks on Cameron's role in promoting the project, affected the premier's popularity; a June poll showed his approval rating had dropped seven points since the start of the year, but he was still the most popular party leader in the province. The Liberals, meanwhile, were making it easy on the Tories. The Grits had turned on leader Vince MacLean, who had come close to winning it all in 1988 but was widely perceived as being more of a street fighter than a future premier. After barely surviving a vote for a leadership review at a bitter annual meeting early in 1992, MacLean resigned. His replacement was John Savage, a physician and longtime mayor of the city of Dartmouth. Savage, who won the leadership in mid-June, was dead last in the polls, adding to the speculation that Cameron would call a snap election and put the rookie opposition leader to the test.

But if the Cameron government went to the polls in the fall, it ran the risk of being upstaged. The Westray inquiry threatened to unleash a daily diet of news stories revealing the government's role in financing and inspecting the mine. Some political observers suggested the government might call an election before the inquiry began its work, to avoid potentially damaging revelations. Then again, that could be seen as an acknowledgment that the Tories had something to hide. Cameron, for his part, scoffed at suggestions that the timing of the inquiry would dictate when he called the election. "I have nothing to fear from the inquiry," he insisted shortly after the temporary injunction was imposed. "I know my involvement, and it was my hope that the inquiry would be going on right now."

Despite Cameron's confidence, the inquiry could deal a serious blow to a government already straining under the weight of fourteen years in power. The inquiry had served notice that it would seek answers to tough questions about the government's dual role as Westray's chief watchdog and biggest cheerleader. Inquiry lawyer John Merrick, in a written submission to Glube, outlined some of the avenues to be explored. Was Westray under political or financial pressure to produce and, if so, did this impair the safe operation of the mine? Did officials of the departments of Labour and Natural Resources exercise proper supervision and control? Why was Curragh, a firm that specialized in lead and zinc mining, allowed to take on the project? Depending on how the inquiry viewed the efforts to regulate Westray, the answers could prove embarrassing in the heat of an election campaign. With the inquiry shelved, Westray could not come back to haunt the Cameron government until well after the next election.

Against that background, the government's motives for requesting a delay were suspect. The move was criticized as an abuse of the legal process, or, at least, an effort at damage control. The attorney general's lawyers took heat for not recognizing the legal pitfalls when the terms of reference were being drawn up. The Liberals and New Democrats challenged the government to halt the inquiry and face the political consequences, rather than have the courts do the dirty work. The inquiry, abandoned by its architects in government, fought back. "To expect the court to make a policy or political decision in these circumstances is much more likely to bring the administration of justice into disrepute," Merrick warned in a rare outburst. But Attorney General Joel Matheson protested that any attempt by government to limit the inquiry would be construed as politically motivated. Since the issue was already in the hands of the courts, he said, the chief justice should decide the inquiry's fate.

It took Glube just over a week to hammer out a decision. She handed down a fifty-five page ruling on November 13, and struck down the inquiry as beyond the scope of the Nova Scotia government's powers. "It is unfortunately my opinion that the terms of reference . . . are so specific to the [May 9] occurrence, and overlap so deeply into the field of criminal law, that as they are presently written, they cannot stand," she wrote. She followed the Supreme Court of Canada's precedent in the Starr case, declaring the inquiry invalid because it intruded into matters of criminal law, which was the preserve of the federal government. Once Glube had reached that conclusion, the fair-trial rights of the accused were a moot point. The bottom line was that the inquiry was out of business. But Glube came out in favour of some form of inquiry into the disaster, and invited the government to draw up a new mandate that avoided issues of criminal law.

The inquiry was left in "legal limbo," as Merrick put it. The ink was barely dry on Glube's ruling when the Steelworkers union, the inquiry's strongest ally, launched an appeal. The government, which had argued only for a delay, was back to the drawing board. Matheson, the attorney general, said the government would either appeal or draft new terms of reference along the lines Glube had suggested. The government eventually opted to let the appeal courts decide whether the inquiry could be salvaged.

In the meantime, four of the Westray officials who had challenged the inquiry had another court date. Gerald Phillips, Roger and Robert Parry, and Glyn Jones, along with lawyers for Curragh, were to appear in a New Glasgow court in December to enter pleas on the Labour Department safety charges. A trial was expected by the spring of 1993. Even with the inquiry out of the picture, perhaps for good, the events leading up to the Westray disaster would soon be aired in a public forum. But the RCMP was hard at work on the Westray file, and criminal charges seemed to be a foregone conclusion. And that raised new legal questions about whether the police and Department of Labour could lay two sets of charges against Westray officials. The Mounties' investigation, one of the weapons the mine managers had used to sink the inquiry, was about to throw a wrench in the Labour prosecution as well.

DAYS OF RECKONING

HE LABOUR DEPARTMENT'S WESTRAY PROSECUTION was too little, too late, to save Leroy Legere's job. For months, Donald Cameron had backed the beleaguered minister, ignoring the chorus of demands that Legere resign or be fired. The minister should not be punished for the mistakes committed by his staff, the premier protested, even though there was abundant evidence of Legere's own shortcomings. But the lack of leadership was not limited to the ministerial level. A consultant's report later revealed that the explosion and the resulting controversy had a paralysing effect on the department's staff and day-to-day operations. And there was still no resolve to get tough on companies violating safety laws. In the fall of 1992, labour inspectors repeatedly ordered Maritime Steel and Foundries to rectify hazardous conditions at its metal-fabricating plant in New Glasgow. But charges under the Occupational Health and Safety Act were laid only after a worker was crushed to death by a steel beam.

Legere's post-Westray reprieve soon ran out. Cameron shuffled his cabinet on November 19, stripping Legere of the labour portfolio but keeping him on as minister of fisheries. Cameron refused to concede that Legere had been demoted, and that Westray was the reason. "I expect that we're going to have a lot of troubled waters in the fisheries," he said, "and I think it's important that we have a minister that can spend his whole time on that issue." Legere's replacement was Tom McInnis, who had put in a lacklustre performance as minister of economic development. But McInnis, the deputy premier, had a reputation as a minister who could turn around a troubled department. He had been attorney general in the late 1980s, when Nova Scotia's justice system was in the throes of the Marshall inquiry. Now he was being handed the challenge of cleaning up the mess at Labour.

McInnis quickly made his presence felt. Just hours after being sworn in as minister on November 25, he announced a comprehensive review of the department's occupational health and safety division. Legere, his credibility all but gone, had promised a similar in-house study a few weeks before he was bounced from the portfolio. His successor, however, promised a thorough, independent examination of the division's management practices, under the direction of Nova Scotia's auditor general. McInnis also bowed to pressure for an overhaul of the province's worker-safety laws. The Advisory Council on Occupational Health and Safety—a body made up of government officials, employers, and union representatives—was directed to recommend improvements to mining and other regulations. He promised to hire more inspectors if they were needed to ensure existing laws were properly enforced. And he made it clear heads would roll. "I'm not a hatchet man," McInnis said, "but there are obviously going to be some changes."

The following day, McInnis brought aboard a new deputy minister to oversee the reform drive. The appointee was Innis Christie, a former dean of Dalhousie Law School, who was held in high regard for his experience in labour law and arbitration. The department's deputy minister, Hugh MacDonald, was shunted into retirement. Jack Noonan, executive director of the occupational health and safety division, was not so lucky. Noonan, the target of criticism from organized labour long before the Westray explosion, was fired the following week from his $77,000-a-year post. "We've been on record for some time as having no confidence in Jack being able to perform that job," said Rick Clarke, president of the Nova Scotia Federation of Labour. The New Democratic Party's labour critic, Robert Chisholm, praised the government for moving to "rid the division of what was a real problem." Noonan, who was not offered a severance package, sued the government for wrongful dismissal. According to documents filed in the lawsuit, the department had referred Noonan to the Nova Scotia Commission on Drug Dependency, an agency that deals with alcohol and drug abuse, in February 1992. He was undergoing treatment at the time of the Westray explosion.

In two weeks, the government had dumped the top three officials responsible for worker safety in Nova Scotia—Legere, MacDonald, and Noonan. McInnis refused to discuss the reasons for Noonan's firing. And the premier had nothing but praise for MacDonald, the outgoing deputy minister. "He has served the people of Nova Scotia well over the years," Cameron said in a news release, "and I wish him an enjoyable retirement." But the purge ended with Noonan; chief mine inspector Claude White, who reported to Noonan, along with supervisor Colin MacDonald, inspectors Albert McLean and John Smith, and other labour officials who dealt with Westray, kept their posts. By implication, the source of the Labour Department's problems had been at the most senior levels. The front-line officials involved with Westray—the group that had borne the brunt of the public outcry unleashed after the explosion—emerged unscathed. No other department employees were disciplined, demoted, or fired over Westray.

* * *

It would be up to the public inquiry to pass judgment on the conduct of the Labour Department's mine inspection branch–assuming there was an inquiry. Its fate was in the hands of the Supreme Court's appeal division, Nova Scotia's highest court. At the request of lawyers for the United Steelworkers of America, the Westray union, the court moved the case to the top of its docket. "The court should move quickly to resolve the uncertainty," said the appeal division judge, who set a hearing for mid-December, less than a month after the inquiry was struck down. The quick service did not sit well with the Nova Scotia government, which was toying with the idea of starting from scratch–a move that would obviate the need for an appeal. A new inquiry, armed with new terms of reference that avoided criminal-law issues, could still probe safety issues raised by the disaster. On the other hand, critics warned that a scaled-down inquiry would not have the power to investigate the causes of the explosion, and how future disasters could be prevented. The government eventually backed off, and left the appeal court to sort out the legal mess.

Besides, there was more at stake than just the Westray inquiry. Chief Justice Constance Glube's ruling, if allowed to stand, would restrict the government's power to set up inquiries in the future. If there were another incident or disaster with criminal overtones, a royal commission or other government-sponsored probe might be precluded. The court ruling, not the public interest, would be the deciding factor in whether an inquiry could be held. Glube's ruling was not binding on courts outside Nova Scotia, but it was a model for other provincial governments that faced calls for public inquiries into controversial incidents or issues. Faced with those high stakes, the government came onside. When the appeal was heard on December 11, the attorney general's lawyers were among the strongest defenders of the Westray inquiry.

The former mine officials who had blocked the inquiry fought back. "There are three intensive investigations focused on the same event," protested one lawyer acting for the group. "They're ploughing over the same ground and coming up with the same allegations." The RCMP proved that point in early December by using a search warrant to seize more documents from the inquiry. This time, the Mounties tapped into the mountain of government files that had been turned over to the inquiry, including the inspection reports of the Labour Department, which was itself under police investigation.

The appeal court handed down a ruling on January 19, 1993, overturning the Glube decision and reinstating the public inquiry. Three appeal court judges took the view that the inquiry's terms of reference did not amount to a substitute criminal investigation, and that the probe could go ahead. But there was still the problem of the competing Labour Department prosecution and the spectre of criminal charges. "There is a great deal of merit in a regime which requires a government to either lay charges or conduct a public inquiry, but not to do both," Mr. Justice Doane Hallett wrote on the appeal court's behalf. With the inquiry back on the rails, the judges had to deal with its potential impact on the constitutional rights of those facing Westray-related charges. The public's right to answers from the inquiry had to be balanced against the fair-trial rights of those accused of safety violations or crimes. The court came down on the side of protecting the rights of the managers, and imposed a ban on public hearings until all trials arising from the disaster were completed.

The managers lost one legal battle, but won the war. "Our clients didn't have any interest in stopping an inquiry for all time," noted Robert Barnes, lawyer for five of the seven former mine officials. "The purpose of doing what we did was to stop this thing before it did damage to their fair-trial rights." There was no time limit on when the inquiry could start the public phase of its work. The Labour Department's charges were expected to clear the courts by the summer of 1993; if criminal charges were laid, however, it would be two years or more before the inquiry heard its first witness. In the meantime, the inquiry's skeleton staff would quietly gather evidence and take part in the department's review of coal-mining laws. The other option, an appeal to the Supreme Court of Canada, was a long shot. Assuming the country's highest court decided to hear an appeal, it could be a year before a ruling was handed down. But if no criminal charges were laid, the Westray inquiry could hold hearings as early as the fall of 1993. The inquiry's timing hinged on whether the RCMP could build a case for criminal charges.

* * *

IN NOVEMBER 1992, with the public inquiry sidetracked in the courts, another legal battle flared up. This time, Westray Coal was pitted against its former miners and the families of the dead miners. Word got out that the company was going to resume pumping water into the partly flooded mine; a passerby noticed workers laying pipe at the mine site in Plymouth. The southwest section, which had been examined by the RCMP in September, would finally be submerged. The Westray Families Group, claiming the move would destroy evidence needed to pinpoint the cause of the explosion, went to court on November 20 and won a temporary injunction halting the work.

On the morning of November 23, a few dozen laid-off miners and relatives of many of those killed in the explosion staged a protest at the mine. "My husband and twenty-five others died in that mine, and I want to know why," said Donna Johnson. If flooding went ahead, "it's all over as far as a full investigation is concerned," claimed Jay Dooley, an underground supervisor at Westray at the time of the blast. "Equipment similar to what was used in the Plymouth mine is being used in mines in Western Canada today," explained Dooley, an experienced draegerman who had accompanied the RCMP into the southwest section in September. "If the Westray explosion can be traced to a failure of the equipment, the same thing can happen in other mines." Another potential clue to the cause of the explosion was the plywood stoppings used to seal up the Southwest One section about a month before the explosion. If those seals were inadequate or had been improperly installed, they might have allowed methane to escape, touching off the blast. The stoppings were within the area slated for flooding.

But the alarm being sounded at Plymouth was shrugged off in other quarters. Westray announced that flooding was necessary to stabilize the mine, and said its consultants advised against re-entering the southwest section. The Mounties were satisfied with the evidence collected from the mine in September, and a spokesman for the force said no further underground search was necessary.

Lawyers for Westray went before a Nova Scotia Supreme Court judge in Halifax on November 24 in a bid to overturn the injunction. Brian Hebert, a lawyer for the Families Group, said relatives of many of the dead miners planned to sue Westray and Curragh for negligence in the operation of the mine. The Nova Scotia government, responsible for mine inspection, and the federal government, which had provided the financial backing, would also be named in the lawsuits. If flooding went ahead, Hebert said, the families feared evidence crucial to the lawsuits would be lost. He pleaded for a one-month extension on the flooding ban, which would give the families a chance to organize their own underground search.

Mr. Justice Hilroy Nathanson, who was clearly sceptical about the legal basis for the injunction, granted Hebert the right to question the mine manager. Dan Currie, hired to run Westray shortly after the explosion, acknowledged that flooding was part of an elaborate plan to recover the eleven bodies still underground and put the mine back in production. After the water was pumped in, concrete seals would be installed in the main tunnels to guard against another explosion. Then, from a point about halfway down the existing slope, new tunnels would be driven into the Foord seam so that mining could resume. Currie said the new tunnels would be used to reach the flooded North Mains, where the bodies of eleven men were entombed. After the water was pumped out, Westray hoped to break into the North Mains to search for remains. If the Nova Scotia government approved the plan, he said, tunnelling could begin within six months.

All this was news to the two dozen miners and relatives who drove to Halifax for the hearing. And their reaction ranged from surprise to anger. "These people are talking about starting up a coal mine again," said Wayne Cheverie, a Westray draegerman who worked the shift before the explosion. "Who in their right mind is going to go into a coal mine that blew up, and nobody knows why it blew up?"

The company's motives for recovering bodies were also suspect. "They just want to go down there and make money," said Joyce Fraser, choking back tears as the TV cameras rolled. "We're going to fight all the way. We'll fight. They took our husbands away from us."

When the hearing resumed the following day, the company did an abrupt about-face. Westray lawyer Bruce MacIntosh announced that flooding would be shelved for at least three weeks, giving company officials a chance to explain their plans to the families and the miners. "It would be a very Pyrrhic victory if the flooding began, and yet all of the widows and all of the family members of the twenty-six men felt that the system had done them in again and evidence was being destroyed," MacIntosh explained. Westray's opponents saw the reprieve as an opportunity to pressure the provincial government to provide the money and expertise needed to search the southwest section. "Hopefully we can get into the mine and . . . fix it up so experts can go look at it and find out what happened to our twenty-six buddies," said Jay Dooley.

But not all Westray miners shared that view. As early as June, many miners and draegermen had come forward to demand that the mine be flooded as a safety measure. Now that flooding was a cornerstone of Westray's bid to resume coal production, it was the miners' best hope for getting back to work. Randy Facette, president of Local 9332 of the Steelworkers, came out publicly against any further search of the mine. Facette argued that underground conditions were too dangerous to put more lives at risk; one spark, he claimed, and "you've got May 9 all over again." Despite Facette's opposition, a majority of union members at a meeting in early December voted in favour of delaying flooding. Facette promptly resigned as president. Dooley brushed off the warnings of Facette and his supporters. "I'm not one to contemplate suicide," he contended. "If I thought [the risks] were that great, I wouldn't bother making the effort."

The flooding controversy was also taking its toll on the Westray families. The company held sporadic discussions in early December with individual families and small groups of relatives to explain why flooding was necessary. But it was impossible to break through the animosity and suspicion of those who blamed the company for the miners' deaths. The lobbying effort left many feeling they were being used, that the company was dangling the possibility of recovering more bodies to win their support. "A lot of wives are dreading Christmas over all other times. It's the first Christmas without their husbands, without fathers for their children. . . . They don't need this," noted Kenton Teasdale. "Some of them call it emotional blackmail," the Families Group spokesman added. "I think they're right." Widows who had left Pictou County, and were out of touch with events since the explosion, were more likely to support flooding. Bonnie Atkings, who had moved to Nanaimo, B.C., after her fiancé, Trevor Jahn, died in the explosion, got a rundown on the flooding proposal from Curragh officials during a visit to Nova Scotia. "If the only thing that can ensure safety for others is to flood the mine," she later wrote in a letter to the *Chronicle-Herald,* "that's what should be done."

As Westray's hopes of negotiating an out-of-court settlement evaporated, the Nova Scotia government stepped in. The final decision on whether to complete the flooding of the mine ended up on the desk of the new labour minister, Tom McInnis. At McInnis's request, Westray agreed to hold off on flooding until January 6, 1993, to give the Labour Department time to review the plan. The Families Group dropped its injunction on December 22, but not before Justice Nathanson chastised the families for not going to the government in the first place. Teasdale countered that the Families Group had been forced to go to court because of government inaction. "Has the Department of Labour shown leadership in this?" he asked reporters. "Has the Attorney General's Department shown any aggressive attitude in trying to find out the real story of what happened down there?"

The flooding debate had erupted just as the department's top officials were being swept aside. But with McInnis and a new deputy minister at the helm, the department was taking a hard line on Westray. The department issued a stop-work order at the end of November, based in part on Dan Currie's testimony during the injunction hearing. Currie, Phillips's replacement as manager, admitted entering the southwest section in mid-November, without the department's authorization. He and three other Westray employees walked in about 250 metres and used a hand-held methanometer to take gas readings, which were at near-explosive levels. The Labour Department chided Currie for putting "yourself and the men who accompanied you at risk," but stopped short of laying charges because of confusion over whether permission was required to enter the area. The department issued a new order forbidding anyone from entering any part of the mine without its authorization.

McInnis's effort to settle the flooding dispute got off to a rocky start when the new minister travelled to Stellarton in early December for a get-acquainted session with Westray miners and members of the Families Group. McInnis heard them out on the need to delay flooding, and promised to seek expert advice before making a decision. But the mood turned testy after talk turned to the department's questionable inspection practices before the explosion. Then someone accused the provincial government of trying to delay the public inquiry until after the next election. McInnis defended the government and accused the group of "playing politics."

McInnis returned to Stellarton on January 6, 1993, to break the news that flooding would be allowed to proceed. The minister said every mining expert consulted by his department, including two American consultants retained by the public inquiry, advised that the mine be flooded as a safety measure. The unstable roof and presence of methane made sending down more investigators "a highly dangerous exercise." There was also the cost—an estimated $500,000—and the slim chance of finding any evidence that was not already in the hands of the RCMP.

The decision vindicated Westray, which had been arguing all along that it was flooding for safety reasons, not to destroy evidence. It was expected to take about a week to pump in enough water to finish the job. Some relatives of the dead miners felt betrayed; they had fought the company for more than a month, only to have the government let them down. Robert Bell of Eureka, Pictou County, whose twenty-five-year-old son perished in the explosion, summed up his reaction in two words: "It stinks."

* * *

CLIFFORD FRAME must have been glad to have 1992 behind him. The Westray disaster had left twenty-six miners dead, and his company's reputation in tatters. Not only was Curragh facing trial on charges of violating Nova Scotia safety laws, it was under the cloud of a criminal investigation. And, partly because of the Westray explosion and its aftermath, the company was taking a financial drubbing. Curragh lost $54.5 million in 1992; there was no comfort in the fact that the company had lost twice as much in 1991. Westray alone drained away $30 million to pay for everything from the rescue effort to legal fees. At a time when Westray should have been in full production, and generating an annual cash flow of roughly $20 million, the project was dragging Curragh ever deeper into the red.

As the losses piled up at Westray, prices for lead and zinc continued their downward spiral during 1992. The recession was still taking its toll, and the newly independent states of the former Soviet Union were dumping their production on the world market. Low demand for the metals forced a month-long shutdown of the Faro and Sä Dena Hes mines in the Yukon over Christmas; for the first time since the mid-1980s, not one Curragh employee was taking anything from the ground. To add insult to injury, Curragh had to cough up an unbudgeted $22 million in 1992 for failing to meet the terms of the bank loan used to finance Sä Dena Hes. That blow, combined with the Westray losses, virtually wiped out the money recovered from the unwinding of the Asturiana de Zinc partnership. By the end of 1992, the $70 million in cash in Curragh's back pocket had dwindled to $12 million. The Bank of Nova Scotia, which had recovered every penny of its $95-million Westray loan from the federal government and Curragh by December 1992, advanced a short-term $10-million loan in January 1993 to help out.

But Curragh needed a lot more than $10 million to stay afloat. At the end of January, the company released a thick financial report that laid bare its dire financial straits. Immediate steps had to be taken to get new money into the company. There were a number of possibilities. Curragh could issue new shares and cross its fingers that someone would be willing to invest in a money-losing company. A piece of the undeveloped Stronsay property in British Columbia was up for sale. And the company continued to press the Yukon and federal governments to provide the $29-million loan guarantee needed to develop a new source of ore for Faro. The report ended with a blunt assessment of Curragh's prospects: "If additional capital or short-term financing cannot be obtained, the assumption that the company will be able to continue as a going concern . . . may not be valid."

The only way to end Westray's drain on the Curragh treasury was to re-establish coal production. Coal from the Wimpey pit, processed at the Plymouth wash plant, would generate a cash flow of about $12 million over three years, but environmental hearings were expected to delay government approval of the strip mine until at least September 1993. There was another potential source of money; in December 1992, Westray's insurers had paid $16 million to settle the policies on the damaged mine. The federal government claimed 85 per cent of the insurance proceeds as partial compensation for repaying $80.75 million of the $95-million Westray loan. But Curragh argued that the money should be used to repair the underground mine and build new tunnels for the recovery effort. Once Westray was back in production, the company could begin repaying its massive debt to Ottawa and the Nova Scotia government. While Curragh and federal officials argued over how to split up the insurance proceeds, the $16 million sat in a trust account, out of the company's reach.

The first step in reviving Westray was to find a replacement for mine manager Dan Currie, who was eager to get out of the spotlight. Curragh's choice for the daunting task was Richard Rouse, a giant of a man with three decades of coal-mining experience in the United States. Rouse, an Illinois-based mining consultant, had formerly been in charge of a dozen open-pit and underground mines for the Ohio-based Consolidation Coal Company. "When I reopen this coal mine, we will have our plans laid out with safety our number-one goal," Rouse vowed when he took over in January 1993. With a union in place and the Department of Labour breathing down the company's neck, Rouse said, there was no way the mistakes of the past could be repeated. Rouse also took over as Westray vice president, replacing Gerald Phillips, the man in charge at the time of the explosion. Phillips, who had become a liability in Nova Scotia, had been transferred to head office in Toronto shortly after the strip mine fell through. He left the company in early 1993 and headed off to Florida, where he put in a bid on a small nursing home. Phillips's reputation was on the rocks, and his chances of finding a new mining job were slim; after twenty years in the coal business, he was ready for a career change.

Westray's mining plan was delivered to the Department of Natural Resources on January 8, two days after flooding was approved. A pair of thousand-metre tunnels would be dug over a two-year period, putting thirty miners and another twenty surface staff back on the payroll. Tunnelling and repairs to the existing mine would cost about $25 million. It was billed as an effort to recover bodies, but Curragh acknowledged that the new tunnels would also be used to resume coal mining. And the company was quick to point out that it might be too dangerous, or technically impossible, to drain the flooded North Mains and send in crews to search for what remained of bodies submerged for close to a year. The plan flew in the face of Clifford Frame's assurances, at the time of the explosion, that the mine would remain closed until an inquiry cleared the air. But with the inquiry on hold, Rouse argued, it was time to put people back to work and repay the government loans that were in default. "It doesn't make sense to let such a significant investment lie idle and deteriorate if it can be mined safely."

Curragh's offer did not wash with the Westray Families Group. "They're outraged. In plain English, they're not buying this," spokesman Kenton Teasdale said, after about one-third of the group's one hundred members met to discuss the proposal. Even those with the body of a father, husband, or brother still underground distrusted the firm's motives. "Westray's commitment to recover bodies means nothing," Teasdale charged. "Westray will say whatever it has to to get a mining permit, but do whatever it wants." The group launched a petition drive demanding that the Nova Scotia government reject the proposal and turn the mine over to Devco or some other "credible operator." The company reacted by attacking Teasdale, questioning whether his views represented those of most of the families. John Forgeron, Westray's personnel director, said there was no opposition when he met with family members to explain the proposal. "Do they want the bodies or not?" he asked one reporter. Curragh was willing to spend millions of dollars to search for bodies; Forgeron's remark, born of frustration, made it clear that the company expected gratitude from the families, not flak.

Frame displayed more tact when he showed up in Stellarton on January 23 to make a personal pitch for the plan to reactivate Westray. It was the first time Frame had been to Nova Scotia since the memorial service to the Westray miners in mid-May. The atmosphere was tense as he faced the families in a question-and-answer session. Since the meeting was behind closed doors, the talk was frank and the accusations flowed freely. "They all want to lay blame and string somebody up," Frame complained to reporters after the gathering. "But what happens if two weeks later they find they hung the wrong guy? . . . We must be careful not to become vigilantes."

Curragh, Frame explained, hoped to finance the recovery effort with the insurance proceeds and cash from the strip mine, if approved. "So you might say that to be self-sufficient in the cash we would need the strip mine, yes," he explained. "But if we don't get the strip mine, we'll have to find the money some other way." Frame denied he would seek government aid if the strip-mine revenue did not materialize. But it was not clear who would bankroll Curragh, a company that had lost $150 million in two years, in a costly and possibly futile effort to recover the remains of eleven miners. The payoff, of course, was that the new tunnels would enable Westray to resume coal production. But Frame bristled at the suggestion that the recovery plan was simply a ploy to reopen the mine. "There is no way I want to mine this coal body if I don't make a real attempt at trying to get those people out," he maintained, reminding reporters that he had lost a brother in a mine accident.

That was not the spin the families put on Frame's message. Teasdale emerged from the meeting to denounce the Curragh chairman for using the recovery effort as a lever to get provincial approval for the strip mine. "The message we got is they will need the insurance money and the funds from the strip mine to recover the bodies," he said. Frame was accused of using blackmail to win the families' support for the strip mine, just as, days after the explosion, he was blamed for offering the widows a slice of the profits from strip-mined coal. Frame, who was back in Toronto by the time Teasdale's comments appeared in print, was outraged. "Let me make it absolutely clear. The recovery operation does not depend on the open pit," Frame shot back in a press release. "Any attempt to link the two is simply not true."

Frame also took advantage of the opportunity to lash out at the media for what he called the "spirit of vigilantism" pervading news reports on Westray. "Predictably," he protested, "those whose interests are served by inflaming passions have accused me of using blackmail to recover the remains." A week later, Frame announced that Curragh was about to release the findings of the experts investigating the explosion. With the public inquiry on hold, the move would be an answer to critics who said the mine should not reopen until the cause of the explosion was known. It also offered relatives and friends of the dead miners an explanation of what went wrong at Westray—even if it was Curragh's version. Frame never followed through on the promise; the findings were never released.

Frame had allies in his fight to reopen the mine—his former employees in Pictou County. As many as one hundred Westray miners and surface workers were facing welfare when their unemployment insurance benefits ran out in June. Some would lose homes unless they could find work, and the recovery effort was their last hope. It was a replay of the debate over the strip mine in the previous summer; men who had lost their livelihood in the explosion were pitted against the families of those who had lost their lives. Local 9332 of the Steelworkers backed the company to the hilt. "We don't give a damn who we work for," said Randy Facette, who had been reinstated as the local's president after resigning during the debate about flooding the mine. "We want to mine this coal seam and we want to get on with the recovery." With the mine flooded, making it safe for repair crews to enter, it was time to put people back to work. Facette and other laid-off miners argued it would be different this time around; the union and the government inspectors would make sure Westray followed safety laws to the letter.

As their benefits ran out, the unemployed miners turned their anger on the Westray Families Group. They accused Teasdale, a prominent Liberal in Antigonish County, of using the Westray controversy to defeat the Tories in the upcoming election. Don Dickson of New Glasgow, who had worked at Westray's wash plant, deluged newspapers with letters questioning whether Teasdale's views were shared by the majority of the families. The recovery plan put enormous strain on the group's unity; relatives who blamed Curragh for their husbands' deaths were squared off against widows who saw the company as their only hope of recovering the remains of their loved ones. Bernadette Feltmate broke ranks, saying she was tired of the bickering within the group and wanted to speak for herself. "We should leave it up to Westray to reopen the mine and retrieve the bodies, and we can get on with our lives," she wrote in a letter to the *Chronicle-Herald*. "If we fail to do this, will our husbands ever be found?"

But dissent within the group's ranks was quelled by an attack from another quarter. In March, about thirty miners staged a sit-in at a Department of Natural Resources building in Stellarton to draw attention to their plight. As the television cameras and tape recorders rolled, Facette put in a call to Donald Cameron. The premier, unaware he could be overheard on a speaker phone, tore into Teasdale. "So who does he represent, except himself and the Liberal party?" asked Cameron. "He's destroyed the lives of the families by playing his games." After the premier's comments were broadcast, a number of relatives leaped to Teasdale's defence. "He's not speaking for himself and he's not speaking for the Liberal party; he's speaking for us," Joe MacKay, whose brother Mike was still entombed in the mine, told reporters called to a press conference in Plymouth. "If anybody destroyed our lives, Curragh and the politicians did." Teasdale declined to enter the debate, but MacKay insisted the chairman had the confidence of at least two-thirds of the twenty-six families. The incident underlined the animosity between Cameron and many of the Westray families. Partisan leanings aside, the Families Group had emerged as a political force to be reckoned with. And its goal, like that of the Liberals, was to bring down the Cameron government.

The final decision on the recovery plan rested with the Department of Natural Resources, which, unlike the government's critics, did not consider the mine-safety charges pending against Curragh and Westray's former managers to be an impediment. "Nobody has been found guilty of anything," Natural Resources Minister John Leefe stressed. "Our adjudication will be solely on the basis of the mining plan and whether it meets the requirements one would lay down for the best use of the resource for Nova Scotians." The proposal was in the hands of the same officials who, before the explosion, had rubber-stamped all Westray's requests to revise its underground layout and mining plans. But those days were gone; Natural Resources rejected the proposal. The company's ideas for retrieving bodies were "deficient," Leefe said; because of the explosion, an entirely new mining plan, covering all facets of the mine's operation, had to be submitted.

With the strip mine in limbo until the fall and the recovery effort nipped in the bud, Frame put Westray on a back burner. Dealing with the financial implications of the disaster had absorbed the attention of Frame and the staff at Curragh's head office for far too long. The biggest crisis facing the company was low metal prices and losses at the Yukon mines. For months, Frame's top executives had been telling him that the company needed to raise new money. But Frame was used to issuing orders, not taking advice. "Infallible," a senior Curragh official later said when asked to describe Frame's managerial style. "He believed that he did not make mistakes." He had built Curragh from nothing, and turned it into one of Canada's largest exporters, and he was determined to run it his way.

Frame finally faced reality at the end of February, when Curragh floated a new share offering in hopes of raising $50 million in cash. The shares, intended for large institutional investors such as insurance companies, would sell for $1.20, even less than the company's existing shares were fetching on the Toronto Stock Exchange—$1.45. The money would be used to pay down some of Curragh's growing debt, plus provide working capital to revive Faro and Sä Dena Hes. Even though the new shares would more than double the number of Curragh shares in circulation, Frame would still wield total control. His only concession was to offer his new investors some seats on the board of directors. With metal prices at an ebb and Curragh's finances in a shambles, it would be a tough sell. "If it weren't for bad luck," noted the *Financial Times of Canada,* Frame "would have no luck at all."

The only bright spot was the Yukon government's decision in March 1993 to provide the long-sought $29-million loan guarantee. The territorial government had little choice—about 200 people had been laid off at Faro, and the mine's remaining 180 workers had been given layoff notices. Another 100 people were out of work at Sä Dena Hes. That represented a major chunk of the Yukon workforce. Unless something was done, Faro would once again become a ghost town, as it had been when Curragh came on the scene in 1985. But the territorial government was not about to let Cliff Frame take one more penny out of the Yukon to prop up his ailing mining empire. To get the loan guarantee, Curragh would have to raise the $50 million in new equity and repay an existing $5-million provincial loan. There was one other condition attached to the package; Curragh would have to relocate its head office from Toronto to the Yukon. The territorial government, apparently, wanted Frame close by, where it could keep an eye on him.

But the $50-million share issue bombed, leaving the Yukon offer in limbo. With no new money on the horizon, it was time to haul in the sails and try to weather out the storm. One precaution had already been taken; in mid-February, $5 million in company money was set aside in a trust fund. A former director who approved the fund says it was intended to ensure that employees would have severance packages and vacation pay if Curragh did not survive. But while company officials were bracing for the worst, that was not the message they wanted to send to potential investors. In financial statements released to the public, Curragh described the fund as a war chest to cover possible liability for breaches of environmental laws and other legal claims. That may not have been the fund's real purpose, but Curragh was sure to face its share of legal actions over Westray. The week the trust was established, relatives of more than half the miners killed in the explosion served notice that they would sue the federal and Nova Scotia governments for negligence. Curragh and its directors were expected to be named in the lawsuits.

Curragh's house of cards came crashing down on the morning of April 5, 1993. The company announced it no longer had enough money to meet its bills, and was seeking court protection to stave off bankruptcy. The debt was staggering: about $220 million, including $130 million owed to large American institutional investors, such as insurance companies and mutual funds, which held notes payable in the year 2000; $24 million borrowed from the Bank of Nova Scotia, including the $10 million advanced in January; and $21 million owed to contractors and suppliers in the Yukon. Under the Ontario Companies' Creditors Arrangement Act, Curragh's creditors would have to wait until the end of June before they could petition the company into bankruptcy. The three-month reprieve gave Curragh one last chance to scrape together the estimated $75 million needed to put its financial house in order. "I remain optimistic that Curragh will emerge from this difficult time as a healthy and successful Canadian mining company with world-class facilities," Frame said in a short press release announcing the insolvency. Eleven months after the Westray explosion dealt a heavy blow to Curragh's pocketbook and credibility, the company was in a high-stakes fight for survival.

* * *

THE PITCHED BATTLES over Westray that convulsed Nova Scotia did not spill over into the Ottawa political scene. The Liberals and NDP periodically called for an inquiry or parliamentary committee hearings into the mine's financing, but the Mulroney government used its majority to squelch an investigation at the federal level. Although Ottawa had the right to seize the mine to recover the $80.75-million loan payout, the government chose to sit back and let Curragh float the scheme to recover bodies and put the mine back in production. Then, at the beginning of 1993, Mulroney announced he would quit as prime minister in June. With an election in the offing by the fall and the Tories preoccupied with the search for a new leader, the name Westray was rarely heard on Parliament Hill.

But the controversy over the mine resurfaced with a vengeance in March 1993. The federal Liberals accused the Mulroney government of withholding evidence that showed the cabinet was warned of safety risks before approving aid to Westray. Cape Breton MP David Dingwall produced a copy of an internal memo to Jake Epp, the minister of energy, mines, and resources; the heading indicated that the subject was federal funding for Westray. It took party researchers six months to pry the document out of the government through a freedom-of-information request, but the effort was next to worthless. Bureaucrats had blanked out all eighteen pages of the memo to protect cabinet secrecy and confidential advice to a government minister. Even the date had been removed. The heavy-handed censoring did not deter Dingwall, who speculated that the document "spelled out in spades" what the safety risks could be in mining Westray coal. "One could reasonably conclude that this is probably one of the smoking guns in the project," said Dingwall, the Liberal's house leader. "The government could alleviate all our fears and all our suspicions by coming clean."

When Dingwall raised the issue in the House of Commons, Tory House Leader Harvie Andre denied the government was withholding anything. Since the explosion, Ottawa had released about six thousand pages of documents to the Nova Scotia public inquiry probing the disaster. These records, most of them financing agreements required for the September 1990 loan guarantee, had already been made public. Andre, the former industry minister, who had been in the thick of the bureaucratic battle over Westray, said that "all of this information" was in the hands of the inquiry. And he assured the opposition that Ottawa was cooperating fully with the inquiry's investigation.

Andre's statement did not sit well with the inquiry's chief counsel, John Merrick. For months, inquiry staff had been trying to convince Ottawa to release its remaining Westray files. But federal officials were reneging on their promise to produce the material, claiming the information could prejudice the RCMP investigation and any criminal trials that resulted. That argument held little weight with the inquiry, which was under a court-ordered stay and could not make the information public until the criminal prosecution was completed. And it did not stack up with Andre's claim of full cooperation. Merrick responded with a press conference revealing the impasse and threatening to haul federal officials into court to get the documents. "If that is not clear, unequivocal evidence of a cover-up, I do not know what is," noted Dingwall.

The Mulroney government scrambled to mend fences with the inquiry. "Nothing has been withheld," Small Business Minister Tom Hockin, who normally fielded questions about Westray, told the Commons on March 30. "It was just getting through acres and acres of pages of paper." The Liberals, however, saw a chance to take a swipe at Kim Campbell, the frontrunner to take over Mulroney's job. Campbell had been justice minister in 1992, when the inquiry served her department with an order to produce all federal documents relating to Westray. The Liberals accused Campbell of withholding the documents in violation of the inquiry's order, a move that drew Mulroney into the debate. Despite Merrick's statements to the contrary, the prime minister said Ottawa was doing everything it could to help the inquiry, "because we want the truth to come out." But the allegations that documents had been withheld were never addressed.

The federal government relented during the first week of April, shipping its remaining Westray files—11,000 pages of material—to the inquiry. Mulroney and his ministers had made much of their cooperation with the inquiry, but all the while the government had been holding back the bulk of its records pertaining to the mine. And when the remaining files arrived in Halifax, they had been carefully vetted to remove references to cabinet discussions, information covered by solicitor-client privilege, and other material the government considered confidential. Inquiry staff reviewed and sorted the documents but, as promised, kept them under wraps.

But some of them made it into the hands of the *Chronicle-Herald*. The newspaper had filed a freedom-of-information request in the fall of 1992, seeking Westray documents that had ended up in the Prime Minister's Office. The documents were heavily censored, but they outlined the role played by Peter White and Stanley Hartt in shepherding the Curragh assistance package through the reluctant bureaucracy. After a series of articles based on the documents appeared in the newspaper in the first week of April, the federal New Democrats cried foul. The NDP accused the Mulroney government of misleading Parliament about the PMO's role in Westray for more than two years. In December 1990, the NDP had formally requested records of any contact between Curragh and the PMO. The government replied in March 1991 that all letters and phone calls to the PMO related to Westray had been "noted and forwarded" to the Industry Department. "Even before the disaster, this government was going out of its way to mislead the public and Parliament over the role of the Prime Minister's Office in all of this," charged the NDP's Rod Murphy, a Manitoba MP. "They've been involved in a long-term cover-up."

The controversy died out as quickly as it surfaced. With Mulroney's departure from 24 Sussex Drive only weeks away, the debate over the PMO and Westray was academic. In June, the Mulroney era ended and Kim Campbell took over a party that was only months away from being wiped out at the polls. Campbell's succession signalled the end of the line for another key player in the Westray deal, Elmer MacKay. MacKay had backed runner-up Jean Charest for the Conservative party leadership, and was left out in the cold when Campbell formed her first cabinet. After twenty-two years as Pictou County's champion in Ottawa, MacKay announced he would not run in the upcoming federal election.

* * *

JOHN PEARSON CAME to Nova Scotia in 1990 with a mandate to clean up a justice system tainted by charges of corruption and racism. In the late 1980s, a royal commission had been set up to investigate the wrongful conviction of Donald Marshall, Jr. A native teenager, Marshall served eleven years in prison before he was cleared of the 1971 murder of another young man in Sydney. The commission, headed by an out-of-province judge, was able to reconstruct how police and prosecutors used flimsy evidence to railroad Marshall while ignoring eyewitness testimony absolving him of the crime. But the Marshall case pointed to deeper problems within the Attorney General's Department. The commission revealed that politicians had intervened to prevent the RCMP from laying criminal charges against Conservative cabinet minister Roland Thornhill, in the early 1980s. To prevent such abuses, the commission recommended that the prosecution of crimes in Nova Scotia be taken out of the government's hands. In 1990, the Buchanan government responded by setting up Canada's first office of public prosecutions. The office's director could be removed from office only by a majority vote in the House of Assembly—which was supposed to make it independent of government control.

An independent committee screened candidates and unanimously recommended Pearson for the job. A career prosecutor in his early forties, Pearson was the son of a Toronto cop who had spent ten years in the Ontario Attorney General's Department. On paper, Pearson seemed as good a candidate as any for the $110,000-a-year job; he had a master's degree in criminal law, courtroom experience as a prosecutor, and administrative experience as deputy director of the criminal branch of Ontario's Crown Law Office. He was something of an expert in the arcane field of motor vehicle law, and he had been part of the prosecution team at the high-profile trial of Ernst Zundel, the Toronto neo-Nazi who faced a criminal charge for publishing anti-Jewish pamphlets.

But within hours of his appointment in early 1990, Pearson was at the centre of a controversy over a case that was not mentioned in his résumé. Pearson had prosecuted Ottawa journalist Doug Small and two others in 1989 over a leak of federal budget documents. Small, a reporter with Global Television, embarrassed the Mulroney government by reporting details of the budget before it was released in the House of Commons. The RCMP was called in and laid theft charges. But when the case came to trial, a senior investigator testified that the criminal charges were laid because of political pressure. Pearson withdrew from the case after accusing the Mountie of perjury; he later apologized to the police officer for the damaging allegation, but his actions did not escape censure. The judge hearing the case criticized the prosecutor's actions as "offensive to the principles of fundamental justice and fair play."

Despite the questions raised by his role in the Small case, Pearson's appointment stood. But the tone was set for his controversial tenure as director of prosecutions. Behind the scenes, Pearson oversaw a staff of about sixty-five Crown attorneys, who handled prosecutions across the province. He brought in a new protocol for dealing with police investigators, designed to eliminate the threat of political influence; while prosecutors would offer legal advice, the final decision to lay charges was left to the police. He also introduced ground-breaking rules for the disclosure of Crown evidence to defence lawyers before trial.

In the public eye, however, Pearson's performance did not inspire confidence. He flunked a qualifying exam for admission to the Nova Scotia bar and had to rewrite it. He personally prosecuted—and lost—a case against a former Tory health minister on a charge of releasing confidential medical information about Michael Zareski, the bureaucrat who blew the whistle on Tory patronage. He prosecuted former cabinet minister Roland Thornhill on criminal charges arising from a ten-year-old debt settlement with the banks; after most of the charges were thrown out at a preliminary hearing for lack of evidence, Pearson withdrew the rest. He won the conviction of a former high-level bureaucrat on a charge of breach of trust, only to have the verdict overturned on appeal. Despite the barbs from editorial cartoonists and newspaper columnists over his poor record in the courtroom, Pearson made a point of publicly defending and explaining his decisions. "If I couldn't take the heat," he once told an interviewer, "I wouldn't be in the kitchen."

The fatal explosion at Westray set in motion one of the most complicated investigations and prosecutions in Nova Scotia history. And Pearson's office, which provided legal advice to the RCMP and to Labour Department investigators, was thrust into the spotlight. For senior lawyers in the prosecution service, many of whom had endured the bad old days before the Marshall commission came to town, Westray was the first real test of Pearson's legal judgment and independence. Some prosecution staff were outraged that government lawyers—including counsel for departments that had been responsible for enforcing safety laws at Westray—sat in on a meeting with RCMP officers to discuss the progress of the investigation. To their minds, the ostensible independence of the public prosecutor's office had been compromised. Pearson, however, later argued that the meeting dealt with delineating the roles and responsibilities of the investigators, but not with the substance of the investigation.

Despite the seriousness of the case—twenty-six men had died in circumstances that could lead to criminal charges—Pearson did not assign prosecutors to handle it full time until September 1992. His choices were Chris Morris, the son of a former Conservative cabinet minister who had about eight years' experience, and a rookie prosecutor, Stephanie Cleary. They were put on the file in time to go over the Labour Department charges and advise the RCMP on the drafting of the search warrants. But assigning staff was one thing; giving them the resources to do the job was another. Morris and Cleary were expected to research the complex issues raised by the case and prepare for the upcoming trials without full-time support staff, offices, desks, computers, or even their own telephones. Until Pearson was able to wrangle additional funding from the Nova Scotia government, they set up shop in a spare office just down the hall from the law library at the prosecution service's main office in Halifax. To show their frustration with the spartan accommodations, they delineated their work space by pasting lines of masking tape on the floor.

Poor working conditions paled in the face of a bigger problem—a possible conflict of interest. Morris and Cleary were advising both the RCMP and the Department of Labour, but the department was itself under criminal investigation by the Mounties for its failure to enforce safety laws at Westray. Cleary was so troubled by the predicament that she asked to be taken off the Department of Labour file. Pearson insisted, at the time and in later interviews, that the prosecution service could not shirk its duty to provide legal advice to a prosecuting agency like the Department of Labour. He solved the dilemma, and helped ease the workload on the two prosecutors, by personally taking over the Department of Labour prosecution.

But those charges were in jeopardy, anyway. The department's investigation and the RCMP probe covered similar ground, so it followed that the fifty-two Occupational Health and Safety Act charges were based on the same allegations of unsafe working conditions as the potential criminal charges. Take, for example, methane and coal dust, the two substances that fuelled the explosion. Labour Department charges included failure to spread stone dust and allowing mining to continue when gas levels were high. It was a safe bet that these charges would also figure prominently in any criminal action. Curragh and any former Westray officials singled out for prosecution could wind up being convicted twice—under the Criminal Code and the labour safety act—for what amounted to the same offence. And that gave defence lawyers a strong argument for having the criminal charges thrown out to ensure that their clients did not face "double jeopardy" for their actions.

Pearson and others within the prosecution service recognized the risk to the criminal prosecution. But under the procedures established in the wake of the Marshall commission, it was up to the Department of Labour, not Pearson's office, to decide whether charges should be laid under the Occupational Health and Safety Act. And the department was under the gun—safety charges had to be laid within six months of the explosion, but the RCMP was still a long way from deciding whether charges would be laid. "Nobody could guarantee whether there would be criminal charges, so the Department of Labour had to make a decision," Pearson contended later. "Our concerns were expressed to them about the consequences of them laying charges, but they made their decision. Then the ball comes into our court. Then we have to decide what's going to happen." In early October, when the department filed the fifty-two charges, the ball was in Pearson's court.

When lawyers for Curragh and the four former Westray officials showed up in New Glasgow provincial court on December 10, Pearson asked that the arraignment be put off. He admitted he was trying to buy time to enable the Mounties to complete their work. Defence lawyers, who were just as eager to get not-guilty pleas on the record, balked at the delay. After some sharp debate, the presiding judge, Clyde MacDonald, turned down the request. Pearson was forced to play his hand. He announced that the Crown would drop thirty-four of the fifty-two Occupational Health and Safety Act charges to forestall any future legal challenge to the criminal prosecution. The dropped charges covered some of the most serious and controversial allegations levelled in the wake of the explosion—tampering with methane detectors, working in high levels of methane, failure to prevent the build-up of coal dust. Those charges, Pearson explained to reporters as he left the courtroom, were the most likely to cause legal headaches once criminal charges were laid.

The surviving charges, eighteen in all, dealt with inadequate training, allowing miners to work in areas lacking sufficient roof support, use of acetylene torches underground, discarded fuel containers, and improper storage of oil. Curragh, which had faced eighteen charges at the beginning of the day, now faced only six. Gerald Phillips and Roger Parry were named in ten, while Glyn Jones and Robert Parry each faced two charges. The former officials were not required to be in court for the arraignment; lawyers for the four men and the company entered pleas of not guilty. The trial was to begin on April 19, 1993. Even with the number of charges slashed by two-thirds, the Crown and defence expected it would take two months to present their cases.

But little more than a month before the trial was set to start, Pearson pulled the plug on the Labour prosecution. He imposed a stay of proceedings on the remaining eighteen charges, which could not be reactivated. "I'm simply not prepared to imperil potential criminal charges in this case," he explained at a March 4 press conference in Halifax. Although he had concluded in December that the eighteen charges could go to trial, the Nova Scotia Court of Appeal had meanwhile handed down its decision halting the Westray public inquiry. In its January ruling, the province's highest court determined that the inquiry's hearings would prejudice any future criminal proceedings. In Pearson's view, that reasoning applied with equal force to the trial on the remaining Occupational Health and Safety Act charges.

"I'm simply not prepared to imperil potential criminal charges." In March 1993, John Pearson, Nova Scotia's director of public prosecutions, dropped all Occupational Health and Safety Act charges against Curragh and four former Westray managers. [Courtesy of the *Chronicle-Herald* and the *Mail-Star*]

The eleventh-hour decision to halt the prosecution did not sit well with some relatives of the Westray victims. First the inquiry had been derailed, then the bulk of the labour charges had been dropped, and now the remaining charges had gone by the wayside. Whatever faith they had left in the justice system was shaken. "It seems as if they're making up the rules as they go along," Joyce Fraser, the widow of miner Robbie Fraser, told one newspaper. "I just don't understand it." The Liberals and New Democrats questioned Pearson's timing. A provincial election call was imminent, and the Cameron government could go to the polls without a trial to dredge up bad news about Westray. Pearson, for his part, said he would rather face criticism for dropping the charges than the public's wrath if the Labour Department prosecution scuttled a future criminal trial. "My view is that, legally, I had no choice."

It was up to the RCMP to decide if there had been any wrongdoing in the deaths of the Westray miners. Pearson had consulted the RCMP before paring down the Labour charges, and his bold decision to clear the decks left no doubt that criminal charges were in the works. By the beginning of 1993, ten officers were working on the file full time, wading through a half-million pages of documents seized from the public inquiry and other sources. Lab reports were filtering in on the dust samples and other items removed from the mine. About two hundred witnesses had been interviewed. And the Mounties said they might charge Curragh and some former managers not only with criminal negligence causing death, but also with manslaughter. Manslaughter, which, like criminal negligence, carries a maximum penalty of life in prison, is defined as causing death by means of some unlawful act. In the Westray case, the unlawful act needed to prove manslaughter could be the violation of a safety law.

As the Mounties continued to build their case, the prosecutors assigned to the case faced indifference. Morris and Cleary sent countless memos to their superiors, asking for the resources needed to do their job. Even though criminal charges had not been laid, they argued, the RCMP deserved to receive "the best advice possible" to ensure "appropriate charges are laid, or not laid, as the circumstances require." By early 1993, neither Morris nor Cleary felt able to offer that advice. "We have been maintaining responsibility for the largest criminal prosecution in the history of the province without offices, desks, telephones or full-time support staff for the last five-and-one-half months," they complained in a February 22 memo to their direct superior, criminal trial director Martin Herschorn. "The people of Nova Scotia and the families of the 26 dead miners deserve better. If we cannot do the job properly, we cannot do it at all." They asked to be taken off the case, and eventually they were reassigned.

Morris and Cleary used the memo to put their concerns on record. The Nova Scotia government, they said, was a "suspect" in the Westray case—an obvious reference to the government's responsibility for worker safety and mine inspection. Yet the prosecutor's office was forced to go "cap in hand" to the government for money to fund the prosecution. "This case may become a watershed for the Public Prosecution Service in terms of its relationship with government," they predicted. "It is a prime example of what can happen when we do not control our own budget." Government control of the purse strings was a serious threat to the service's much-vaunted independence. Months later, after the memo was leaked to the press, Pearson acknowledged that the government had been slow to approve his request for an additional $600,000 to conduct the Westray prosecution. But he contended that the prosecution service, like any other publicly funded agency, had to be fiscally responsible.

While Pearson made his pitch for more money, the RCMP was without "the best advice possible." And that led to a slip-up that left police and prosecutors redfaced. Under the Criminal Code, police can keep items seized under a search warrant for only three months without laying charges. A court order is required to retain such evidence for a longer period, and the owner of the items must be notified. But no notice was sent to Curragh enabling the police to keep the methanometer, dust samples, and other items removed from the mine during the risky underground search in September 1992.

On March 26, Curragh's lawyers went to court to have the materials returned for analysis by the company's experts. The judge gave the Mounties one month to comply. The RCMP tried to downplay the oversight. "We haven't dropped the ball, and I don't want the public to think that," insisted Superintendent Robert Tramley of the Truro subdivision, where the investigation was based. "It's not the end of the world." For months, the RCMP had refused to set a deadline for their decision on laying charges. Now the force had until the last week of April to either lay charges or hand over evidence that could be crucial to the criminal case. And it would have to make the decision in consultation with a new team of prosecutors.

* * *

JUST BEFORE NINE O'CLOCK on the morning of April 15, 1993, a half-dozen unexpected visitors turned up at Curragh's Toronto offices. It had been ten days since the company sought court protection from its creditors, but this was not another delegation of lawyers or bankers; the men in the dark suits were RCMP officers. Among them was a tall, lanky Mountie from Nova Scotia—Staff Sergeant Ches MacDonald, head of the Westray investigation. The Mounties produced a search warrant and spent the rest of the day leafing through file cabinets and desk drawers, as Curragh executives and their secretaries looked on. They searched Clifford Frame's desk and the hard drives of computers, looking for anything relating to the Westray mine. They settled on about thirty documents and, after poking around for about eight hours, they left.

The Mounties left a calling card of sorts. Curragh lawyer Peter Atkinson, who was called in to oversee the search, was told to expect criminal charges to be filed against the company within a week. That piece of information was guaranteed to lower the value of Curragh's shares—already beaten down to twenty-six cents per share—and the company was bound by stock-exchange rules to go public with the news. After the Mounties had finished their work and the Toronto Stock Exchange had closed for the day, Curragh announced that charges were imminent. That was not all: the company was abandoning the Westray mine, and would "turn over the keys" to the federal and Nova Scotia governments. Criminal charges erased any chance that Curragh could rehabilitate the coal mine and recover the eleven miners' bodies. "Since we see no prospect of Curragh being able to complete the task, we believe the people of Pictou County will be best served if we stand aside," Frame said in a statement released to the press. "We trust this will assist in the transition to another operator who can recover the bodies and secure the future of the mine, for the benefit of everyone."

It was exactly what the Westray Families Group had asked for back in August 1992—criminal charges, and Curragh out of Nova Scotia. "If there is a trace of justice left in the system, it is shining through tonight," a jubilant Kenton Teasdale said when told of the announcement. "We are really gratified." But Curragh's decision to wash its hands of Westray caught government officials off guard. Neither Ottawa nor the provincial government had any intention of repairing and operating the damaged colliery. Premier Donald Cameron said his officials were trying to sort out the ramifications of the company's decision to pack up and move on. The Nova Scotia government did not have the expertise to recover the miners' remains, he said, and he doubted private companies were interested in taking over a mine that had generated so much controversy.

All that remained was to make the criminal charges official. The Mounties called a press conference for the afternoon of April 20—nineteen days shy of the anniversary of the disaster—to make the announcement. The location was the Plymouth Fire Hall, where relatives and friends of the victims had held their marathon vigil during the rescue operation. As many as one hundred people showed up; for most, it was the first time they had set foot in the fire hall since those five grim days of waiting and praying. As Staff Sergeant Ches MacDonald looked on, Superintendent Robert Tramley announced that Curragh, Gerald Phillips, and Roger Parry had been formally charged with manslaughter and criminal negligence in the deaths of all twenty-six miners. The charges spanned an eight-month period, from September 10, 1991—the day before the mine officially opened—to May 10, 1992, a day after it exploded.

The onlookers signalled their approval with a standing ovation. "This is like lifting a weight off our shoulders," Joe MacKay, whose brother Mike died in the explosion, said to the reporters hovering around as tearful relatives exchanged hugs. "It's going to be a little lighter to bear." At Curragh's Toronto offices, a few employees gathered in front of a television set to watch the live broadcast of the RCMP's announcement. The only bright spot, Curragh lawyer Peter Atkinson noted, was that none of the company's officers, directors, or head-office staff had been charged.

The corporation and top officials responsible for safety at Westray were facing the music, but not the government officials who acted as their watchdogs. Tramley announced that the investigation had found "no evidence of criminality" on the part of government officials responsible for safety at the mine. In a 1994 interview, the officer in charge of the investigation, Staff Sergeant MacDonald, defended that decision. "During the investigation, we looked at their function in inspecting the coal mine," he said of the Department of Labour official. "We had to look at when they visited, and what they did when they visited, and the contact they'd had with the officials running the mine, and the duties imposed on them and how they carried that out. . . . We haven't, in my view, found any evidence to support the laying of any criminal charges against anyone in the Department of Labour, or the Department of Labour itself."

While the criminal charges came as no surprise, the timing of the first court appearance raised eyebrows. The company and the two former managers were to be arraigned before a judge in New Glasgow on May 25; Cameron had just called a provincial election for the same day. Tramley, who looked startled when reporters pointed out the matching dates, insisted the timing was a coincidence. But it was bad news for a government eager to distance itself from the events of May 9, 1992. The Westray explosion had dogged the ruling Tories throughout the final year of their mandate. The spectre of twenty-six dead miners would pursue the Cameron government all the way to the polls.

* * *

THE ROLE THE WESTRAY mine disaster would play in the provincial election campaign was anyone's guess, as politicians took to the hustings. A poll taken at the beginning of 1993 had suggested that Nova Scotians were more likely to blame the mine's owners for the explosion. "We asked questions about who they most blamed for the disaster, and the provincial government did get some blame," said a pollster with Halifax-based Corporate Research Associates, "but the premier got virtually none." Donald Cameron, whose program of fiscal restraint and government reform was being lost in the din over Westray, was frustrated with the finger-pointing. "You'll never find another example where the deaths of people have been exploited in such a vicious manner," the premier told a reporter for the Canadian Press in January, refusing an interview on the subject. "I think it's sickening. . . . You keep milking this story–you're going to milk it without Don Cameron."

Cameron had done his best to put Westray behind him and get on with the business of governing. In the two years since he had taken over from John Buchanan, he had pushed through the multimillion-dollar privatization of the Nova Scotia Power Corporation and introduced new tendering policies for government contracts. To project a no-nonsense, businesslike image, Cameron had dispensed with the ceremonial Speech from the Throne to introduce his government legislative package at the opening of House of Assembly sessions. In the weeks leading up to the election call on April 16, the government unleashed a torrent of good-news announcements–tax cuts for business, more money for victims of child abuse, and the sale of government-owned resorts and a money-losing amusement park. In an attempt to attract female voters, Cameron catapulted two women–Tory candidates who had yet to win elected office–directly into cabinet.

Whether those changes were enough to persuade voters to support a party that had been in power for fifteen years was unclear. Cameron tried to distance himself from the taint of the Buchanan era; his campaign literature focused on the man, proclaiming the theme: "Leadership that's making a difference." But the effort to downplay the party only alienated dyed-in-the-wool Tories, many of whom still preferred John Buchanan's folksy ways to Cameron's icy aloofness. The Liberals, meanwhile, looked like a shoe-in to form the next government. With new leader John Savage at the helm, the Grits promised action on the issue every poll identified as number one in the minds of Nova Scotians–jobs. The New Democrats, Nova Scotia's perennial third party, led by Alexa McDonough, flailed away at the old-style politics of the other two parties.

For the most part, the opposition parties shied away from the Westray controversy during the campaign; to do otherwise would have invited more comments from the ruling Tories about how reprehensible it was to use the disaster for political gain. And that was the tack Cameron took when pressed on the Westray issue during a televised leaders' debate early in the campaign. "Politics has reached a real low level here," the premier said. "I'm sure there's people that believe the very worst about it–the bottom line is that because of the exploitation of this disaster in the province, we put families through living hell." McDonough refused to let Cameron turn the tables on his critics. "Are you ever going to acknowledge that the human tragedy of Westray was not an act of God, but was the result of a number of things about which your government could have done something?" she asked. Cameron acknowledged that mistakes had been made at Westray, but again accused his opponents of exploiting the disaster.

Other events conspired to keep Westray in the public eye. The first anniversary of the explosion fell at about the mid-point of the campaign, unleashing a torrent of news stories chronicling the human cost of the disaster and criticizing the long wait for justice. It was also the deadline for filing lawsuits seeking damages in the miners' deaths; virtually every family brought an action against Curragh, the federal and provincial governments, and the manufacturers of equipment used in the mine, alleging varying degrees of negligence. The lawsuits also named Clifford Frame, Gerald Phillips, Roger Parry, and four long-time Curragh directors: James Hunt, Ralph Sultan, Walter Bowen, and John Mitchell. On May 9, relatives and friends of the twenty-six killed, and miners spared by the explosion, gathered for a private ceremony at a newly erected memorial bearing the names of the Westray dead, located directly above the last resting place of the eleven men whose bodies had not been recovered.

They never have been. At about the time of the ceremony, a high-tech plan was being drawn up to locate, and possibly even recover, the mens' remains. A

Pictou County–based consultant specializing in the use of remote-controlled underwater cameras and probes was coordinating the effort. The idea was to drill a hole from the surface to tap the flooded North Mains; if the tunnels were clear, small probes powered by jets of water would be lowered to search for bodies. The idea seemed far-fetched, but similar equipment had been used for delicate operations such as capping offshore oil wells. The Nova Scotia government, through the Department of Natural Resources, offered up to $200,000 to finance the effort. But a disagreement arose between the consultant and the government over the terms of a contract to cover the work. The plan fell through shortly before polling day, creating more bad press for the government.

The Cameron Tories did not entirely manage to avoid talking about Westray. During the campaign, Labour Minister Tom McInnis released the report of the consultants hired to review practices and procedures within his department. It was not a flattering picture: the department had been mismanaged for years, record-keeping was shoddy, and inspection reports did not contain enough information to properly assess which companies were violating safety laws. "Management skills and work processes had been allowed to deteriorate to an unacceptable level," admitted McInnis, who noted that steps had already been taken to improve the way the department was run. Cameron pointed to the report as proof that his government was committed to cleaning up the mess he had inherited from the previous Tory regime. "We went to outside consultants to do an audit on the department, and the deficiencies will be taken care of," he said. "That's leadership."

One powerful union was not impressed with the provincial government's newfound interest in worker safety. The Canadian Union of Public Employees, already smarting over civil-service cutbacks, circulated leaflets and posters bearing a stark silhouette of the mine site and admonishing voters to "remember Westray" at the ballot box. Cameron did have the support of Local 9332 of the United Steelworkers of America, the union representing Westray miners. "I think he's the only hope we've got of getting anything happening here in fairly short order," said the local's president, Randy Facette. The target of the union's wrath was the Liberals, who were opposed to any move to reopen the mine until after a public inquiry. That could take years–too long for Westray employees facing the prospect of going on welfare. When Savage drove into Westville on May 21 to deliver a speech at the local Legion hall, a mob of angry union members and their supporters used a huge dump truck to block the road in front of his campaign bus. "We want some jobs, you bastard," one protester shouted. Savage stood toe-to-toe with the demonstrators and refused to back down. "Twenty-six people have died and we don't know why," he shouted over the hecklers. "We're not prepared to compromise the safety of Nova Scotian workers."

As decision day approached, it became clear that a Savage government, not the Cameron Tories, would decide Westray's future. A poll released in the final week of the campaign gave the Liberals a two-to-one lead over the Conservatives, more than enough to form a majority. On May 25, as the Westray criminal charges went to court in New Glasgow, Nova Scotians trooped to the polls. The outcome was never in doubt. The Liberals swept forty of the province's fifty-two seats, leaving the demoralized Tories with nine. The NDP kept the three seats the party had held before the election. More than a half-dozen cabinet ministers went down to defeat, including former labour minister Leroy Legere. Despite the Westray disaster, Pictou County was one of the few bright spots for the Conservatives; Cameron and two other Tories held onto the area's three ridings with comfortable margins. Cameron outpolled his Liberal rival by 720 votes, a result almost identical to his margin of victory in 1988.

Cameron shouldered full responsibility for the party's drubbing. "I blame no-one else–the federal Tories or the Buchanan government," he told a sombre gathering at his campaign headquarters in New Glasgow. "I would like to have finished the job, but the people have spoken and we must accept that." Then, as the men and women who had worked to gain Cameron a sixth straight term looked on in stunned silence, he announced he was quitting as party leader and as MLA for Pictou East. "After twenty years of public life, it is time for me to leave," Cameron explained, declining the consolation prize of opposition leader. "I'm a doer, not a hanger-oner." Donald Cameron, the man who had worked as hard as anyone to make Westray a reality, was leaving the political game for good.

BITTER LEGACY

HE OCTOBER DAY IS AS COLD and lifeless as the bare trees that stretch their bony fingers towards the sky. Dark clouds scud overhead in the brisk wind, threatening an early snow. The weather and the season have conspired to make Pictou County look much as it did a year and a half before, when the Westray explosion wiped out the men of B-crew and tore apart the lives of their wives and children. In the spotless kitchen of the Johnson bungalow in Westville, two women talk about broken promises and shattered dreams. They talk about their husbands, their ordeal in the days following the explosion, their ongoing struggle to rebuild their lives. Occasionally, a happy memory will flood back, bringing a chuckle or a wry smile. But the events of May 9, 1992, have left an indelible stamp on those left to carry on.

"He was a great father and husband. He was my best friend. I told his mother I had the best, and I'll never, ever, have it again." That's Donna Johnson, describing her husband of twelve years, Eugene. She's petite, with curly blonde hair and a quiet manner–and she has two rambunctious sons to raise alone. At the other end of the kitchen table is Joyce Fraser. She's only twenty-six, and she is a widow with four children under the age of ten. Her youngest, Alexander, was born in August 1992, three months after Robbie Fraser died in the Westray explosion. "I envied her," says Johnson. "She had that last thing of Rob's." The dark-haired Fraser, a feisty woman with a sharp wit, refuses to let the comment slip by unchallenged. "*Now* you don't, do you, honey?" she jokes, a reminder, mother-to-mother, of the headaches of looking after a toddler. Johnson laughs and shakes her head. "Nope."

The boys have kept Donna Johnson going these past months. School's out for the day, and they descend on the kitchen, all noise and excitement. The conversation turns briefly to the logistics of getting Michael to hockey practice. David, the youngest, raids the candy bowl in the centre of the table, which is by now the domain of empty soft-drink cans and half-filled ashtrays. Michael is twelve, and to this day Johnson has never seen him cry about his father's death. But his anger is palpable. More than once, he has wished out loud that he could blow up the buildings at the mine site, obliterating that constant reminder of the explosion. Westray Coal had sponsored his baseball team, but Michael asked his mother to cover up the company's logo by sewing on his own name, taken from a hockey jersey. David, now eight, refused to talk about his father for a long time; if Eugene's name came up in conversation, he would cover his ears and complain that the subject was boring. Until Curragh became insolvent, the company was paying a psychologist to help the widows and children. Johnson and Fraser were among the few who regularly attended the sessions. Counselling helped her and the boys cope, Johnson says, but she knows the emotional scars run deep.

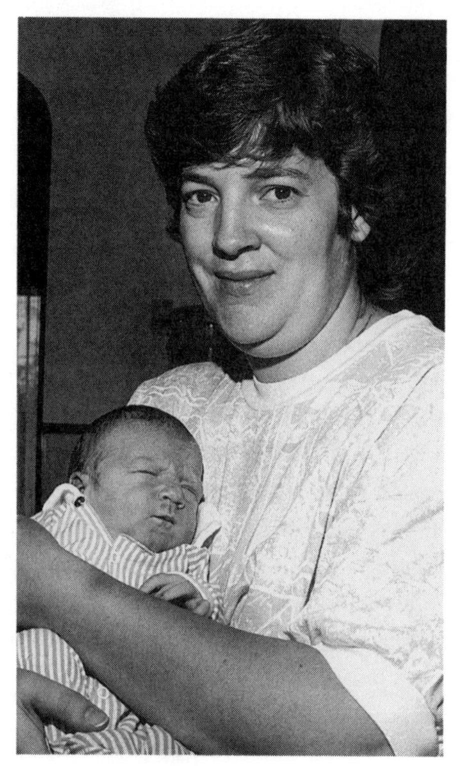

Joyce Fraser cradles her son Alexander, born three months after his father, Robbie, died in the Westray mine. [Photo by Steve Harder, courtesy of the *Chronicle-Herald* and the *Mail-Star*]

Those who lost husbands, fathers, or brothers also had to deal with frustration as the Westray disaster turned into a legal and political football. Promises of timely investigations and swift justice have been broken; the name Westray conjures up images of endless legal wrangling and delay stacked upon delay. "I'm hoping to see some justice done," says Fraser. She is confident that those responsible for her husband's death will be punished by their own consciences, if not by the courts. At the very least, she hopes the disaster will lead to improved mining laws, so the mistakes made at Westray will not be repeated. Johnson, for her part, is bitter that the men in the pit did not stand up to management and demand better working conditions. "They knew what it was like, and they should have done something," she notes. "I blame everyone, even Eugene."

None of those widowed by the explosion has money worries. Westray employees were covered by a group life insurance plan, and mortgage insurance took care of the house payments. Nova Scotia's workers compensation program will pay an estimated $15 million in the years to come to look after the dependents of the miners killed at Westray. Each widow received a $15,000 lump-sum payment, and will be paid 75 per cent of their husband's salary—up to a maximum of $36,000—every year for the rest of their lives. An additional $2,300 will be paid annually to support the children of each miner until they reach age eighteen. The families have also benefited from the generosity of strangers. Donations from across Canada and beyond had swelled the Plymouth Mine Disaster Fund to $1.8 million by the fall of 1992. The family of each miner received about $50,000; the remaining money was used to create $10,000 trust funds to pay for the education of the forty-three children left fatherless.

Relatives of twenty-three of the twenty-six men have launched legal actions seeking damages for negligence. The suits name Curragh, Clifford Frame and other company directors, Gerald Phillips and Roger Parry, the federal and Nova Scotia governments, and manufacturers of mining and safety equipment. The case is a massive one, perhaps the most complex civil action in Nova Scotia history. There will be a flurry of pre-trial motions as defendants try to have themselves exempted from proceedings. It will likely be 1995 before lawyers sit down to examine witnesses and sift through the evidence; if the lawsuits get to court, it will be the late 1990s before trials are held. Time has already become a factor. One of the former Curragh directors named in the suits, John Mitchell, has died. Curragh has been carved up by its creditors, its assets sold off to cover a portion of the mountain of debt left behind. Curragh's $5-million trust fund, if not used to compensate former employees, could be the only source of money if the company is found negligent.

The only comparable lawsuits in recent years arose after the 1982 sinking of the *Ocean Ranger* off Newfoundland. Eighty-four men died when the giant oil rig capsized in a fierce North Atlantic storm. The rig was not fitted with proper life-saving equipment, and the men operating the ballast controls had no formal training. But the Ocean Drilling and Exploration Company of Canada, the rig's owner, and the company leasing it at the time of the disaster, Mobil Oil of Canada, never faced criminal charges. The only court action came from the families of the victims who launched lawsuits in Canada and the United States, where both companies had head offices. After years of negotiation and legal manoeuvring, the cases were settled out of court for amounts ranging from $25,000 to $300,000 per family. No dollar figure has been placed on the damages sought in the Westray cases; Ray Wagner, a Halifax lawyer acting for most of the families, will only say the amount will be large. But, he adds in the same breath, the lawsuits are not just about money. They are about corporate responsibility, about making sure employers and government officials are held accountable if they fail to ensure the safety of workers.

For Donna Johnson and Joyce Fraser and the other women widowed at a young age, the donations and other forms of compensation have brought financial security. In a small town like Westville, the support has also brought unwanted attention, even jealousy. "We're always going to be the Westray widows," Johnson says, her voice heavy with resignation. "Everybody's watching every move that you make," Fraser chimes in. "We'll always have that label." People have told them to their faces that they have it easy. Even some of their relatives have criticized them for buying new cars, or dirt bikes for the kids—or, in Fraser's case, the new home that she and Robbie had dreamed of owning together. "It's jealousy," she says bluntly, "and they've got nothing to be jealous about." No amount of money can replace a husband or a father. And they know that Robbie and Eugene would have wanted them to get on with their lives. "People just don't understand what you go through, and how you do change," Johnson says. There is a long pause. "But we'll survive."

* * *

DURING THE SPRING and summer of 1993, Clifford Frame and his cohorts worked frantically to keep Curragh afloat. The Ontario Supreme Court kept the company's creditors at bay, while trustees Ernst & Young doled out money to maintain Westray and the inactive mines in Western Canada. The only offer on the table was from South Korea. Samsung Corporation and Korea Zinc Company, one of Curragh's Pacific Rim customers, were agreeable to investing a combined $50 million in the ailing company. In return, the Koreans demanded a fifty-fifty partnership in a revamped and refinanced Curragh. But before the Koreans would part with their money, Frame had to find another $25 million and persuade the Yukon government to guarantee the $29-million loan needed to provide new ore for the Faro mine.

The deal was still in the works when Curragh held its annual meeting in Toronto on July 28, 1993. It was a sombre affair, a far cry from the upbeat gatherings over which Frame had presided in previous years. "This is not a time for speeches," the chairman said in his brief report to shareholders. "It's a major understatement to say 1992 was a difficult year." Curragh's lawyers were on hand to field questions about the Westray criminal charges and the restructuring effort. Even as Curragh teetered on the brink of bankruptcy, Frame was optimistic that there were better days ahead. If the Koreans came on board and metal prices rebounded, he said, Faro could be reactivated as early as January 1994. He was even rethinking his decision to wash his hands of Westray. "There's nothing wrong with that coal body," he told reporters after the meeting. "It will be produced by us or somebody else." Three weeks later, Curragh announced that it had lost another $30 million in the second quarter of 1993.

The Korean proposal was only part of the solution to Frame's problems. There was still the Yukon government to contend with, but being the territory's largest employer put him in a strong bargaining position. The real challenge was finding the extra $25 million he needed to make the Korean deal fly. Curragh's brokers were beating the bushes in search of potential investors; the company's executives were racking their brains, trying to come up with the white knight who could save the day. Then, out of the blue, someone purporting to represent some well-heeled Middle East investors contacted Curragh's London-based European marketing agent, offering to help. The lead looked promising; these investors might be able to provide the extra money the Koreans were demanding. They might even have enough cash to solve all of Curragh's money woes.

Frame, working under a cloak of secrecy, took charge of the negotiations. Nothing strange about that–Frame had always operated on a "need to know" basis. He was forever going off on his own to work on new projects or acquisitions, confiding only in a few of his most trusted advisers. Once a deal was done, the chairman would let the rest of the staff in on the secret. The Middle East offer was no exception. It was months before some of the firm's senior executives had any inkling that a second potential deal was in the works. And even then, the details were sketchy. All that was known was Frame was busily running up a $40,000 hotel bill at the posh Beau Rivage in Geneva, where he was locked in negotiations with a woman who was fronting for the Middle East investors, and who presented herself as a private banker for wealthy families in Kuwait, Egypt, and the United Arab Emirates. Fortyish and Egyptian by birth, she had fled Kuwait when Iraq invaded that country in 1990. Frame kept her name to himself.

The battle to save Curragh was fought on two fronts–Korea and Europe. Behind the scenes, Frame was spending most of his time in Europe and devoting all of his energies to wooing his potential Middle Eastern partners. He would check in with the Toronto office by phone to report his progress, taking pains not to give away the identity of the people he was trying to bring on board. Curragh officials, meanwhile, shuttled to Seoul for talks with the Koreans, and the Koreans sent engineers to take a look at Curragh's mines and ore properties. As far as the company's creditors and bankers were concerned, the Koreans were the best hope for Curragh's salvation. And they were agreeable to the company's periodic requests that the courts allow them more time to work out a final deal. There were rumours of Frame's effort to court Arab money, but no-one was holding their breath waiting for the money to show up.

When the Koreans presented their proposal on August 20, there were a few surprises. The tentative deal struck back in May envisioned a more-or-less equal partnership; the Koreans would snap up some vacant seats on Curragh's board of directors in return for their money, but Frame would still control the company. That prospect did not sit well with the Koreans. Frame was a loner; he had already gone through one set of partners, the Australians, in the late 1980s. And there were still question marks about the reasons for a more-recent parting of the ways—the dissolution of Curragh's marketing arrangement with Asturiana de Zinc, the Spanish smelter. In their final offer, the Koreans demanded a say in how the company was run—who was hired and fired, the whole works. "The company would carry on," as one Curragh official later put it, "but Mr. Frame would not carry on with anything near the same role that he was expecting to have." The Koreans wanted Curragh, but not its chairman.

Besides giving up absolute control, Frame stood to take a big financial loss. The Koreans wanted to create a privately held company out of Curragh's ashes, offering Frame, Ralph Sultan, and James Hunt—the triumvirate that originally controlled Curragh—options on stock in the new company. Frame and his partners expected to cash in their old stock for a good-sized chunk of the new Curragh—something in the range of forty-five million shares. The Koreans were offering a fraction of that, about three million. The final sticking point was Westray. To insulate themselves from the criminal and civil liabilities arising from the explosion, the Koreans wanted to dump Westray and transfer only the lead-zinc properties to the new company.

It was bitter medicine, and not just for Frame. Under the initial agreement, Curragh's bondholders, who held notes worth $130 million U.S., would have recouped 40 per cent of their investment in cash and the remaining 60 per cent as stock in the new company. But under the August offer, the bondholders, most of them American and Japanese institutional investors, would have to be content with $56 million U.S. Worse, the money would be repaid in instalments stretched out over ten years. The Koreans made in clear through their Toronto lawyers that the package was open to negotiation. But time was running out. Curragh had been in a state of suspended animation for five months, and there was no more money for the court-appointed trustee to keep the operation afloat. The Koreans had come back late in the day with an offer no-one could stomach; the white knight was willing to save Curragh, but only at a bargain-basement price.

Frame was given two weeks to make up his mind. He could either take his lumps and accept the terms, or try to hammer out a more-palatable deal. There was a third option—the Middle East investors. Only Frame and a small circle of insiders knew the details, but the negotiations in Switzerland were having some tangible results. Frame had an escrow account set up to handle the Middle East money, about $80 million U.S., if a deal was worked out. Lenz & Staehelin, a Zurich law firm that worked for Curragh's Swiss marketing subsidiary, Seltique Holdings SA, handled the paperwork. At least $1.5 million U.S. went into the account, presumably to cover the mystery woman's commission once the deal was completed. How Frame managed to scrape together the money was unclear, but it did not come from Curragh.

He was still keeping the identity of the Middle East investors to himself. Frame was in frequent telephone contact with the Toronto office, but the lack of hard information made it tough for anyone else at Curragh to get a handle on the chances of success—or, for that matter, whether the whole Middle East thing was just a wild-goose chase. "You didn't know whether it was a mirage or real," says a Curragh official who was immersed in the restructuring negotiations back in Canada. "You didn't know whether or not he knew who was behind the scenes and couldn't tell us, or whether, in fact, he'd ever been told who was behind the scenes. . . . And you have to take two hypotheses. One, that there was somebody very important and very wealthy behind the scenes, and we weren't being told. Or two, there wasn't." Whoever the shadowy investors were, they were considering an investment in Curragh, sight unseen. No engineering firms or consultants ever came forward on behalf of anyone in the Middle East to conduct the customary check of the company's assets.

It was Frame's call. And, in the words of the Curragh official, he chose "to bet on the desert, to bet on the Middle East." On September 3, 1993, Frame rejected the Korean offer outright. The reason, according to a terse press statement issued under Frame's name, was the Koreans' failure to meet the "investment criteria" contained in the May agreement. "That's a pretty heavy step," the official says. "So he must have had a strong belief at the time that the other alternative was viable." It was a heavy step indeed. The deadline for presenting a restructuring proposal to the courts was 6:00 p.m. on September 10; Frame had seven days to come up with a new proposal that would keep his creditors from pulling the plug. On that score, he was as cryptic with the public as he was with his own staff. Frame's press statement said only that Curragh was in "advance discussions with other investor groups," and that an announcement would be made "as soon as it is appropriate."

But Curragh's bankers were losing their patience. The Bank of Nova Scotia had stood behind Frame for years, from the early days, when Curragh was rolling in dough, through the lean years of the early 1990s, when low metal prices burned a hole in Frame's pocket. There were no hard feelings over Westray; thanks to the federal guarantee of the $100-million loan advanced to build the colliery, the bank had emerged from the disaster unscathed. All through the spring and summer of 1993, the bank remained patient, giving Frame room to work out a restructuring deal. At one point during August, Frame spent a half-hour on the phone from his bunker in Geneva, briefing bank officials about his negotiations for Middle East money. Unlike Frame, however, the Bank of Nova Scotia was not willing to bet on the desert. Late on September 10, with Frame still chasing money in Europe, an

Ontario Supreme Court judge granted a motion from the bank to put Curragh's Sä Dena Hes and Stronsay properties into receivership.

Ten days later, Curragh's remaining creditors secured a receiver for Curragh's most valuable asset, the Faro mine. And the following day, Mr. Justice James Farley put the mothballed Westray mine in the hands of a receiver. Westray was of little use to the creditors, since the federal and Nova Scotia governments had first call on its assets. Ottawa and Curragh had agreed in June 1993 to divide the insurance money, with Ottawa getting $13.6 million of the $16-million settlement. These proceeds reduced the government's loss to $67.15 million of the $80.75-million loan payout. The only way to further reduce the loss would be to sell the damaged mine; in the unlikely event that the property fetched more than $67.15 million, the Nova Scotia government could claim the next $12 million to recover its Westray loan. Curragh, shorn of all its holdings, was reduced to a shell company. But if Frame could find a backer, he still had the right to make a proposal to win back his company. The chances were slim, but it was hard to count out Clifford Frame, the comeback king of Canadian mining. "Mr. Frame has pulled rabbits out of the hat in the past," noted Dan Dowdall, a Toronto lawyer representing the Curragh bondholders. "We'll see what he can do here."

It was the end of the line for an eight-year-old company that Frame had built from nothing into one of Canada's leading exporters. At its peak, Curragh employed one thousand people, sold its lead and zinc concentrates worldwide, and was about to become a Canadian-owned mining giant. But by September 1993, Frame's hard work and dreams of grandeur lay in ruins. The Faro mine had made enormous profits—about $250 million in five years, during an upswing in the boom-and-bust cycle of metal prices. Frame had poured that money into building the Curragh empire, rather than building up a nest egg for the inevitable downturn in prices. The money and corporate credibility lost in the wake of the Westray disaster exacerbated the problem. For Curragh and for Clifford Frame, the failure to put aside money for the tough times was a fatal mistake. And the fortunes of the man and his company were intertwined: "It was not possible to separate the company from Cliff Frame," concedes a Curragh official who hung on until the bitter end. "And the problem is, there was a viable company there."

At the end of September, the receiver laid off Curragh's remaining 380 employees, including the entire head-office staff, from secretaries to top executives. When a former Curragh executive later complained that employees had been left high and dry, without vacation pay or a severance package, Frame expressed little sympathy. "Well, I've given you a job for seven-and-a-half years," he replied. "What's wrong with you?" Almost a year later, the $5 million set aside for those benefits remained in limbo. The only senior official to land on his feet was Colin

Benner, the man who had put a human face on the company during the futile search for Westray survivors. He jumped ship in mid-September 1993, joining Vancouver-based Royal Oak Mines as vice president of operations. Out of the frying pan, into the fire–Royal Oak was no stranger to controversy. One of the four gold mines put under his control was Giant Yellowknife, where nine miners had died in an explosion a year before. That blast, like the one at Westray, sparked an RCMP investigation, which led to murder charges against a striking miner; at Westray, however, the criminal case against the company and its on-site managers focused on safety violations. Benner's role as Curragh's chief apologist in the days after the Westray explosion, and his brief involvement with the mine before the disaster, did not phase his new employer. "He's a very capable guy and has a good reputation," a Royal Oak spokesman said when the appointment was announced.

Curragh's bones have been picked clean. As secured creditors, the Bank of Nova Scotia and Curragh's suppliers in the Yukon were in a position to recoup most of their losses. The company's bondholders, by comparison, stood little chance of recovering their money. Frame was forced to watch from the sidelines as his mining empire was carved up and sold off. Teck Corporation and Cominco–the zinc-smelting giant Frame worked for as a teenager in Trail, B.C.–snapped up the Sä Dena Hes and Stronsay properties for the fire-sale price of $34 million. Korea Zinc, one of the firms involved in the aborted Korean deal, agreed to buy 50 per cent of both properties. Curragh's prime asset, the Faro mine, was sold for $27 million in mid-1994. The successful bidder was Anvil Range Mining Corporation, a fledgling Toronto-based company whose backers included Bill McKnight, a former federal energy minister; Frame's former partner Ralph Sultan; and Adrian White, Curragh's former chief financial officer. The new company aimed to put Faro back in operation by the late summer of 1995.

Frame lost a bundle when Curragh went down the drain–on paper, at least. The huge block of shares that gave him control of the firm had a book value of almost $100 million when the company went public in 1990. But Frame could not sell his shares without relinquishing control of the company; he was forced to hang on, as the stock plummeted to the 20-cent range, then was taken off the market. As Curragh was going through its death throes, Frame and his partners chipped in about $500,000 from some holding companies to cover part of the company's expenses. But Curragh's demise did not leave Frame in the poorhouse. He still had the country estate north of Toronto, his herd of purebred Black Angus cattle, his stable of Jaguars, and whatever remained of the $4 million he had earned before Curragh went public in 1990. Frame and his wife own two homes, a $1.2-million townhouse in Rosedale and an old farmhouse near Uxbridge, which has been extensively renovated and expanded. The farm occupies about one hundred hectares, and the whole works, complete with tennis courts, has a market value of about $500,000. The property was held in the names of Frame and his wife until August 1993; about the time Curragh went under, Frame transferred ownership to a couple of companies with familiar-sounding names—Curraghdale Farms and Curraghdale Cattle Corporation.

Frame's number at the farm is listed in the phone book, and a persistent caller can get him on the line. It's the spring of 1994, just days before the second anniversary of the Westray explosion, and Frame picks up the phone. He's had it with the investigations, the allegations, and the foot-dragging. "This thing should have been brought out in the open two or three months after the accident," he complains. "It's two years now, and they're still looking for someone to hang. . . . They've ruined a company; they've done nothing about the cause of the accident, nothing about solving the problem of continuing mining, nothing about the jobs, nothing about the mining act, nothing about improving the regulations, nothing about anything. Absolutely nothing. Except have a court case and look for somebody to hang."

Were it not for the upcoming criminal trial, Frame would consider giving his side of the whole affair. "You know," he says, "nothing is totally black and white." He takes comfort in the fact that Gerald Phillips and Roger Parry, the men he entrusted to run his Westray mine, were underground on a regular basis; if the mine was so unsafe, he asks, why would they endanger their own lives? But the charges against Phillips, Parry, and his defunct company, he maintains, make it impossible for him to talk about what went wrong at Westray. Curragh's experts spent the better part of a year investigating the cause of the explosion; their findings will remain under wraps until the criminal case is over. That's not the only reason he prefers to hold his tongue. Every time he has spoken out since the explosion, he grumbles, he's been "crucified" in the media. He's tired of the "hatchet jobs" that have sullied his reputation.

"I've worked in the mines and smelters since I was sixteen. At twenty-four, I had one hundred and fifty men working for me. I've developed twenty-some mines around the world, done exploration and mine development to the tune of three or four billion dollars." People are pointing fingers at him over what happened at Westray, he says, simply because he had the misfortune of being both the chairman of Curragh and a mining engineer. "If I'd been a goddamned lawyer, they wouldn't have asked me anything . . . but because I'm a mining engineer and a big developer, I was supposed to know everything and be responsible for everything."

Westray is not the only subject Frame would rather avoid. He deftly ducks questions about a possible comeback in the mining business. Toronto brokers say Frame would have big trouble raising money in Canada; he's already had two kicks at the can, and it will take more than two citations as the *Northern Miner's* "Mining Man of the Year" to make bankers and investors forget about Quintette and Westray. The government well that supported Frame's dreams for so many years has run dry. Then there are the questions about a couple of his past dealings. Some creditors would like to know the whereabouts of $7 million syphoned off into a Frame-owned numbered company, 717077 Ontario Corporation, during a corporate restructuring in 1987. Government officials in the Yukon have taken Frame to task for awarding construction contracts on the Sä Dena Hes mine to companies owned by his son-in-law, Scott LaPrairie.

But Frame says that finding cash to start over is not a problem. "I started with zero in '85, so I can start with zero again," he told a reporter in the fall of 1993, just as Curragh was going under. Neither is assembling a new team of mining executives; "staff," he noted at the time, "are easy to get." In the early months of 1994, former Curragh officials heard rumblings that Frame was still commuting to Europe to negotiate with his mysterious Middle East investors. He was rumoured to be looking at everything from a new farm to a Kentucky coal mine. So what lies ahead for Clifford Frame? On this spring day, two years after the mine explosion that killed twenty-six of his employees and toppled his mining empire, Frame has no interest in showing his cards. "That's a military secret," he says. And then he lets out a hearty laugh, says a quick goodbye, and hangs up.

* * *

THE LONG-AWAITED Westray criminal prosecution got off to a rocky start. The ink on the charges against Curragh, Gerald Phillips, and Roger Parry was barely dry when the defence filed a motion in May 1993 to have all counts quashed. Peter Atkinson, a Toronto lawyer acting for Curragh, contended that the wording of the manslaughter and criminal negligence counts was too vague. "In neither charge is any hint given as to the specific acts or conduct which the Crown asserts caused the deaths," he argued. "How can Curragh Inc. identify what it is alleged to have done wrong so that it can prepare its case adequately?" Lawyers for Phillips and Parry, not surprisingly, endorsed the company's stance. The criminal case was put on hold while a judge considered whether the charges were flawed.

Meanwhile, the prosecution came under fire from another quarter. The memo written by Chris Morris and Stephanie Cleary in February 1993, criticizing the lack of resources devoted to the Westray case, surfaced during the Nova Scotia election campaign, putting Director of Public Prosecutions John Pearson on the defensive. The Westray case "has been handled on a principled basis from day one," Pearson said, suggesting that people within the prosecution service were leaking confidential information as part of "a campaign of distortion and innuendo" against him.

The memo was out of date, Pearson said, and its concerns had been addressed. The Nova Scotia government was devoting $600,000 a year to the case, and the prosecution service had assembled a three-member team to conduct the Crown's case. The new lead prosecutor was Herman Felderhof, who had twelve years of courtroom experience, and whose geology degree would stand him in good stead. The others, Arthur Theuerkauf and Peter Rosinski, had been Crown attorneys for a combined ten years. Pearson's choices were based as much on location as expertise—two of the three were already working out of the prosecution service's New Glasgow office.

But not everyone was pleased with the way the prosecution service was handling the Westray file. Staff Sergeant Ches MacDonald and his team of Mounties had been left high and dry at a crucial point in their investigation. "We were getting a case together and we had a team in place," he told *Saturday Night* magazine in May 1994. "And then we didn't have a team and I wasn't very happy. It was such a setback. I was so disappointed. We had to bring a whole new team up to speed." And Pearson was still under attack from within the ranks of the prosecution service. In June 1993, Robert Hagell, a senior prosecutor who specialized in appeal cases, tore a strip off Pearson in a letter that received wide circulation among his colleagues. "I have on numerous occasions given my opinion that the Westray Case was the first substantial test of the Prosecution Service," Hagell wrote. "The Premier, Cabinet and many Government officials had a direct interest in this file. . . . The failure to provide proper legal and support staff to this file is, in my opinion, a disgrace." Hagell also offered his unsolicited opinion that Pearson's failure to provide proper staff and funding had "fatally flawed" the Westray prosecution. Pearson, for his part, rejected the notion that he somehow imperilled the criminal case, or compromised his independence. The decision to drop the Occupational Health and Safety Act charges, he pointed out in a 1994 interview, was politically unpopular but had a sound basis in law. "As far as I'm aware," he said, "nobody has suggested there was any political interference in either the advice provided to the RCMP or in the way the prosecution is being conducted."

Pearson and his new Westray prosecution team soon had bigger fish to fry. In July 1993, provincial court judge Patrick Curran accepted Curragh's arguments and ruled that the charges were "fundamentally deficient." The operation of the Westray mine over an eight-month period "would no doubt involve tens of thousands of activities and decisions of many kinds," the judge said, and the accused were entitled to know which ones had landed them in the prisoner's dock. Pearson took the rebuke in stride, saying an appeal would be launched, or new charges laid. But the ruling caused an uproar. Shaken relatives of the Westray victims wondered aloud why the Mounties, and the prosecutors advising them, did not draft proper charges in the first place. There were calls for Pearson's head, and his performance was discussed at the cabinet table in Halifax. "No incident in recent provincial history has suggested a more urgent need for a comprehensive and timely public accounting," said a *Chronicle-Herald* editorial. "And fourteen months later, justice has neither been done, nor been seen to have been done."

Within seventy-two hours, the RCMP came to Pearson's rescue with a new set of charges drafted in concert with the prosecution service. The alleged offences were the same—manslaughter and criminal negligence in connection with all twenty-six deaths, encompassing the eight months leading up to the explosion. But there was no lack of detail. The manslaughter count alleged Curragh, Phillips, and Parry caused the deaths by failing to keep coal dust in check. The second charge alleged all three had been criminally negligent in more than a dozen aspects of the mine's operation. Among the allegations: inadequate training, use of unsafe equipment, lack of stone-dusting, failure to deal with the build-up of methane, and straying from government-approved mine plans. Pearson maintained the new charges struck "the appropriate balance," outlining the substance of the Crown's allegations without limiting the scope of the prosecution. The Liberal government displayed its lack of confidence in Pearson by announcing plans to conduct an independent review of his office.

Atkinson mounted a challenge to the new charges, causing a further delay. The prosecution team led by Felderhof responded with a battery of arguments and legal precedents that supported the charges as drafted. In late October, Judge Curran gave the revised charges his stamp of approval. Six months after the RCMP wrapped up its investigation, the criminal case finally reached first base. In April 1994, long after the Westray controversy had cooled down, Pearson announced he was resigning at the end of the summer to take a prosecutor's job with his former employer, the Ontario Attorney General's Department. Pearson said he was heading back to Ontario for family reasons, and even joked about wanting to see the Toronto Blue Jays make a run at a third straight World Series title. He denied that criticism over his handling of the Westray case had convinced him it was time to move on. Within a month, the Nova Scotia government named Joe Ghiz, the former premier of Prince Edward Island, to examine the operation of the public prosecutor's office during Pearson's four-year tenure. Among the areas to be studied were budgeting procedures, independence from government, and public accountability.

No sooner had Curran upheld the charges than a new problem arose with the prosecution. While the legal battle raged, Curragh slipped from insolvency into receivership. With it went legal funding for the defence of the company and its former managers. When the case was called for arraignment in November 1993, no lawyer showed up at the New Glasgow courtroom to represent the company.

Felderhof was nonplussed, saying the Crown would continue to prosecute Curragh in absentia. This was tantamount to pursuing charges against a dead person—even if the company were convicted, it would not be possible to levy a fine as punishment. There was no company left to foot the legal bills, and Curragh's creditors, already out more than $200 million, rejected the notion of paying more money to defend the company's actions at Westray. So did Clifford Frame. Tracked down at his Uxbridge farm by a newspaper reporter, Frame weighed in with the observation that Curragh's financial predicament would be a boon to the prosecution. "They certainly have a better chance of getting a scapegoat, haven't they?" he noted.

Phillips and Parry were left to their own devices. Parry, who was living in Hinton, Alberta, was only able to find sporadic work after being laid off at Westray. He worked odd jobs at a sawmill, then drove a school bus for a time. Estimates of the fees for lawyers and expert witnesses needed to fight the charges were astronomical—anywhere from $750,000 to more than $1 million. Parry applied for a Legal Aid lawyer, but was rejected; houses in New Glasgow and Hinton, plus retirement savings, gave him a nest egg of $220,000, too much to qualify for assistance. The Nova Scotia Legal Aid Commission offered to take the case as long as Parry put up some of his own money. Parry, however, balked at spending "one penny" to defend himself against what he called the "politically motivated" charges. He finally agreed to ante up $50,000, and was assigned a team of Legal Aid lawyers at government expense.

Phillips was in better financial shape to fight the charges. But, like Parry, he had no chance of finding a new job in the coal industry. About a year after the explosion, the former Westray manager headed to Florida in a bid to start over. His goal was to get in on the ground floor of the Sunbelt's booming retirement industry. By late 1993, he was leasing the Plantation Retirement Home, a twenty-bed home for the elderly located in Clearwater, a haven for retired Canadians on the Gulf Coast. Phillips and his wife, Catherine, planned to buy the half-million-dollar facility and expand it to eighty beds. They incorporated a company to operate the home, applied for zoning approval for the expansion, and even printed up lime-green business cards listing them as the owners. But Florida health officials rejected Phillips's application for an operator's licence in March 1994, on the grounds that the leasing arrangement had not been approved in advance. Phillips had also failed to disclose the criminal charges he faced in Canada, even though applicants were required to reveal any past criminal involvement.

Pretrial motions and defence manoeuvres slowed the Westray prosecution to a crawl. A three-month preliminary hearing on the charges, set to begin in late March 1994, was postponed because Parry was having trouble finding a lawyer. Then Phillips waived his right to a preliminary hearing and asked to be sent directly to trial. Parry, however, was still entitled to a preliminary hearing. To avoid separate trials for the two managers, Pearson responded in May 1994 by issuing a preferred indictment, bypassing the preliminary-hearing stage altogether. As of this writing, the trial was set to begin in February 1995 in Pictou. There will be no jury: both Phillips and Parry asked to stand trial before a Nova Scotia Supreme Court judge. The trial will last six months, maybe more, and will hear testimony from scores of Westray employees and mining experts.

Legal battles over the wording of the charges have provided a glimpse of what lies ahead. The prosecution's case will focus on the failure to prevent the build-up of coal dust, the menace that gave the Westray explosion its destructive power. "The Crown intends to lead [with] evidence that the defendants assumed control of a dangerous activity, mined coal day after day and created hazardous accumulations of explosive coal dust without taking reasonable systematic steps to clean it up or to prevent it from exploding," the prosecution noted in a brief filed in support of the charges. "They exposed the twenty-six men to the risk by requiring them to work in hazardous conditions." Whether evidence crucial to the criminal case was lost when the mine was flooded, or when documents were shredded in the days following the explosion, will only become clear when the evidence unfolds in court. Lawyers for Phillips and Parry have yet to play their hand. But one line of defence has already been advanced—since the two managers were out of the country until a few hours before the explosion, they may not be responsible for what was done in their absence.

The case will turn on the role that coal dust played in the explosion, and who was ultimately responsible for safety at the mine. Phillips and Parry had statutory duties under the Coal Mines Regulation Act to ensure that the men working underground were not at risk. To secure a conviction, the Crown will have to convince the courts that the two men failed to live up to those obligations. Curragh's guilt or innocence will be determined by the actions of its employees, from the mine site in Plymouth all the way to the chairman's office in Toronto. A corporation can be held criminally liable for crimes committed by its employees or managers, even if the executives and members of the board of directors had no knowledge of what was happening. Under Canadian law, the crimes of company officials become the crimes of the company. The law of corporate criminal responsibility was set down by the Supreme Court of Canada in a 1985 ruling on fraud charges filed against three Ontario dredging firms. One of the lawyers who argued the winning side, on behalf of the Ontario Attorney General's Department, was John Pearson.

Cases of companies and employees being charged with crimes arising from workplace fatalities in Canada are few. Convictions are fewer still. Syncrude Canada was charged with criminal negligence causing death after a 1981 nitrogen gas leak killed two workmen at the oil sands project in northern Alberta; the company was acquitted. An employee at an Inco mine in Ontario was acquitted of the same offence in 1988, after an ore spill killed four workers. Inco did not face criminal charges, but the company was convicted under Ontario's Occupational Health and Safety Act and fined $60,000.

At least two other Canadian coal-mining companies have faced homicide charges. In 1898, a coal train running on the Union Colliery Company's line near Nanaimo, British Columbia, plunged thirty metres through a trestle bridge, killing seven passengers and crew. The bridge's timbers were rotten, and the company was charged under the criminal law of the day for putting lives at risk by failing to keep the structure in a safe condition. Union Colliery was convicted after a jury trial and fined $5,000. The other is Brazeau Collieries, which operated a mine in the village of Nordegg in the Alberta Rockies. In a case eerily like that of Westray, Brazeau was charged after twenty-nine men died in an October 1941 methane explosion. The charge was the same as in the Union Colliery case—failure to take proper safety precautions over a three-month period leading up to the explosion, an offence now covered under the criminal negligence provisions of the Criminal Code. Lawyers for Brazeau Collieries asked a judge to quash the charge as too vague, but were turned down. Unlike the Westray case, however, none of the mine managers was singled out in the charges.

The evidence against Brazeau Collieries was damning. Miners had complained repeatedly to their bosses about poor ventilation and high levels of methane. An Alberta government inspector had detected explosive levels of gas during an August 1941 tour, in the area where the blast occurred three months later. He reported his findings to the mine's general manager, but nothing was done to correct the problem. Coal was a crucial commodity needed for the war effort, and the emphasis of Brazeau's management was to produce as much as possible. In the words of C. S. Blanchard, the lawyer who prosecuted the case, "the consideration of getting the coal out came first and the interests of the safety of the men came next." In January 1943, after a two-week trial before a judge of the Alberta Supreme Court, the company was convicted. The sentence was a slap on the wrist, even by World War II standards—a $5,000 fine, or roughly $172 for each man who died.

* * *

WHILE THE CRIMINAL PROSECUTION sputtered, the Westray Mine Public Inquiry languished in a legal no-man's land. The inquiry remains the best chance for a comprehensive stock-taking of what happened at Westray, and why. The trial will delve into events and safety problems leading up to the explosion, but the focus will be narrow; the courts will have to determine whether the actions of Curragh and its officials amounted to a crime. Mr. Justice Peter Richard, on the other hand, has the mandate to look beyond the cause of the explosion and who, if anyone, is to blame. Inquiry officials have promised a wide-ranging probe of the disaster and the business, political, and technical decisions behind the mine's development. The inquiry will provide a forum for geologists and engineers to debate whether proper mining methods and procedures were used at Westray, and whether the gassy Foord seam can be mined safely. And it should produce new mining regulations and safety laws to help prevent such a disaster from happening again. The international mining community will closely study the inquiry's evidence and findings, to make sure the mistakes of Westray are not repeated at other mines.

The inquiry should also explain why Nova Scotia's safety officials and mine inspectors failed to prevent the tragedy of May 9, 1992. The provincial bureaucrats responsible for mine safety have yet to publicly defend or explain their actions. Patrick Phalen, who was promoted to executive director of the mines and minerals branch of the Department of Natural Resources around the time of the explosion, will entertain questions about his department's approval of changes in the mine design and layout at Westray. But he is less comfortable discussing whether the company received special treatment. "We're getting into things that are really inquiry-type stuff," he said when the question was raised in a 1994 interview. "That's where I think it's more appropriate to answer those questions." Robert Naylor, the provincial government geologist who criticized Curragh's reading of the Foord seam's structure, also begs off. The inquiry, he says, is the proper place for debating the merits of the geological work that went into the Westray mine.

Tracked down at his Glace Bay office, Albert McLean says he would welcome the chance to discuss his crucial role as the Labour Department's eyes and ears at Westray, but at the inquiry, not for a book. "There's been a lot of things said," he notes. "I'll have my day, that's all I can say." John V. Smith retired as the electrical/mechanical inspector at the end of May 1994, at age sixty-one. He's a gregarious fellow, knowledgeable about all facets of coal mining. Now that he's retired, he's willing to talk about the shortage of inspection staff and the deficiencies of the Coal Mines Regulation Act. But he was not privy to the decisions made by his superiors in Halifax, and he claims to be as in the dark as anyone about why the department did not move faster to deal with the coal-dust menace at Westray. All he knows is that, just hours before the explosion, Claude White told him the department was on the verge of closing down Westray.

White, Nova Scotia's chief mine inspector and the man who presumably can answer such questions, politely declines to be interviewed; he expects to be a prosecution witness at the upcoming criminal trial, and Labour Department lawyers have advised him to say nothing in the meantime. Most of the bureaucrats who dealt with Westray say the inspection reports and other records made public since the explosion speak for themselves. But the mound of paperwork does not answer one nagging question: why did government officials allow Westray to put the lives of twenty-six men on the line?

Even without an inquiry, the Westray disaster has prompted changes. The Department of Labour has been put under new management. An internal review has led to improvements in the way the department and its inspectors do business. Nova Scotia's outdated coal-mining laws and regulations are being redrafted, with the help of inquiry officials. The only thing lacking is any sense of urgency, because there is little risk the Labour Department will repeat the mistakes made at Westray, or that a flaw in the Coal Mines Regulation Act will put another miner's life in jeopardy. An underground fire has permanently closed the Evans mine on Cape Breton's west coast. The only underground coal mines still operating in Nova Scotia are Devco's collieries on Cape Breton Island, and those mines are monitored by federal inspectors enforcing federal safety laws.

Westray will remain closed, at least for the foreseeable future. That has left Nova Scotia Power, the former Crown corporation privatized during Donald Cameron's period in power, without a supply of low-sulphur coal for its Trenton generating stations. The utility considered buying low-sulphur coal from the United States or South America, but the idea of importing coal to coal-rich Nova Scotia was rejected by the province's Liberal government. Nova Scotia Power ended up going back to its old supplier, Devco, to keep the Trenton plants running. The move was a $25-million shot in the arm for Devco, and created 150 jobs in the Cape Breton collieries. Despite all the pre-explosion posturing about the environmental importance of Westray coal, the utility said it could burn the higher-sulphur Cape Breton coal without exceeding federal or provincial restrictions on sulphur dioxide emissions. NSPC president Louis Comeau, who told federal officials in 1988 that the utility desperately needed Westray coal, had softened his stance five years later. "There is no question that the lower sulphur coal you get, the better off we are from an environmental perspective," he noted in the fall of 1993. "But we also have an economy to run in Nova Scotia, and I think we can marry the two."

The mine site in Plymouth sits idle, and the ghostly silence that has fallen over the well-maintained buildings is the only hint of the calamity that occurred hundreds of metres below. Curragh has disappeared from the scene; the federal government took over the property from the company's receivers in early 1994; the mineral rights to the Foord seam have reverted to the province. The Nova Scotia government assumed the $35,000-a-month cost of keeping security guards on the site and making sure that settling ponds meet environmental standards.

In early 1994, the federal Industry Department retained a Pittsburgh-based consulting firm, John T. Boyd Company, to assess whether the mine could be rehabilitated. Boyd handed in its report in June 1994, and concluded the mine could be put back into production. The existing shafts could be repaired and extended to bypass the explosion-damaged workings, but the cost was exorbitant. Boyd estimated it would cost $76 million to resume coal production, a figure approaching the $120 million Curragh spent to develop the mine in the first place. The coal would have to sell for about $100 per tonne—double the 1994 price on the world market—for even a modest return on investment.

No-one was about to question those findings. Boyd was one of the world's leading mining and geological consultants, a specialist in coal properties. Based on the report's bleak outlook, any effort to sell Westray to a new operator would be futile. When the report was released, Industry Minister John Manley announced that Ottawa would swallow its $67.15 million loss, along with the $3.9 million already paid out to ease the interest-rate burden on the original loan to Westray. The federal government would also turn the mine over to the province, which was absorbing all $12 million of its loan on the project. In other words, the potential $96.6-million loss to taxpayers had been "whittled down" to $83 million. "It's not very good news for anybody," said Don Downe, Nova Scotia's minister of natural resources, who suggested that "something" would be done to replace the 160 jobs lost when the mine exploded. But Westray miners still in the area were more interested in getting their old jobs back. "You're sitting there today telling us we can't work any more," a former Westray foreman, Jay Dooley, protested as the study's results were revealed at a press conference in Stellarton. There was one obvious oversight; Boyd had not been asked to take into account the possibility of strip-mining coal from the Wimpey pit to generate the cash needed to put the Plymouth mine back into production.

The Westray Families Group was no happier with the findings. Relatives of eleven Westray victims had a lot riding on the study; unless the mine was put back in production, there was little hope of recovering the remains of the men left behind when rescue operations ceased. The idea of using robotic equipment to search the tunnels had been abandoned. It would cost $12 million to drain and re-enter the existing workings to search for the miners' remains, on top of the $76 million needed to resume production. If the mine did not reopen, the cost could be as much as $30 million. Either way, it was a risky, time-consuming proposition. Chris Martin, who knew the report probably represented the last chance of bringing his brother, Glenn, to the surface, dismissed it as "a study of convenience."

The report also rejected much of the work done at Westray during Curragh's tenure—the exploration data, the mine layout, the safety procedures. Boyd concluded that a new operator would have to begin almost from scratch. The $76-million cost of reopening the mine included millions of dollars for new ventilation shafts and safety equipment designed to prevent a repeat of the 1992 explosion. In Boyd's opinion, Westray's ventilation system was flawed from the outset. Fresh air was forced into the mine through the main slope travelled by men and machinery, and returned via the adjacent tunnel used to bring coal to the surface on a conveyor belt. Methane was being flushed out, but the air blown over the freshly dug coal on the belt line stirred up dust. Something had to be done to reduce the air velocity in the conveyor tunnel, and keep the dust down, while still removing the gas. Boyd recommended drilling dual vertical shafts, each 450 metres long and 7.3 metres in diameter, into the heart of the mine workings, at a cost of $12 million.

That was only part of Boyd's plan for dealing with coal dust. Roadways would be graded with gravel and doused in calcium chloride, as country roads are, to prevent heavy machinery from kicking up dust. But the main attack on dust was an elaborate system to ensure that stone dust was spread throughout the mine. Stone dust would be stored on the surface and delivered underground through a hole bored adjacent to the new ventilation shaft. A half-million dollars would be spent to buy a dozen machines to spray the dust on the roof, walls, and floor. Stone dust would be spread continuously as coal was mined—not the way Westray did it, expecting miners to work overtime after a gruelling twelve-hour shift.

Boyd also took a swipe at the conclusions of the earlier feasibility studies on the Westray property. Westray's estimates of the coal available to be mined, and its quality, were grossly exaggerated. Curragh and Westray had hoped to produce more than one million tonnes of coal per year for fifteen years, and speculated that there was enough coal in adjacent, unexplored areas to keep the mine going another fifteen years. Boyd said a new operator would be lucky to get a fraction of that, maybe five million tonnes of good-quality coal over ten years. The lower recovery rate was due in part to Boyd's insistence on taking no more than six metres of coal out of the seam. The top two metres of the coal seam would be left behind to keep the roof in place, and prevent a repetition of the cave-ins that plagued Westray. But Boyd also suggested that much of the coal was high in ash and better left in the ground. Before mining resumed, the report recommended spending $500,000 to collect core samples, in order to target the best areas for mining. This plan sounded suspiciously like the drilling program that geologists working for the federal and Nova Scotia governments had espoused in 1988, while Westray was still on the drawing boards.

Mining methods would remain largely unchanged. Despite all the seismic work done in the mid-1980s, Boyd felt the location of faults remained "uncertain." It was clear, however, that there were a lot of faults, and not much distance between them. For that reason, Boyd rejected the thought of using longwall equipment, which required large areas of uninterrupted seam. Employing room-and-pillar methods, Boyd said, would give mining crews the flexibility to work through the faults. Curragh had used this method, and it was one of the few approaches of the previous regime that won Boyd's endorsement. That was not the case with procedures for sealing off mined-out areas, which were prone to methane build-up and spontaneous combustion. When Westray abandoned part of the southwest section a few weeks before the disaster, plywood sheets were used to block off the area. Boyd said such areas should be cordoned off with explosion-proof seals, which would keep the gas in and cut off the oxygen needed to feed a fire.

And there was no sense trying once again to mine the tricky Foord seam without experienced, properly trained miners. The Boyd report envisaged a $2.5-million training program "to develop prudent safe-work practices." Miners would have to be given time to master mining techniques, and to learn to work as a team. The amount of coal dug would rise as employees became more familiar with their jobs; under Boyd's scenario, it would take three years to reach full production. Westray's miners did not have the luxury of time to develop a sense of how to mine the seam. The company was locked in a constant struggle to meet deadlines and reach production targets. Inexperienced men were pressed into service in order to get the project up to speed. As early as 1988, a mining engineer with the federal agency CANMET warned that Westray's miners would face a steep "learning curve." If the mine ever reopened, employees would finally be given the time they needed to reach the top of the curve.

Boyd's harsh conclusions will not be the final word on how Westray should have been operated. That will be the task of Mr. Justice Peter Richard and his public inquiry. But with the inquiry on hold, the reports of consultants hired to study the operation of the mine, and suggest what caused the explosion, remain under wraps. A half-million pages of Westray records gather dust under lock and key at the inquiry's offices. With each day that passes, memories fade and potential witnesses leave Pictou County in search of work. The endless legal wrangling has been a boon for lawyers—legal fees account for about half the $1.8 million the inquiry spent by mid-1994. Nova Scotians have watched from the sidelines with growing cynicism. Some relatives of those killed—the people whose lives were torn apart by the explosion—wonder whether the inquiry will ever answer their questions. Disputes over legal funding for the Westray Families Group and the inquiry's decision to go along with flooding of the mine have bred hostility and suspicion. The families, who long ago pinned their hopes on the RCMP, view the criminal prosecution as their last hope for justice.

The legal mess surrounding the inquiry is a tragedy in its own right. The Westray probe has the dubious distinction of being the first public inquiry shut down to prevent a conflict with the constitutional rights of individuals. The Court of Appeal ruling applies only within Nova Scotia, but it has influenced the way public inquiries are conducted across the country. The New Brunswick government followed the precedent when it launched an inquiry into the sexual abuse of inmates at a correctional facility for boys. The military investigation into the beating death of a Somalia prisoner was delayed until Canadian peacekeepers stood trial on murder charges. In British Columbia, justice officials suggested that nine public inquiries set up between 1985 and 1992 could have been vulnerable to the same legal challenge that derailed the Westray probe.

The outcome of the two-year legal battle over the Westray inquiry is unknown as of this writing. The inquiry's future rests on a Supreme Court of Canada ruling expected in the latter half of 1994. The United Steelworkers of America and Justice Richard have asked the country's highest court to lift the ban and allow the inquiry to do its work. The case could have far-reaching implications for inquiries in all provinces. If the court endorses the Nova Scotia ruling, no provincial government wil be able to hold an inquiry into an incident or event that is the subject of a criminal investigation or prosecution. The Steelworkers union has proposed a compromise: that the Westray inquiry proceed without testimony from Gerald Phillips and Roger Parry, the former managers facing criminal charges. But the governments of Ontario, Quebec, British Columbia, and Saskatchewan have joined the appeal to argue against restrictions on who can be called as a witness at inquiries. Whatever the Supreme Court decides, the Westray case will set new ground rules for Canadian public inquiries. And it could prevent a repetition of the legal complications that have become part of Westray's bitter legacy.

* * *

THE CRIMINAL CASE has been to court more than a half-dozen times since charges were laid in April 1993. Members of Glenn Martin's family have dutifully shown up at each hearing, and the delays and endless legal wrangling have left them frustrated and disillusioned. "It just seems like high-priced lawyers are just having a game with it all," Chris Martin complained on the eve of the second anniversary of the explosion that killed his younger brother. "It's been two years, and we haven't even got to court yet. You're embarrassed at the system. It's not right, is it? It's not working for the victims at all." But he still pins his hopes on the criminal case. "We know it's taken a long time and it's a long road, but I guess our hopes aren't dampered because we feel someday it will get to court, and justice will be served."

Martin's desire to see justice done spurred him to become active in the Westray Families Group. He took over from his father, Bert, whose health was suffering under the strain of the running battles with government and inquiry officials. Bert turned his attention to siding his son's house—one of the projects Glenn wanted to finish before he quit Westray—before it was sold. Chris Martin would still like to see an inquiry held; he hopes that politicians and government officials will finally be forced to justify their actions. He has talked to a lot of the miners who worked at Westray, and he would like to know if their versions of what happened will be contradicted. "I'm hearing one side of it. I'd like to see the other side of it, and then I can draw a better conclusion."

Like Martin, Isabel Gillis has poured her heart and soul into the quest for justice. In the months following the explosion, she fought tirelessly to fulfil the promise she made to her husband, Myles—that she push for a full inquiry if he died at Westray. She was at the forefront of the families group, headed by her father, Kenton Teasdale. "I want every one of my questions answered. I have that right," Gillis explains. "Somebody took away my husband; somebody ripped apart my whole life and broke the hearts of my three children. Nobody does that, and doesn't at least answer to it." Like Donna Johnson and Joyce Fraser, she is incensed that people think the Westray widows are set for life. "You can have my money if you give my husband back," she says. "I'm not bragging, but we had what I would consider a perfect family. The Brady Bunch, whatever. We had it. And I lost that." No amount of money will bring back the husband and father who went to work on May 8, 1992, and never returned. "When my kids are in bed at night and I'm sitting alone on the couch, and the whole house is so cold and dreary—what does money do for you then?"

The struggle for answers has forged a strong bond among many of the Westray families. The wives and children of miners who came to Westray from other provinces have left the area. But parents, widows, and siblings of men from the Pictou County area have been brought together by a common desire to make sure their loved ones are not forgotten. "Some of these families that we've been in contact with over these two years are as close as family," notes Martin. "I guess you'd say the disaster started a new family, because you know their children and what they do and their hobbies. It has really developed a friendship."

The focal point of their effort has been a memorial erected just outside an industrial park in New Glasgow. It is located about one-and-a-half kilometres northeast of the mine site, directly above the North Mains, the resting place of eleven men. The polished black-granite memorial bears the image of an old-style miner's lamp. Thirteen rays of light shoot out from either side, each inscribed with the name and age of a man who worked and died in the Westray mine. Below the lamp are the words: "Their light shall always shine." Joe MacKay, who lost his brother Mike in the explosion, was the driving force behind raising money for the memorial; he even managed to wring a $10,000 donation out of Westray Coal before Curragh went under. It has become a labour of love; on some Saturdays, dozens of family members and assorted tradespeople and volunteers have showed up to work on the memorial or landscape the surrounding park.

For the Martins and ten other families, the memorial is the only link to their loved ones, who lie 350 metres below. On May 9, 1993, relatives and former Westray miners gathered there for a private service. Reporters and camera operators honoured a request that they stay away. When the second anniversary rolled around, families living in the area decided against a formal service; many felt it would only darken yet another Mother's Day weekend. So each family marked the day in their own way. The fire hall in Plymouth, where the families held vigil for five terrible days, once again became a gathering place.

On May 9, 1994, the Martin family went to church, then visited the memorial. It was a day to remember Glenn, who loved to hunt and fish, and who died two days shy of his thirty-sixth birthday. And when another May 9 had come and gone, it was time for Chris Martin to get on with the task of making sure his broth-

"Their light shall always shine." Joe MacKay wipes away a tear as he surveys the memorial to the Westray victims. The remains of his brother, Mike, lie about 350 metres below. [Photo by Andrew Vaughan, courtesy of the Canadian Press]

er did not die in vain. "I want the system cleaned up to be sure nobody else gets hurt the way we've been hurt, and that the people who caused it answer for it," he says. "I don't think it's too much to ask."

* * *

THE LAST THING Donald Cameron wants to talk about is the Westray coal mine. "No-one wants to hear the truth," he says over the phone from his office in Boston, his voice tinged with anger. "I've had my fill of it, and I have nothing to be ashamed of." It's early 1994, and Cameron is right where he wants to be—far away from the Nova Scotia political scene. Within weeks of his disastrous defeat in the 1993 election, Cameron landed on his feet; Brian Mulroney appointed him Canada's consul-general to New England. The $110,000-a-year post, complete with a house in the posh Boston suburb of Cambridge, is intended to foster trade links between the United States and Canada. Its last occupant was a defeated Mulroney cabinet minister, and no less an authority than John Buchanan has called it a patronage job. But Cameron, Nova Scotia's former trade minister, bristles at the suggestion that he sullied his record as an opponent of patronage by taking the job.

In the two years since the explosion, Cameron has been besieged with requests to be interviewed about Westray. He has declined them all. Other than occasional comments, in the House of Assembly or when confronted by reporters on specific issues, he has held his tongue. But he has not forgotten the way his critics tried relentlessly to lay responsibility for the disaster at his doorstep. He resents the way his political opponents used the issue to bring down his government. And he's not about to let all the accomplishments of his long political career be wiped out by the Westray tragedy and its aftermath. "I have a lot of bitter memories about all this," he says. "No-one is happy that there was an explosion and loss of life. I'm sure as hell not happy with the way the media's dealt with it. I'm not happy with the way the politicians have dealt with it."

And Cameron is not interested in being interviewed for a book about Westray. He's been pilloried in the press for supporting the mine's development. Even the fact that he sold his old power-boat to Westray manager Gerald Phillips in the summer of 1991 became the stuff of headlines and innuendo. He's wary. "I'm not going to have any long interview that's going to be a major part of that book," he says. But he would like to set the record straight on a number of things. "I guess what I want to see is for someone to maybe, just maybe, after all the pain and agony people went through, the families went through, that something objective could be written. These dark schemes that everyone seems to have, that you know, this was something cooked up by Don Cameron so he could get himself elected in 1988, just a political decision, there's a lot of evidence that just won't support that."

Cameron finally agrees to field questions about specific aspects of his dealings with Westray. For an hour and a half, he breaks his long-imposed silence and offers his views on everything from the cause of the explosion to the "sick" politics of those who have blamed him for what happened at Westray. He relives his battles with the federal bureaucrats who blocked the project to protect Devco. His government drove a hard bargain with Curragh, he argues, scoffing at claims that a $12-million, full-interest loan for a $130-million project amounted to a "sweetheart deal." And he takes the opportunity to point out that the Pictou County coal project was in the works long before Curragh came on the scene – long before he was industry minister, let alone premier. "I didn't know Cliff Frame. I didn't choose them as the company to come to Pictou County," he says.

Cameron is particularly incensed by the way the disaster was used to undermine his integrity. The Liberals easily won the 1993 by-election held to fill his old Pictou East seat, and he claims party supporters are bragging about how they used Westray to defeat the Tories. He has dug out yellowed newspaper clippings and transcripts of political debates about the 1979 methane explosion at Devco's No. 26 colliery–the one that killed twelve miners–and he smells a double standard. "How many politicians supported Devco over the years?" he asks. "Have they paid any price for any loss of life there? Were there any allegations against Allan J. [MacEachen, the Cape Breton Liberal MP] because he poured billions into Devco?" The answer, of course, is no. "There'll be losses of life at Devco again some day, and I'm just wondering if they're going to have the same standards." A miner died in a cave-in at Devco after the Westray explosion, he notes, causing barely a ripple in the media. "I suppose if you kill them one at a time, it's all right."

The point, in Cameron's view, is that politicians who support development projects should not become instant targets if something goes wrong. "What kind of standards do you want from now on? . . . If you support an industry, and anybody gets killed in it, it's your fault? If that's the standard everyone's going to have from now on, it's a pretty sick standard." That was the standard applied after the Westray explosion, he says, and it's wrong. "Whether it's in the fishing industry, or whether it's in the forest industry, we are going to have these accidents. And the families are going to be touched by the people who were lost in those industries. And we've got to make sure the rules are such that every precaution is taken. Don't take one company, and just turn it into a political snowball, and ignore what happened in other places."

Were the proper precautions taken at Westray? Ask the people at the Department of Labour and the Department of Natural Resources, Cameron says. "That was their job, not mine." What he resents most of all is the implication that he or members of his government pressured bureaucrats and mine inspectors to go easy on

Westray, or that his government's stake in Westray's success prompted the officials responsible for mine safety to treat the company with kid gloves. "There better not be anyone in that bureaucracy that tries to slough off on me their responsibility for doing their job," he says angrily. "Any weasel that wants to try to get out of the responsibility of carrying out his own or her own job, by saying, 'Well, you know, we didn't want to offend anyone in high places . . . so we overlooked some of that,' those are the people that you should be going after. . . . I purposely never called them, not ever once, because I didn't want to interfere." He challenges anyone to produce evidence of a letter or phone call from him that would prove otherwise.

On the other hand, Cameron admits, Westray officials complained to him "on a couple occasions" about the Department of Labour taking too long to approve equipment needed in the mine. "I just absolutely ignored that," he insists. "People ask you to do things in politics. You just use your own judgment." When safety concerns turned up in the newspapers, or when his sparring partner, Bernie Boudreau, complained about cave-ins, Cameron got on the phone to the mine site, by-passing the Department of Labour. "Lots of times I would call the company, and say, 'Look, is there any truth to this?' " Most of the time he put his concerns directly to the mine manager, Gerald Phillips. "Gerald believed that they had enough equipment there that would take care of the safety of the people," Cameron says. And he was in no position to dispute the company's claims that safety problems were being dealt with. "I wasn't doing the inspections. I know nothing about coal mining. In fact, I was never in a coal mine in my life before. . . . I don't know what to believe or not believe."

To Cameron, the real issue is what set off the explosion, not whether Westray Coal failed to live up to the letter of safety laws. Cameron has made it his business to find out what happened in the southwest section on the morning of May 9, 1992. He says he has it on good authority that the explosion was touched off by the continuous miner working in the area. The machine's methanometer had been tampered with to keep it running in high levels of methane. The gas ignited when the machine's cutting head struck a band of pyrite, setting off a spark. When the RCMP searched the mine, he says, the pyrite could be seen embedded in the coalface. "People want to blame it on the company, and the company certainly wasn't perfect, as we found out. Bottom line was, that explosion took place at the continuous miner, . . . and we hide the fact that machine was tampered with only a few hours before the explosion took place."

He has said his piece, but already Cameron is having second thoughts. "I did something I said I wouldn't do at the start—give you a long interview," he says, slightly irritated with himself. "And all you'll do is just use a whole lot of this in the book." Three times during the wide-ranging interview, he has threatened to strike back at anyone who prints anything he feels is inaccurate about him or his dealings with Westray. "When you're in politics, you have to accept a lot of things that you normally wouldn't. I'm not in politics any more, and I never plan to be at any level in the future. And I'm going to be very careful about what people say about me from now on. I'm not going to overlook things that are not true," he says. "I have tried very hard in my public life to do what was right and to change the system, and I'm not going to allow anyone to leave an impression in the minds of people reading any book or any article, anything otherwise."

"That's not trying to intimidate you; you write what you have to," he adds, almost as an afterthought. "But you better make sure it's true about me."

* * *

IT WAS A LONG TIME before Lenny Bonner could close his eyes without seeing the faces. "It was at least six or seven months before I could even sleep," he says, his booming voice suddenly subdued. "I'd lay down, especially for the first couple of months, and close my eyes, and all I'd see was just [myself], marching by all the bodies, looking at them. And I'd just see their face. Just face after face after face, every time I'd close my eyes." They were the faces of the first eleven men pulled from the southwest section of the Westray mine, two days after the explosion. Bonner was dispatched to the New Glasgow arena to help identify the bodies, and the horrible images of their lifeless faces are forever stamped on his memory.

The man at the other end of the couch has been tormented by his own demons. While Lenny Bonner was too shaken to return to the rescue effort, his buddy Shaun Comish soldiered on for three more days. All told, Comish went underground as a draegerman a half-dozen times in the risky search for survivors. He groped through the darkness and debris, and dodged falling rock—put his own life on the line. But by May 13, the day before the rescue effort was called off, he had given up hope. "I knew then that everyone was dead. And I just said, cold as it sounds, 'I'm not going down, and I'm not dying looking for dead bodies.'"

Listening to the two men talk, over coffee at Shaun Comish's duplex on the outskirts of Halifax, is like eavesdropping on two veterans having flashbacks about some long-ago war. "It's an awful thing, bodies lying around like that," Comish says, shaking his head as the memories flood back. "Exactly like you see on those newsreels from Bosnia." Bonner jumps in and finishes his friend's sentence without missing a beat: "Only they're your buddies, they're not strangers." Comish vividly recalls his last task as a draegerman—putting the badly burned bodies of two men into bags so they could be taken to the surface: "It wasn't real. You just functioned," he explains. "You just put it out of your mind, did what you had to do." He pauses for a moment. "And then it hit you later."

Bonner and Comish are both in their mid-thirties. Between them they have more than twenty years' experience working in hard-rock mines in Northern Ontario and Nova Scotia. Westray was their first coal mine, and it will be their last. The nightmare they survived was their ticket to a new life. Doctors diagnosed Bonner and Comish and at least a half-dozen other Westray miners as suffering from post-traumatic stress syndrome. They were unable to work, unable to sleep, unable to hold back the tears; and they needed counselling to deal with what they had seen and done in the desperate days after the explosion. They are on workers' compensation while they retrain for new jobs. "It's a brand-new chance, just like if you were young again," offers Bonner, who immersed himself in courses at a technical school in Halifax.

Comish entered a computer-programming course, with the promise of a job working for a friend when he graduates. He has already made his mark as a writer. Night after night, when the images of Westray kept him awake, Comish sat at his word processor and began to write. He wrote about the poor working conditions at Westray. He expounded on his theory of what caused the explosion. He described his horrible experiences as a draegerman. He wrote about the toll the disaster has taken on the wives and families of all Westray miners, living and dead. Comish ended up with a book, and Fernwood Publishing, a small Halifax-based publishing house, released it in the fall of 1993. *The Westray Tragedy: A Miner's Story* was greeted with a splash of national media attention, and has sold a respectable eight thousand copies.

"It was therapy, just getting it out of my system," Comish says of the book. It helped put Westray behind him and allowed him to get on with his life. By the time the first anniversary of the explosion rolled around, he was still having trouble sleeping. This time, however, he was up nights with his newborn daughter, his third child. His wife, Shirley, has spearheaded a petition drive to convince the governor general to award medals of bravery to the scores of draegermen who risked their lives at Westray. Comish, for his part, thinks the unsung heroes of the Westray tragedy are the wives and families of the draegermen, who knew the risks the rescue crews were taking. But most of all, Comish is grateful that he lived to tell the truth about his brush with death. "We went through a rough time," he says, exchanging a glance with Bonner, "but we're the lucky ones."

Mike Wrice, the man who let John Turner know what he thought of Westray's roof problems, counts himself lucky that he got out of the mine when he did. He quit in the fall of 1991, citing safety reasons, but got no support from the Nova Scotia Department of Labour or the bureaucrats who handle unemployment insurance claims. Wrice lost several thousand dollars in benefits when unemployment officials ruled he had no grounds for quitting. Despite the explosion, despite the documented safety problems at the mine, and despite the politicians who have championed his case, Wrice has yet to receive his retroactive benefits.

It was a year and a half before he once again set foot in a mine. In May 1993, after drawing pogey and taking a course in diesel mechanics in his home town of Glace Bay, Wrice landed a job in a gold mine in the Northern Ontario town of Marathon. He's back at the work he knows best, the work that brought him to Westray in the first place–carving out development tunnels in advance of mining. But his harrowing few months digging coal in Pictou County are never far from his thoughts. And he knows the torment of the miners who stayed to the bitter end. "It's over, but it's not done with, because there are still people living with it, sometimes on a daily basis," he says. "Hopefully, they're all getting their acts together and their heads together."

"Ten or twenty years from now, they're just going to be twenty-six men who died at Westray," says Steven Cyr, who waited in vain for the call back to work that never came. "They were all people, not just a number." [Photo by Len Wagg, courtesy of the *Chronicle-Herald* and the *Mail-Star*]

Wrice was not the only former Westray miner who was forced to leave Nova Scotia to find work. Alberta miners lured to Pictou County by the promise of fifteen years of steady work returned home; some got back on at the mine in Grande Cache. A dozen others were hired at Quinsam, an underground coal mine in Campbell River, British Columbia. Most of the miners and surface employees from the Pictou County area stayed put, hoping the underground mine would be reactivated or the strip mine approved. That hope never materialized, and a few ended up losing their homes and cars. Those lucky enough to find other jobs could only dream of the steady work and big bucks they had enjoyed at Westray.

Steven Cyr, born and raised in Westville, decided to stick around. His wife had a part-time job and, combined with his unemployment insurance cheques, there was enough money to support the couple and their four-year-old daughter. Cyr was among the union members who backed Curragh's bid to reactivate the mine and mount an effort to recover the bodies trapped below. To Cyr, the recovery plan was more than just a way of getting his old job back. It was the only way he could finish the task of finding the eleven men left behind when the rescue operation was called off. "You know, ten or twenty years from now, they're just going to be twenty-six men who died in Westray," Cyr says. "I'd like to see all the guys remembered, other than just as a statistic. They're friends of mine or they're friends of somebody else, and they all had families. They were all people, not just a number."

Westray did not cost Carl Guptill his life, but it cost him his livelihood. The back injury he suffered during his stint at the mine brought his mining days to an end. He won disability benefits after a long fight with the Workers Compensation Board, and he headed back to school for an upgrading course. By the fall of 1994, he was ready to begin a community college program, in either small business administration or aquaculture. He moved away from Pictou County about a year after the explosion, to a small village on Nova Scotia's rugged Eastern Shore. He can look out his window and see the ocean, and it reminds him of the times he worked on the fishing boats, or filled in as a lighthouse keeper. It's a lot better than waking up each morning and seeing the towers of the Westray mine looming in the distance.

Guptill took some heat for appearing on the CBC National News within two days of the explosion to condemn Westray's safety practices. A few of the draegermen complained he had spoken too soon, that he should have waited until the rescue effort was completed before criticizing the company. He was accused of seeking revenge for his firing; when reporters asked about his allegations, Westray officials went out of their way to discredit Guptill and his knowledge of mining. In the months after he went public, people in Pictou County would either shake his hand for what he did, or give him the cold shoulder. Moving away meant he no longer had to guess what people were thinking as he passed them on the street.

But Guptill has no regrets about blowing the whistle on Westray when he did. The way he sees it, he paved the way for Shaun Comish, Lenny Bonner, and other miners who eventually came forward to back up his allegations. If he and the others had not had the courage to come forward, Westray might have been allowed to reopen the mine as if nothing had happened. It was the hue and cry over safety that prompted the RCMP to step in, and led to the criminal charges. Guptill has been interviewed by RCMP investigators on a couple of occasions, and he is among the scores of Crown witnesses on tap for the criminal trial. He will describe what he saw at Westray. He will probably be asked about his dealing with the officials at the Department of Labour, who ignored his long list of accusations about unsafe conditions. And he will once again keep the promise he made to the men of B-crew that spring morning in Roy Feltmate's kitchen—to speak on their behalf.

* * *

SITTING AT HER KITCHEN TABLE, Donna Johnson is amazed at her own resiliency. "You do learn to go on," she says as she talks about facing each day without Eugene. "I mean, they're a part of your life, you never forget them. But life does go on. You don't think it does at the time. But you do go on." She pulls out a photo of Eugene taken in Montreal when he accepted the safety award on Westray's behalf. He looks uncomfortable as he stands with Gerald Phillips and other mining executives, clutching the plaque that should have meant he was working in Canada's safest mine. Little more than a week later, Eugene and twenty-five of his fellow workers were dead.

On the kitchen counter sits a bottle of white wine. The label bears a regal-looking lion set against a shield—the Curragh emblem. Westray had a local cottage winery do up dozens of cases with special labels to mark the mine's grand opening in September 1991. Eugene managed to get his hands on a bottle as compensation for missing the opening celebrations. He put it aside, telling Donna he wanted to save it for the day he retired. The Westray mine was supposed to be in operation for fifteen years, and that was how long he planned to work. Eugene would be forty-seven by then, the house would have been paid off long before, the kids would be grown. Then, and only then, would he open the bottle and drink a toast.

In her grief and frustration, Donna has thought many times about taking the bottle over to Plymouth and smashing it against the mine gate. But she has a better idea. When the fifteen years are up, she'll open it, just as Eugene would have if he were still alive. There's no way of knowing if the wine will still be drinkable by then. There's no telling if all the questions raised by the Westray disaster will have been put to rest. There's no guarantee Pictou County's gassy seams will not have claimed more lives in the quest for coal. But there will be no shortage of bitter memories or shattered lives. And there will always be the broken promise of the Westray mine. That's not what Donna Johnson will drink to, though. She'll be raising a glass to the hope that such promises will never be broken again.

INDEX

D

U

V

W

Z